Training, Education and Research in COVID-19 Times: Innovative Methodological Approaches, Best Practices, and Case Studies

About the Editors

Miltiadis D. Lytras is an expert in advanced computer science and management, an editor, lecturer, and research consultant, with extensive experience in academia and the business sector in Europe and Asia. He is a Research Professor at Deree College, The American College of Greece, and a Distinguished Scientist at the King Abdulaziz University of Jeddah, Saudi Arabia. He is a world-class expert in the fields of cognitive computing, information systems, technology-enabled innovation, social networks, computers in human behavior, and knowledge management.

Abdulrahman Housawi is a practicing Nephrologist and Transplant Physician at King Fahad Specialist Hospital, Dammam (KFSH-D), the Director of Privatization Affairs at the Ministry of Health (MoH), and the Chief Officer for Planning and Organizational Excellence at the Saudi Commission for Health Specialties (SCFHS). He joined the faculty at Umm Al-Qura University in 2006 as an Assistant Professor in the Department of Medicine. In 2008, he joined King Fahad Specialist Hospital, Dammam (KFSHD), where he participated in the establishing team of the Multi-Organ Transplant Center (MOTC). He was the founding Chairman of the Kidney/Pancreas Transplant (KPTx) Department. In addition, he led a project for establishing the MOTC Registry and MOTC Information System at KFSH-D. He was the Director of Academic Affairs Administration (2008–2013) and the Director of Training Administration (2013–2014). He was a member of KFSH-D's Center for Health Research's establishing team and its Research Steering Committee (2009–2015). Currently, he is responsible for the development and implementation of the Saudi Commission's strategy, including its transformation to a data-driven organization (2016–present).

Basim Alsaywid is a Pediatric Urology Surgeon with extensive clinical experience and an extensive research output. He is the current Research and Education Skills Director at the Saudi National Institute of Health. He also directs the Research and Development Department at the Saudi Commission for Health Specialties (SCFHS) within the Planning and Organizational Excellence Administration division of the Saudi Commission for Health Specialties, Riyadh, Saudi Arabia. Other affiliations related to his professional and educational practice include King Abdulaziz Medical City, Ministry of National Guard, Jeddah, Saudi Arabia and College of Medicine, King Saud Bin Abdulaziz University for Health Sciences, Ministry of National Guard, Jeddah, Saudi Arabia.

Contents

Editors
Miltiadis D. Lytras
Effat University
Saudi Arabia

Abdulrahman Housawi
Saudi Commission for Health Specialties
Saudi Arabia

Basim Alsaywid
Saudi Commission for Health Specialties
Saudi Arabia

Editorial Office
MDPI
St. Alban-Anlage 66
4052 Basel, Switzerland

This is a reprint of articles from the Special Issue published online in the open access journal *Sustainability* (ISSN 2071-1050) (available at: https://www.mdpi.com/journal/sustainability/special_issues/Training_Covid-19).

For citation purposes, cite each article independently as indicated on the article page online and as indicated below:

LastName, A.A.; LastName, B.B.; LastName, C.C. Article Title. *Journal Name* **Year**, *Volume Number*, Page Range.

ISBN 978-3-0365-2101-5 (Hbk)
ISBN 978-3-0365-2102-2 (PDF)

Training, Education and Research in COVID-19 Times: Innovative Methodological Approaches, Best Practices, and Case Studies

Editors

Miltiadis D. Lytras
Abdulrahman Housawi
Basim Alsaywid

MDPI • Basel • Beijing • Wuhan • Barcelona • Belgrade • Manchester • Tokyo • Cluj • Tianjin

Preface to "Training, Education and Research in COVID-19 Times: Innovative Methodological Approaches, Best Practices, and Case Studies"

Being a global pandemic, COVID-19 posed a major challenge in all aspects of life, including how graduate training of healthcare practitioners is conducted. In Saudi Arabia, there were over 14,000 graduate health professional trainees in different stages of their training in various specialties distributed in many healthcare facilities across the country. The vast geographical distribution and the diversity of health specialties training programs and activities have remarkably magnified the challenge posed by the pandemic. However, recently the SCFHS has implemented a health training governance reform that granted more autonomy to accredited training facilities in supervising training activities according to preset policies. This autonomy was crucial for mitigating various risks imposed by the pandemic, especially during the extended periods of strict lockdown.

The challenges faced during this period were numerous. However, several common themes constituted most of the challenges and warranted significant focus and deliberate attention to mitigate. First, trainees' safety and wellness were the primary concern for all those involved in the immediate supervision and oversight of training processes in all programs. During the early phase of the pandemic, uncertainty was a common denominator in a multitude of professional circumstances; training was not an exception. As a result of the uncertainty related to the extent of the disease virulence and spread early on, coupled with the variability in the preparedness of many health organizations for combating such a devastating public health catastrophe, hospital and training supervisors reacted quite variably. Trainees and training supervisors from different programs raised concerns about their state of health. Second, training continuity was severely threatened as many hospitals were overwhelmed with the increasing numbers of COVID-19 cases; some hospitals have been converted completely to accommodate the growing numbers of COVID-19 cases. As a result, many trainees were either forced to practice in training rotations that were not initially scheduled for, were transferred to other training locations, or were forced to withhold their training till further notice. The lucky ones whose training rotations continued faced various logistical difficulties and distractions from issues related to the pandemic. Third, as the pandemic progressed, training quality emerged as a central concern because of the limitation of oversight over training that resulted from training interruptions, unplanned training structural and content update, and redirecting resources to combat the pandemic and its consequences. Trainees in surgical specialties had their training interrupted due to the sensation or limitation in the elective and subacute surgeries during the pandemic.

These challenges called for immediate interventions to mitigate and control the risks posed by this devastating crisis. The response came from different levels of training governance entities at the Saudi commission, including its executive leadership scientific counselors and committees, and supervisor training bodies in training institutions around the country, all joining forces to address the crisis systematically and collaboratively. Their motto focused on a common goal of sailing together through this rampant storm to alleviate his impact and minimize future consequences on training outcomes. As a result, several initiatives were designed and implemented was successful immediate and short-term outcomes. In addition, plans were put together to monitor and address long-term outcomes related to disturbances inflicted by the pandemic on the healthcare system. Long term impacts of this crisis on training outcomes are the main challenge that is still uncertain and, therefore, unaddressed. Therefore, in the long run, coalitions of training leadership must continue to manage this challenge as an integral component of their activities. They also need to restructure the training system to continuously evaluate the pandemic's long-term effects and adjust the training system accordingly to achieve the best outcomes.

Lessons learned from this experience:

- The importance of planning, anticipating risks, and sitting proper controls minimize unpredictable crises;
- Navigating through a major crisis such as COVID-19 requires thoughtful, innovative, and compassionate leadership;
- Frequent and effective communication is invaluable to manage crises and promptly adapt as circumstances change properly;
- Strong data analytics infrastructure has been proven to be a critical component of successful crisis management plans.

Despite all the difficulties mentioned above and the challenges caused by this pandemic, there was a silver lining of opportunities for innovation and transformation in the graduate health training systems. A key example is the effective utilization of available and emerging technologies in training (e.g., webinars and teleconferencing). Undoubtedly, this accelerated adoption of technology has created an ample window of opportunity for the successful digital transformation of health training environments and practices.

Finally, navigating through the tough COVID-19 challenges was exhausting, but perseverance, hope, and foresight will eventually make us stronger and more agile.

Abdulrahman Housawi

Foreword: Integrating experiences, lessons learnt and know-how in Training, Education, and Research for Post-COVID-19 Times

As a global pandemic, COVID-19 has posed a major challenge in all aspects of life, including how graduate training of healthcare practitioners is conducted. In Saudi Arabia, over 14,000 graduate health professional trainees are in different stages of their training in various specialties distributed in many healthcare facilities across the country. The vast geographical distribution and the diversity of health specialty training programs and activities have remarkably magnified the challenge posed by the pandemic. Among other roles, the Saudi Commission for Health Specialties (SCFHS) is the country's regulatory organization responsible for governing health professionals' training and accreditation and graduate health training. Given its complex mandate and wide stakeholders base, the commission plays an integral role in the healthcare ecosystem.

Over the past three years, the SCFHS has implemented a health training governance reform that granted more autonomy to accredited training facilities in supervising training activities according to preset policies. This autonomy was crucial for mitigating various risks imposed by the pandemic, especially during the extended periods of strict lockdown.

The challenges faced during this period were numerous. However, several common themes constituted most of the challenges and warranted significant focus and deliberate attention to mitigate. First, trainees' safety and wellness were the primary concern for all those involved in the immediate supervision and oversight of training processes in all programs. During the early phase of the pandemic, uncertainty was a common denominator in a multitude of professional circumstances; training was not an exception. As a result of the uncertainty related to the extent of the disease virulence and spread early on, coupled with the variability in the preparedness of many health organizations for combating such a devastating public health catastrophe, hospital and training supervisors reacted quite variably. Trainees and training supervisors from different programs raised concerns about their state of health. Second, training continuity was severely threatened, as many hospitals were overwhelmed with the increasing numbers of COVID-19 cases; some hospitals have been converted completely to accommodate the growing numbers of COVID-19 cases. As a result, many trainees were either forced to practice in training rotations that were not initially scheduled for, were transferred to other training locations, or were forced to withhold their training until further notice. The lucky ones whose training rotations continued faced various logistical difficulties and distractions from issues related to the pandemic. Third, as the pandemic progressed, training quality emerged as a central concern because of the limitation of oversight over training that resulted from training interruptions, unplanned training structural and content update, and redirecting resources to combat the pandemic and its consequences. Trainees in surgical specialties had their training interrupted due to the cessation or limitation in the elective and subacute surgeries during the pandemic.

These challenges called for immediate interventions to mitigate and control the risks posed by this devastating crisis. The response came from different levels of training governance entities at the Saudi commission, including its executive leadership scientific counselors and committees, and supervisor training bodies in training institutions around the country, all joining forces to address the crisis systematically and collaboratively. Their motto focused on a common goal of sailing together through this rampant storm to alleviate its impact and to minimize future consequences on training outcomes. As a result, several initiatives were designed and implemented to successfully achieve immediate and short-term outcomes. In addition, plans were put together to monitor and address long-term outcomes related to disturbances inflicted by the pandemic on the healthcare system. Long-term impacts of this crisis on training outcomes represent the main challenge that is still uncertain and, therefore, unaddressed. Therefore, in the long run, coalitions of training leadership must continue to manage this challenge as an integral component of their activities. They also need to restructure the training system to continuously evaluate the pandemic's long-term effects and adjust the training system accordingly to achieve the best outcomes.

Lessons learned from this experience:

- The importance of planning, anticipating risks, and sitting proper controls minimize unpredictable crises;
- Navigating through a major crisis such as COVID-19 requires thoughtful, innovative, and compassionate leadership;
- Frequent and effective communication is invaluable to manage crises and promptly adapt as circumstances change properly;
- Strong data analytics infrastructure has been proven to be a critical component of successful crisis management plans.

Despite all the difficulties mentioned above and the challenges caused by this pandemic, there was a silver lining of opportunities for innovation and transformation in the graduate health training systems. A key example is the effective utilization of available and emerging technologies in training (e.g., webinars and teleconferencing). Undoubtedly, this accelerated adoption of technology has created an ample window of opportunity for the successful digital transformation of health training environments and practices.

Finally, navigating through the tough COVID-19 challenges was exhausting, but perseverance, hope, and foresight will eventually make us stronger and more agile.

Guest Editors

sustainability MDPI

Editorial

Training, Education, and Research in COVID-19 Times: Innovative Methodological Approaches, Best Practices, and Case Studies

Miltiadis Lytras [1,2,3,*], Basim Alsaywid [1] and Abdulrahman Housawi [1]

1 Saudi Commission for Healthcare Specialties, Riyadh 11614, Saudi Arabia; b.alsaywid@scfhs.org (B.A.); a.housawi@scfhs.org (A.H.)
2 Distinguished Scientists Program, Faculty of Computing and Information Technology, King Abdulaziz University, Jeddah 22332, Saudi Arabia
3 College of Engineering, Effat University, Jeddah 22332, Saudi Arabia
* Correspondence: mlytras@acg.edu

check for updates

Citation: Lytras, M.; Alsaywid, B.; Housawi, A. Training, Education, and Research in COVID-19 Times: Innovative Methodological Approaches, Best Practices, and Case Studies. *Sustainability* **2021**, *13*, 9650. https://doi.org/10.3390/su13179650

Received: 23 August 2021
Accepted: 25 August 2021
Published: 27 August 2021

Publisher's Note: MDPI stays neutral with regard to jurisdictional claims in published maps and institutional affiliations.

1. Editorial

The recent pandemic of COVID-19 challenges the delivery of training and education worldwide. In parallel, research in the times of COVID-19 has become more demanding in terms of research methodologies and the delivery of sound scientific contributions. In this Special Issue, we analyze the diversified training, education, and research context in the times of the COVID-19 pandemic.

The preparedness of the educational systems for a rapid shift in the delivery mode of the training was challenged by the diverse aspects of the COVID19 pandemic. Within this multifaced context new innovative methodologies, frameworks and approaches seem to provide a bold response to the global need for timely, high-quality education, training, and research on a remote and distant basis [1–6].

The widespread nature of the phenomenon proved to initiate global actions and local strategies. Our special issue communicates not only different approaches but also communicates the key lessons learnt. We believe that the know-how and the best practices cultivated during the COVID-19 period will lead future interventions in similar circumstances. From this point of view, a continuous quest of resilience and sustainability in training and education is a great gain from the pandemic. On the other hand, the psychological aspects and the impact on the behavior of trainees and trainers, as well as academics and researchers, need to be analyzed further. The short-, medium-, and long-term impact of COVID-19 on educational practices and delivery will require more time to be fully understood.

Our intention is to promote the scientific debate for the key implications of the modified online mode of delivery as well as the components of strategies and policies that enhance the social value and the social impact of training, education, and research. Our intended contribution is multifold:

- To promote recent sound research on effective training, education, and research strategies and methodologies.
- To contribute to the body of knowledge by promoting sound methodological approaches for effective training, education, and research in COVID-19 times.
- To communicate best practices and key lessons learned related to training, education, and research in the COVID-19 period.
- To invite relevant scientific debate from diverse communities from different domains of human activity including medical training and research, business training and education, information systems and computer sciences education, etc.
- To contribute to the discipline of training and education by synthesizing complementary approaches, limitations, and key findings.

Topics covered in the twelve accepted papers of our Special Issue include but are not limited to the following:

- Training strategies in the COVID-19 period.
- Effective technology for enhanced learning interventions in the COVID-19 period.
- Transforming education from an off-line to an on-line paradigm in the COVID-19 period.
- Research excellence and strategies in the COVID-19 period.
- Enhancing the skills, competencies, and knowledge of trainees in the COVID-19 period.
- Engagement, interaction, and active training and learning in the COVID-19 period.
- Higher education best practices for delivering top-quality education in the COVID-19 period.
- Case studies in medical training and research.
- Business training and education.
- Information systems and computer sciences education.
- Strategic frameworks for training, education, and research efficiency in COVID-19 times.
- Key performance indicators and learning analytics for measuring the effectiveness of online training in the times of COVID-19.
- Integration of academia and industry for training, education, and research.
- Social networks research for enhanced training, education, and research in COVID-19 times.
- Quality initiatives for effective training, education, and research in COVID-19 times.

Some of the key conclusions of our special issue debate are summarized as follows:

- COVID-19 challenged all the educational systems worldwide and also forced a radical adoption of the digital transformation of education.
- Several lessons learnt from the adopted educational process emphasize in a bold way the need for collaboration among diverse stakeholders in the training, education, policy making, and research communities.
- The various technological means and the new generation of learning management tools and online lectures prove their capacity to support, in the long term and in the post-COVID era, a new era of blended learning in academia [7–9]. This trend will require additional research related to the equivalent of online training and the capacity of faculty to support this multidimensional role of the online tutor.
- COVID-19 proves the impact of diverse psychological factors on the efficiency of training, education, and research. From this point of view, additional research needs to focus on the design of sustainable counseling and psychological support of trainees and trainers during times of crisis.
- In terms of policy making and possible adjustment at the strategic planning of educational, training, and research management in times of crisis, it is evident that new flexible procedures for time management, significant rewards and incentives for skills and competencies management, and allocation of required resources to the various stakeholder are a few of the measures that need to be considered.
- One of the most important challenges in the post-COVID time is to capitalize on the know-how gained and to design sustainable policies, processes, and socio-technical systems capable of supporting education, training, and research in crisis times of the future.
- Last but not least, the research production about the phenomenon of COVID-19 should initiate a scientific debate on how local contributions can be integrated with similar ones in the global context in order to promote the idea of a global scientific collective action.

2. List of Contributions

A more focused overview of the collection of articles published in our special issue is summarized below:

1. Factors Influencing Students' Behavior and Attitude towards Online Education during COVID-19
2. Residents' Training in COVID-19 Pandemic Times: An Integrated Survey of Educational Process, Institutional Support, Anxiety and Depression
3. A Progressive Model for Quality Benchmarks of Trainees' Satisfaction in Medical Education: Towards Strategic Enhancement of Residency Training Programs at Saudi Commission for Health Specialties (SCFHS)
4. University Students' Perception, Evaluation, and Spaces of Distance Learning during the COVID-19 Pandemic in Austria: What Can We Learn for Post-Pandemic Educational Futures?
5. Multilevel Antecedents of Organizational Speed: The Exemplary Case of a Small Italian R&D Organization
6. Impact of COVID-19 on the Educational Process in Saudi Arabia: A Technology–Organization–Environment Framework
7. Factors Influencing the Adoption of COVID-19 Preventive Behaviors in Chile
8. The Effectiveness of Online Education during COVID-19 Pandemic—A Comparative Analysis between the Perceptions of Academic Students and High School Students from Romania
9. Conducting Population Health Research during the COVID-19 Pandemic: Impacts and Recommendations
10. The Critical Factors of Student Performance in MOOCs for Sustainable Education: A Case of Chinese Universities
11. Cyber-Physical System of Psychophysiological Support of Professional Self-Realization in Professions of the 'Man-Nature' Type in the Formation of Specialists for Sustainable Development
12. Building on Strategic eLearning Initiatives of Hybrid Graduate Education a Case Study Approach: MHEI-ME Erasmus + Project

Conflicts of Interest: The authors declare no conflict of interest.

References

1. Daniela, L.; Visvizi, A.; Gutiérrez-Braojos, C.; Lytras, M.D. Sustainable Higher Education and Technology-Enhanced Learning (TEL). *Sustainability* **2018**, *10*, 3883. [CrossRef]
2. Visvizi, A.; Daniela, L.; Chen, C.W. Beyond the ICT- and sustainability hypes: A case for quality education. *Comput. Hum. Behav.* **2020**, *107*, 106304. [CrossRef]
3. Zhuhadar, L.; Yang, R.; Lytras, M.D. The impact of Social Multimedia Systems on cyberlearners. *Comput. Hum. Behav.* **2013**, *29*, 378–385. [CrossRef]
4. Lytras, M.D.; Mathkour, H.I.; Abdalla, H.; Al-Halabi, W.; Yanez-Marquez, C.; Siqueira, S.W.M. An emerging social- and emerging computing-enabled philosophical paradigm for collaborative learning systems: Toward high effective next generation learning systems for the knowledge society. *Comput. Hum. Behav.* **2015**, *5*, 557–561. [CrossRef]
5. Sicilia, M.A.; Lytras, M. On the representation of change according to different ontologies of learning. *Int. J. Learn. Chang.* **2005**, *1*, 66–79. [CrossRef]
6. Vargas-Vera, M.; Lytras, M. Exploiting semantic web and ontologies for personalised learning services: Towards semantic web-enabled learning portals for real learning experiences. *Int. J. Knowl. Learn.* **2008**, *4*, 1–17. [CrossRef]
7. Visvizi, A.; Lytras, M.D. Editorial: Policy Making for Smart Cities: Innovation and Social Inclusive Economic Growth for Sustainability. *J. Sci. Technol. Policy Mak.* **2018**, *9*, 1–10.
8. Naeve, A.; Yli-Luoma, P.; Kravcik, M.; Lytras, M.D. A modelling approach to study learning processes with a focus on knowledge creation. *Int. J. Technol. Enhanc. Learn.* **2018**, *1*, 1–34. [CrossRef]
9. Vossen, G.; Lytras, M.; Koudas, N. Revisiting the (machine) Semantic Web: The missing layers for the human Semantic Web. *IEEE Trans. Knowl. Data Eng.* **2007**, *19*, 145. [CrossRef]

sustainability

MDPI

Article

Factors Influencing Students' Behavior and Attitude towards Online Education during COVID-19

Gratiela Dana Boca

Department of Economics and Physics, Faculty of Sciences, Technical University Cluj-Napoca, 430122 Baia-Mare, Romania; bocagratiela@cunbm.utcluj.ro

Abstract: Universities around the world have faced a new pandemic, forcing the closure of campuses that are now conducting educational activities on online platforms. The paper presents a survey about students behavior and attitudes towards online education in the pandemic period from the Technical University of Cluj Napoca, Romania. A group of 300 students participated. The questionnaire was structured in four parts to determine student's individual characteristics, student's needs, students' knowledge in using virtual platforms and students' quality preferences for online education. The students said that online education in a pandemic situation is beneficial for 78% of them. A total of 41.7% percent of students appreciated the teachers' teaching skills and the quality of online courses since the beginning of the pandemic, and 18.7% percent of the students appreciated the additional online materials for study to support their education. However, students found online education stressful, but preferred online assessment for evaluation. This pandemic has led to the new stage of Education 4.0, online education, and the need to harmonize methods of education with the requirements of new generations.

Keywords: digital education; management change; student behavior; student attitude

Citation: Boca, G.D. Factors Influencing Students' Behavior and Attitude towards Online Education during COVID-19. *Sustainability* **2021**, *13*, 7469. https://doi.org/10.3390/su13137469

Academic Editors: Miltiadis D. Lytras, Abdulrahman Housawi and Basim Alsaywid

Received: 18 May 2021
Accepted: 30 June 2021
Published: 4 July 2021

1. Introduction

The concept of education has changed dramatically over the last few years, with many questions being raised as to what the best mode of instruction is with the advent of technology and the Internet. The waves of the evolution of education in history begin in the 1780s, with the first wave concerning the individual context of learning and memorization, known as Education 1. The second wave of mass learning appears around the 1900s, known as Education 2. The Internet that allows learning known as Education 3, begin from the 1970s, and has the addition of computers, but only as an interface with students which produces knowledge. Distance learning was first introduced in the 18th century in parallel with the postal service, but it did not pick up steam until communications technology evolved in the 1990s [1]. If we look in time at the stages of the evolution of education, we can see that from a traditional system that focused on books and teaching on the blackboard, over time the use of technology induced a new stage known as Education 4.0, when the computer and the Internet changed the concept of education and the new digital generation offered more possibilities for education.

The future belongs to Education 4.0, as a part of the evolution of education but with a very high impact of digital technology. Empowering education to improve innovation, the transition to the new stage requires the development and harmonization of education systems by employing the new relationship that must be established: student-teacher-technology = smart education and the use of e-education (online, electronic tools). Zhu et al. [2] are supporters of smart education for an environment in which students work as close as possible to reality, which is the reason the education system must combine reality with the virtual world. Zhu et al. [2] and Hartono et al. [3], set out the needs for hybrid education and the term smart learning for students to adapt education to the digital age.

The pandemic contributed to the faster transition to towards the new stage of Education 4.0. Under the imposed conditions, the use of online education was the tool to save and implement digitalization as a beneficial alternative. Education has changed, and online learning is the next big transformation, as Frecker and Bieniarz [1] suggest. One of these advantages is the diversity of educational possibilities and the multiple ways of placing the content. There is also great diversity in terms of assessment, with the teacher having the opportunity to place continuous or summative assessment tests. An assessment by Bond and Lockee [4] and Jackson [5] identifies online needs and considers online professional development courses mandatory in online teaching for higher education faculties. In terms of the impact upon students, it was beneficial to take into consideration their interests and the desire for teachers to post more online courses for the future (Elzainy et al. [6]). Since 2016, a vision for higher education has been designed [7]: a concept put forward by McGee et al. [8], who stress that the training of instructors for online teaching and the preparation of online teaching faculties must also be taken into account. The vision went further, and Rhode and Krishnamurthi [9] developed the concept of self paced training for academic person. During the pandemic, Iwai [10] researched the effects of virtual classroom learning through adaptive learning and virtual reality by use of technology, considering the satisfaction and improvement of skills of the staff and students.

If we take a look in recent time to the evolution concept of online education, we can identify different shapes in the authors' visions, as presented in Table 1.

Table 1. Authors different shapes of online education.

Author	Year	Shape
Dandara, O. [11]	2013	electronic platforms, a means of modernizing educational technologies
Allen and Seaman [12]	2017	digital learning compass: distance education
Espiritu and Budhrani [13]	2019	showed the importance of e-learning as a culture
Dhawan, S [14]	2020	a panacea in the time of COVID-19 crisis
Jæger and Blaabæk [15]	2020	importance of library because of inequality in learning opportunities during COVID-19

Following the European Commission's public consultation [16] in 2020 to develop an action plan for the digitalization of education at EU level, it has been shown that there are still students who have not used distance and online learning tools, but who before the crisis were willing to improve their digital skills.

The idea of improvement and specialization in the field of online education was brought up in the paper by Schmidt et al. [17], and also involvement in blended learning. According to recent studies of Mishra et al. [18] and Bojovic et al. [19], the effects of education during the COVID-19 crisis led to the need for a rapid transition to distance learning. Moskal et al. [20], Moore et al. [21], and Mohr and Shelton [22] established that online education also presents good opportunities for good practices that are necessary for professional development. Based on the action plan of the European Commission for the action of digital education and teaching in higher education [16], the theme of online courses was addressed by Baran and Correia [23] and Baran et al. [24]. At the same time, Espiritu and Budhrani [13] showed the importance of e-learning as a culture. Brinthapt et al. [25] and Elliot et al. [26] considered online learning important for the professional development of staff, following a strategy in [7,16]. The sustainability of online learning is an opportunity, because it is a flexible option that allows the development and improvement of skills.

Along with current digital tools, wider access to the Internet offers different people the opportunity to access higher education. In addition, we can also reduce the carbon footprint by reducing travel and provide a more personalized learning.

2. Literature Review

2.1. Online Learning and Education

The use of the Internet and state-of-the-art technologies to obtain information for fast communication has become extremely important in the communication and promotion strategy of any university [27]. Communication in the university environment is one of the basic elements on which the student-teacher-university relationship is built. The motivation to approach the communication made by universities starts from the premise that most of the times the students' performances in the learning process and in the integration of the university environment are determined by the way in which the information is made by the universities. Moreover, the COVID-19 pandemic has demonstrated the usefulness of these platforms, as more and more schools move to the red scenario, which means that virtually the entire educational process moves to the online system on educational teaching and learning platforms.

Electronic platforms have a number of advantages over traditional teaching (Elzainy et al. [6]), exploring the impact of e-learning and assessments on students and having observed important changes in improving student's technological skills during the pandemic period. Martin et al. [27] noted the use of traditional assessments to assess the students and course templates, and the processes of quality assurance and surveys, learning analyses and intermediate assessments. Timely response and feedback, availability and regular presence and communication were some of the facilitation strategies used by the award-winning instructors.

The use of educational platforms has allowed finding solutions in the imposed situation and innovating teaching methods and tools in various fields such as geography by Cazacu [28], and medicine by Chatterjee and Chakraborty [29] and Elzainv et al. [6]. Additionally, the use of information and communication with the help of technology has been useful in the medical field, as noted by Grishchenko [30] and Hasan and Bao [31].

The quality of the platform used in the educational process has a favorable effect on the performance of students in online education (Ionescu et al. [32]). According to [33], we can consider that in 2020 the sustainability of online learning offers professionals a flexible option in accordance with their schedule, contact with university staff and platforms for advice and information. Others, such as Singh et al. [34] present the importance of platforms in education but in the same time Diaz and Walsh [35] became advocates for telesimulation based education during COVID-19.

Becker et al. [36] sustain that electronic platforms allow the storage and management of an unlimited number of courses, as well as the storage and management of an unlimited volume of content within a course.

The use of online educational platforms has become a necessity and has spread rapidly since 2020, being the only tools that could be used during the break for online teaching. The influence of smart learning was presented by Budharani et al. [37], and Bojovic et al. [19], like an education in times of crisis: rapid transition to distance learning. However, the use of these educational platforms also has disadvantages, among which we mention: it requires experience in the field of computer use, both by teachers and students, and involves high design and maintenance costs.

The environment in which the students carry out their activity must be as close as possible to reality, which is why it is between the real and the virtual world, and why the educational system must combine reality and the virtual world. Zhu et al. [1]. Martin et and Bolliger [38] agree with the term smart learning to adapt to the digital age Hartono et al. [3], establish the needs for the smart hybrid education.

The COVID-19 crisis has brought to light digital inequalities among students, which is a major risk factor for social vulnerability. Additionally, the inequalities were identified in the research by Beaunoyer et al. [39], because not all students have the same social conditions or lifestyle, and not all have access to the internet or have high-performance digital equipment or have the necessary skills.

Following the study by Jæger and Blaabæk [13], we notice that another inequality is the learning opportunity during and after COVID-19. They highlighted the importance of the library in the learning process, because the backgrounds of students and families are different, as are education and income of the parents.

Based on the action plan proposed by the European Commission for digital education for the period 2021–2027 [16], Zhu and Liu [40], and Hasan and Bao [31] mention that in the pandemic context the innovation in the educational field allowed the identification of the niche elements in digital and post digital education.

The European pandemic COVID-19 has led nationally to the development and taking of rapid and effective measures that have caused significant disruption to education systems, training for both students and teachers but also educators, who at the same time had to adapt to online courses, as Ursan et al. [41] observed.

2.2. Online Education and Teachers

Universities and teachers were not completely taken aback by online courses and activities, and Windes and Lesht [42] highlighted the effects of the online courses and their impact on education.

There are currently few studies on the effectiveness of online courses, the teacher-student relationship, and the effectiveness of online assessments. Among those who approached the new topic were Chakraborty et al. [43] and Aguilera-Hermida [44], who noticed that students believe that online education helped them to continue their training and studies during the pandemic with digital platforms, but at the same time to have access to faculty libraries.

Online education for teachers requires time to identify and build the platforms and materials needed, according to Hodges et al. [45]. Bojovic et al. [19] and Chakraborty et al. [43] noted that teachers still lack confidence on online assessment techniques. Aguilera- Hermida [44], argues that teachers' experience can also be closely related with the students' learning experiences. In his opinion, Chakraborty et al. [43] students prefer face-to-face interaction with teachers because teachers do not trust online assessment techniques.

Si et al. [46] stressed the importance of online teaching skills of teachers but also of students who were not prepared for online courses. Teaching platforms have a number of advantages, including real-time access to education but also to the resources placed on these platforms, but also a number of obvious disadvantages, such as the necessary experience and the appropriate means that can involve considerable costs. Lundsford et al. [47], Martin and Bolliger [38] in their research establish the connection between students and professor and the importance of adaptation of methods and strategies in the online learning environment, and also student's involvement [48].

At the same time, the teachers were unprepared for the online activities, and the students also felt unprepared. However, the online platforms allowed at the same time the monitoring of the activities, by visualizing the frequency of the entries and establishing the result of the activities, allowing evaluations on the course but also the efficient final evaluation. The influence of the pandemic after Carroll and Conboy [49], led to new practices that emerged under the pressure of the pandemic "big bang" introduction of technology and "tech driven" practices. Based on the interviews, Martin et al. [27] found that online instructors create the reverse method through a design taking into account the needs of students.

Polly et al. [50] examined the barriers in the use of digital technologies and the necessary support for academics staff. The barriers identified were the time required to learn new technologies and the time required to learn how to use them in the teaching process. Another factor would be the conflict between the focus on teaching and other service responsibilities, including research. The entire teaching and learning methodology must be transferred online, requiring a systematic reorganization of the learning process through the computer.

The pandemic period provided the opportunity for universities to identify the optimal solutions and to adapt the educational act by opting for the best solution in the given situation. The educational act can be influenced by other factors, such as the communication between students and teachers, identified by Coman et al. [51] or the emotions in the use of digitalization and twitter addiction during the blockade imposed by COVID-19, not only in India, as pointed out by Arora et al. [52].

Perception depending on the field of specialization and the student-teacher relationship were addressed by Trammell and LaForge [53] in their paper, but also the continuous changes necessary for the faculty in the future development of faculties was noted by Stark and Smith [54]. Sheffield et al. [55] instead explored teachers' competencies and attitudes regarding online courses and adapting to students' needs, also Schmidt et al. [17], Trust and Krutka [56] and Rhode et al. [57] identifying what is needed to improve teaching activities and their personalized adaptation.

Other factors were analyzed by Ionescu et al. [32] from three perspectives—teachers, students and parents—which led to the identification of possible psychological effects on students, resulting from the corroboration of social isolation with the online continuation of the educational process.

Electronic platforms have a number of advantages over traditional teaching. One of these advantages is the diversity of educational possibilities and the multiple ways of placing the content. There is also great diversity in terms of assessment, with the teacher having the opportunity to place continuous or summative assessment tests. On assessments used in faculties, Martin et al. [27,38] noted the use of traditional assessments to assess students, course templates and the process of quality assurance and surveys, learning analyzes and intermediate assessments. Timely response and feedback, availability and regular presence and communication were some of the facilitation strategies used by the award-winning instructors. Aguilera-Hermida [44] pointed out that at the same time not only the faculty, but also the students and administrators faced unexpected situations with repercussions in the teaching and learning processes. Some online educational tools facilitate teacher-student collaborative learning as argued by Chakraborty et al. [43] and Adhikary et al. [58]. Important issues are the lack of access to a printer and consumables needed to print worksheets or materials received, children being left alone at home; stress generated by the risk of illness and unemployment.

We can observe that COVID-19 in fact give a push for the acceleration of digitalization in universities all over the world, with researchers trying to identify and diagnose the new provocation, such as Arora et al. [52] in India, Ionescu et al. [32] in Romania, Bojovic et al. [19] in Serbia, Skulmoski and Rey [59] in Germany, as well in the Philippines by Toquero and Talidong [60], in China by Zhang et al. [61] and Islam et al. [62] in Bangadesh.

2.3. Students Behavior and Attitude toward Online Education

Since 1986 when the first Technology Acceptance Model (TAM) appeared to identify the factors that affect students behavior and intention to use technology, in time the model was improved and new factors and a more complex investigation was developed, such as in Table 2.

However, today's generations of students were born and started school in the age of the Internet and online browsers, the Google search engine and social media platforms. The digital world is part of the lives of young students, from their first years of life, and online education, through digital applications, is a language of learning that they have been using for a long time.

The means of information used in the university environment are in the process of reconfiguration and development. In the near future, the means of access to information transmitted by universities will be multimedia, mobile and miniaturized.

Table 2. Factors influencing student's behavior to use technology.

Authors	Factors	Results
Davis [63] 1993	"Perceived Usefulness" and "Perceived Ease of Use" are the two key factors that affect an individual's intention to use a technology	to investigate the impact of technology on user behavior
Liu et al. [64] 2010	namely Online Course Design, User-interface Design, Previous Online Learning Experience, and Perceived Interaction	to investigate the impact of technology on user behavior
Al Kurdi et al. [65] 2020	involved E-learning Computer Self-Efficacy, Social Influence, Enjoyment, System Interactivity, Computer Anxiety, Technical support, Perceived Usefulness, Perceived Ease of Use, Attitude, and Behavioral Intention to Use e-learning	a suitable theoretical tool to comprehend the acceptance of e-learning by users
Al Kurdi et al. [66] 2020	"social influence, perceived enjoyment, self-efficacy perceived usefulness, and perceived ease of use" are the strongest and most important predictors in the a virtual E-learning atmosphere intention of and students towards E-learning systems	to improve ongoing interests and activities of university students in a virtual E-learning atmosphere
Mailizar et al. [67] 2020	The model consists of six constructs: system quality, e-learning experience, perceived ease of use, perceived usefulness, attitude toward use, and behavioral intention.	to improve the understanding of students' intention to adopt e-learning.

More and more universities are adopting coherent strategies for integrating technology into the educational process and the media used in both internal and external communication tend to migrate to online. Student's opinion regarding the digital learning and the impact in their daily life was investigated by Martin et al. [27], Bao [68] and Chakraborty et al. [43], but also, they take in consideration the student's opinion about online education in the pandemic period.

The transition to online education and students' intention to use online education was a challenge during the pandemic, and the studies provided information that will underpin future strategies for developing education and improving the quality of online education use and involvement of both actors involved in the system—students and teachers.

There is a wide range of factors that have been taken into account in order to identify student behavior and attitudes, as shown in Table 3.

Table 3. Online education factors influencing the student's behavior and attitude.

Authors	Year	Factors
Lee, J.W. [69]	2010	online support service quality, online learning acceptance, and student satisfaction
Hung and Jeng [70]	2013	age, online teaching experience, implications of the findings were discussed the lecturer's competence, the lecturer's attitude towards learning, and the nature of the subject knowledge's attitude and practices,
Hatabu et al. [71]	2020	the frequency and activities of information acquisition, the correct explanation of the information and willingness to collect anxiety information
Adil Zia [72]	2020	attitude, curriculum, motivation and technology training
Alzahrani et al. [73]	2021	service quality, information quality and self-efficacy, satisfaction
Yunus et al. [74]	2021	the effort expectation, the performance expectation, social influence and facilitating conditions in using the online education

Studies on e-learning and the impact on students were conducted recently in 2020 by Bao [68], Islam et al. [62], Essadek and Rabeyron [75] and Paea et al. [76] showing that the new impact of online education among students has fostered depression and anxiety in the pandemic period. Cao et al. [77], suggested that the mental health of college students should be monitored during epidemics because of the pressure and stress. Mishra et al. [18] found that due to the limiting of travel, i.e., the academic exchange programs of students and staff between universities, there was also a deterioration of academic research actions and activities in education.

Students' behavior and attitudes toward online education and the use of digital platforms during the pandemic have led many researchers to conduct new studies to identify the new environment and these factors.

3. Materials and Methods

The pandemic did not find the university and teachers totally unprepared, because a platform (edu.utcluj.ro) was created long before for the students from the low frequency system, where the courses, seminar materials and other information necessary for those who work and attend the faculty were posted. The platform was also for full-time students to be able to send homework proposed by teachers or for teacher-student information.

During the pandemic, the platform was improved, and another one, the Knowledge Base (kb.cunbm.utcluj.ro) platform was created, allowing the students from the economic specialization to access them and to build the teacher-student bridge.

In addition to the digital platform for seminars, homework, additional materials, systematic or periodic evaluation, some teachers have chosen other teaching tools, such as ZOOM or Microsoft Team, for a more attentive and beneficial communication. In this way digital technology has brought a plus to the educational act.

The transition from the traditional classroom teaching system to online education, keeping to the schedule, was made gradually at the beginning, and the courses were held online and with face-to-face seminars, but with the lockdown during the pandemic everything moved to the total online system. The academic staff put in practice Hrastinski [78] and Flora Amiti [79] suggestions, by keeping the teaching systems but using actual modern digital tools, all three modes of online learning: asynchronous, synchronous and hybrid.

The questionnaire was applied between October–November 2020 when the pandemic restriction allowed students' to participate in specific situations in activities at the university. The sample of 300 respondents consisted of graduate students in the final grades and master students from the department of economics, because they could compare the two methods of education before and during the pandemic. The items were established to determine the factors which are influencing the students behavior and attitudes concerning online education.

The study wanted to identify students' behavior and their attitudes in the new context of online education. The structure of the questionnaire and factors influencing their behavior are presented in Table 4.

In order to obtain the necessary data, the target group was selected on a voluntary basis, and the period was targeted respectively when students chose their place, the period of the internship or the coordinators of the license or dissertation.

All students voluntarily consented and confirmed their participation in the study after being previously informed of the purpose of the study.

The study allowed the analysis of the behavior and association with student's attitude towards online education.

The first part of questionnaire tried to identify the socio-demographic characteristics of the respondents, three questions were used regarding gender, age and education level. The education level includes two categories: bachelor and master students. The age groups were established between 18–26 years old, 26–32 age, 32–38 age and 38–42 age.

The second part of the questionnaires were focused on students behavior related to their needs: time spend using online education tools. Additionally, the questionnaire establishes and identifies the frequency of hours of consumption of virtual tools used in education, which was measured by asking respondents: "How often do you enter weekly online?". For the response categories for their time spent online, read Supplementary Materials (1 h, 10 h, more than 10 h) and the emphasis they put on the quality of courses and how this affects students behavior.

Table 4. Questionnaire structure.

	Questions	Items	Factor
1	Age	I1	Individual Characteristic
2	Gender	I2	
3	Education level	I3	
4	How many hours are you spending weekly online?	H	Needs
5	Do you find digital learning and platforms useful?	U	
6	How many hours do you devote to individual study?	S	
14	How often you enter online	F	
8	What kind of examination do you prefer (online)	E1	Knowledge
9	What kind of examination do you prefer (writing)	E2	
10	What kind of examination do you prefer (test)	E3	
11	What kind of examination do you prefer synthesis	E4	
12	What kind of examination do you prefer portfolio	E5	
7	Do you read the specialty materials?	R	Quality
13	How you appreciate the online courses quality	Q	
15	How do you consider the learning activity	QE1	

In the third section of the questionnaire, students' knowledge was established, to determine students culture on virtual media, their abilities to use the modern tools, the benefits of education online, how important for students their education is, and their orientation between the traditional and new types of education. The answer categories for the open question about type of examination preferred by students offered the respondents the possibility of stating their favorite one. The key element was the item regarding the final evaluation of their activity, respectively the way of conducting the final evaluation exam. They were able to choose from several types of assessment so as to see which type of assessment is considered the most optimal assessment of their knowledge. Regarding the students preferences type of examination, it was possible to choose between online, writing, multiple choice test, syntheses and portfolio. A Likert scale was used for the study, starting from a score of 1 representing the student disagreement—"Not at all" and up to a score of 5—"Very well", representing the student's strong agreement.

The last part of the questionnaire included three questions that identify students satisfaction about the quality of online courses, the quality of learning activity on the pandemic period and the quality of specialty materials. From research the article wanted to obtain and know students interaction with the professor through an online platform quality with the item "*How you appreciate the online courses quality*", how useful online education iswas identified with the item "*Do you find digital learning and teaching platforms useful?*"

3.1. Sample and Measurement Tool

To determine the factors that influence students' behavior and attitudes during the pandemic due to the use of online education, a statistical analysis was performed using SPSS software applied to the given database, as well as the independent simple t tests and hypothesis testing. The proposed model and the correlations between the items of the questionnaire were made using the solutions of the Lisrel 8.7 program.

3.2. Purpose of the Study

The study presents the following four factors that influence students behavior that were taken into consideration:

1. Student's individual characteristics (age, gender, education level);

2. Student's needs: frequency of entering the online platform (F), hours spent on the virtual platform (H), hours to study and learn from materials proposed by teachers (S), and how useful are digital platforms for their needs and for their better understanding (U);
3. Student's knowledge regarding the type of knowledge, evaluation portfolio, syntheses, test, written examination and online examination (E1, E2, E3, E4 and E5);
4. Student's perception about quality of online education: courses quality (Q), education and learning quality (QE1) and quality of materials presented and information (R).

The research model from Figure 1 is based on the research objectives and hypothesis.

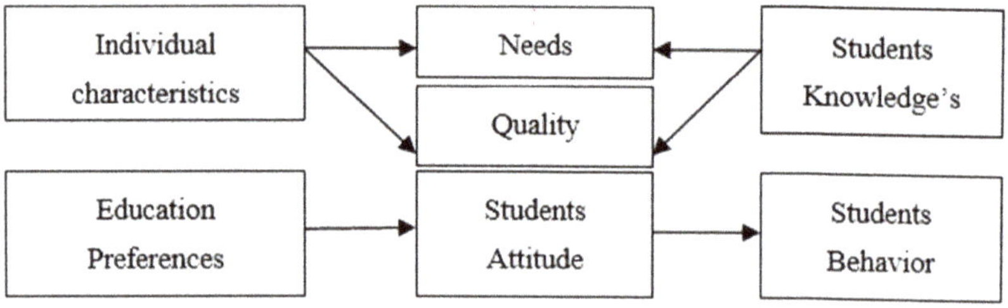

Figure 1. Research model to examine the students behavior and attitude. Source: by author.

The hypotheses tested on the attitude of students in the present study are:

Hypothesis 1 (H1). *Students preferences have a significant effect on their attitudes towards their online education.*
Hypothesis 2 (H2). *Students preferences have a significant effect on their attitudes towards needs.*
Hypothesis 3 (H3). *Students preferences have a significant effect on their attitudes towards knowledge.*
Hypothesis 4 (H4). *Attitudes of students concerning their needs of evaluation has a significant effect on their behavior.*
Hypothesis 5 (H5). *Attitudes of students towards online education have a significant effect on their behavior.*

Taking into consideration Saraçli et al. [80] and Gumus et al. [81], the Cronbach's alpha were calculated as a quality instrument to analyze the items. The total Cronbach's alpha value of scale was calculated as 0.72 which is statistically one of the indicators that shows that the reliability of the scale.

With α value in interval $0.8 \leq \alpha \geq 0.7$ interval following the statistical rule [82,83] we can consider acceptable and good enough the model. After reliability analysis, exploratory factor analysis (EFA) was applied over the four factors: Individual characteristics of students, Needs, Knowledge and Quality.

The root mean square error (RMEA) of approximation value was 0.63: normed fit index (NFI) was 0.93, non-normed fit index (NNFI) was 0.96, comparative fit index (CFI) was 0.90, goodness of fit index (GFI) was 0.90 and adjusted goodness of fit index (AGFI) was equal to 0.88. If we compare the goodness of fit, the model is in acceptable fitness.

Using the Schermelleh-Engel and Moosbrugger structural model [84], the next step was to compare with the program solution, and for the structural equation model we obtained the values shown in Table 5.

Table 5. Structural equation model fit.

Criteria	Acceptable Fitness	Model
RMSEA	$0.05 \leq \text{RMSEA} \leq 0.10$	0.63
NFI	$0.90 \leq \text{NFI} \leq 0.95$	0.93
NNFI	$0.95 \leq \text{NNFI} \leq 0.97$	0.96
CFI	$0.95 \leq \text{CFI} \leq 0.97$	0.90
GFI	$0.90 \leq \text{GFI} \leq 0.95$	0.90
AGFI	$0.85 \leq \text{AGFI} \leq 0.90$	0.88

Source: adaptation after Schermelleh-Engel and Moosbrugger [84].

4. Results

Based on the data obtained by applying the questionnaires following the data analysis, we can draw the following results regarding the factors that influence the behavior and attitude of students towards online courses. For the first part of questionnaire regarding the students characteristic in an equal percentage there are from master's and bachelor's degree. The target group was made up of 56.66% females and 43.34% males, aged between (18–26) (66.33%), ages (26–32) (17%), ages (32–38) (11.34%) and ages (38–42) (5.33%) (Table 6).

Table 6. Distribution according to respondents education by age and gender.

		Age				Cumulative Percent
		18–26	26–32	32–38	38–42	
Gender	female	29	13.33	9	5.33	56.66
	male	37.33	3.67	2.34	0	43.34
Total		66.33	17	11.34	5.33	100
Education license master		22.33	13.67	8.67	5.33	50
		44	3.33	2.67	0	50
Total		66.33	17	11.34	5.33	100

Regarding students attitude about online education, 43.3% of respondents consider that being face-to-face and physical presence in amphitheaters is beneficial, and only 33% of students considered that online education is better. A total of 23% of respondents considered that online education can be an annexes or complementary education together with face-to-face lectures to improve education.

At the same time 41.7% of students believe that the pandemic situation has led to improved teachers' skills through new teaching methods and techniques adapted to the online environment since the beginning of the pandemic, and 34.7% of respondents consider satisfactory the new tools used by professors in their online lectures. However, 23.7% of students consider online courses, seminars and evaluation as the only beneficial solution in the pandemic situation created.

Regarding the items about the reading materials from platforms and online education tools used by professors, such as slideshows, notes, problem solving, student's perception is positive at 63.3%, feeling that the volume of information is adequate and is available online.

Another barometer that come to support the students' behavior is regarding the attention paid to the additional materials, we found that professors use different platforms such as ZOOM, Knowledge Base, together with classical tools such as Power-Point and Prezi to follow, which make sessions more interactive, as appreciated by the students. At the same time, some courses of specialization require supporting quality interaction (Singh et al. [34]).

Table 7 presents that 58.66% of respondents spend weekly between 1–10 h on the platforms, and 24.7% more than 20 h.

Table 7. Time spend by students online.

	Female	Male	Cumulative Percent
How many hours are you spending weekly online			
Non	14	36	16.67
1–10 h	98	78	58.66
10–20 h	58	16	24.67
Total	170	130	100

As a result, 49.3% percent of students read additional materials proposed by teachers if the information provided through online courses is sufficient, which shows that students are aware of the situation and the new context. A total of 16.7% percent of students do not actually participate in online courses when they are scheduled, due to the fact that they either work or cannot access the Internet with connection problems, but have access to the information on the platforms appreciating this in online courses. A total of 56.66% of female respondents are more active on platforms in comparison with males at 43.33%, maybe because they are more curious and conscientious.

Factors Influencing Students Behavior

In Table 8, we can see the correlation between respondent's age and their needs and preferences regarding the type of evaluation.

Table 8. Correlation between respondents age and examination needs.

	Age				Cumulative Percent
	18–26	26–32	32–38	38–42	
What kind of examination do you prefer					
Portfolio	19	5	2	0	8.67
Synthesis	17	4	3	1	8.33
Test (multiple choice)	71	20	15	9	38.33
Written exam	17	1	2	3	7.67
On-line exam	65	20	10	2	32.33
Total	189	50	32	15	95.33

The students prefer the grid test in a percentage of 38.33%, online assessment in a percentage of 32.33%, and portfolios in a percentage of 8.67%. We can conclude that the student's behavior was influenced by the pandemic isolation and also the communication, they prefer the short answer, i.e., fast communication which does not involve too much of their participation.

The results present a young generation accustomed to the Internet and who prefer online examination for 28.3% and verification in the form of tests for 30.33%. For a generation known as native to the digital life, which grow up with new technology the Internet, it is part of their everyday life, so the impact of online education was something normal. For the mature generation, the new technology is a problem because they do not have the necessary skills to use the latest generation technology.

From the gender point of view, the results from Table 9 show a highest percent of feminine preferences for online examination at 17% and a preference for multiple choice tests evaluation at 23.33%.

For male respondents, the preferences are 15.33% for online and multiple choice tests and 4% for synthesis and written exam for evaluation. We can conclude that students attitude towards digital platforms and online education influence their behavior in the function of individual characteristics of age, gender and their abilities and skills with the new technology generation.

Table 9. Correlation between respondent's gender and kind of examination.

	Gender		Cumulative Percent
	Female	Male	
What kind of examination do you prefer			
On-line exam	51	46	32.33
Written exam	11	12	7.67
Test (multiple choice)	70	45	38.33
Synthesis	13	12	8.33
Portfolio	16	10	8.67
Total	161	125	95.33

5. Discussion

5.1. Variables Correlation

To determine the correlation between the variables and to obtain the degree to which online education influences students' behavior, the classification and regression tree (CRT) analysis was used.

Figure 2 shows the classification of students' behavior regarding the information they obtain using supplementary materials for a better understanding of the information received but also the benefits of online education.

Even if the courses and seminars took place online, the teaching staff followed the same steps of the traditional system, with the recommendation of some additional materials for the students.

A total of 18.7% of students apply these additional materials for a good understanding or for information. For a percentage, the technology was a form of rescue in the pandemic situation, for 18% of students this was out of curiosity, and 63.3% do not apply because they consider that they receive enough information based on the assessment and materials on the platform provided by the teacher.

The online courses are considered beneficial by 41.3% of the students, appreciated especially by the working segment, as they have part time jobs, due to the flexibility of the program and the possibility to access the information when they have time.

On the other hand, 64.7% of those who are pursuing a bachelor's degree appreciate this type of course, but for those at the master's level, 39.3% find it as beneficial.

The data of this exploratory study highlights the fact that students go through a process of adaptation and learning, constituting an indirect but favorable argument.

Figure 3 represents the classification for students' behavior using online education. It is a weekly or daily routine for 24.7% of respondents to spend more than 10 h on the platform, and 58.7% of students are spending between 1–10 h daily.

Additionally, 26% of students are male and 32.7 % are female. Taking into account the time spent by students online on educational platforms, we notice that a percentage of 57.8% of students spend between 11–0 h per week, of which 43.2% are at the bachelor's level and 56.8% are at the master's level, we note that they use the information on the platforms put on display by the teachers.

The highest value was obtained for undergraduate students who spend more than 10 h on the university platform, respectively for the field of management and economics, at 24.4%.

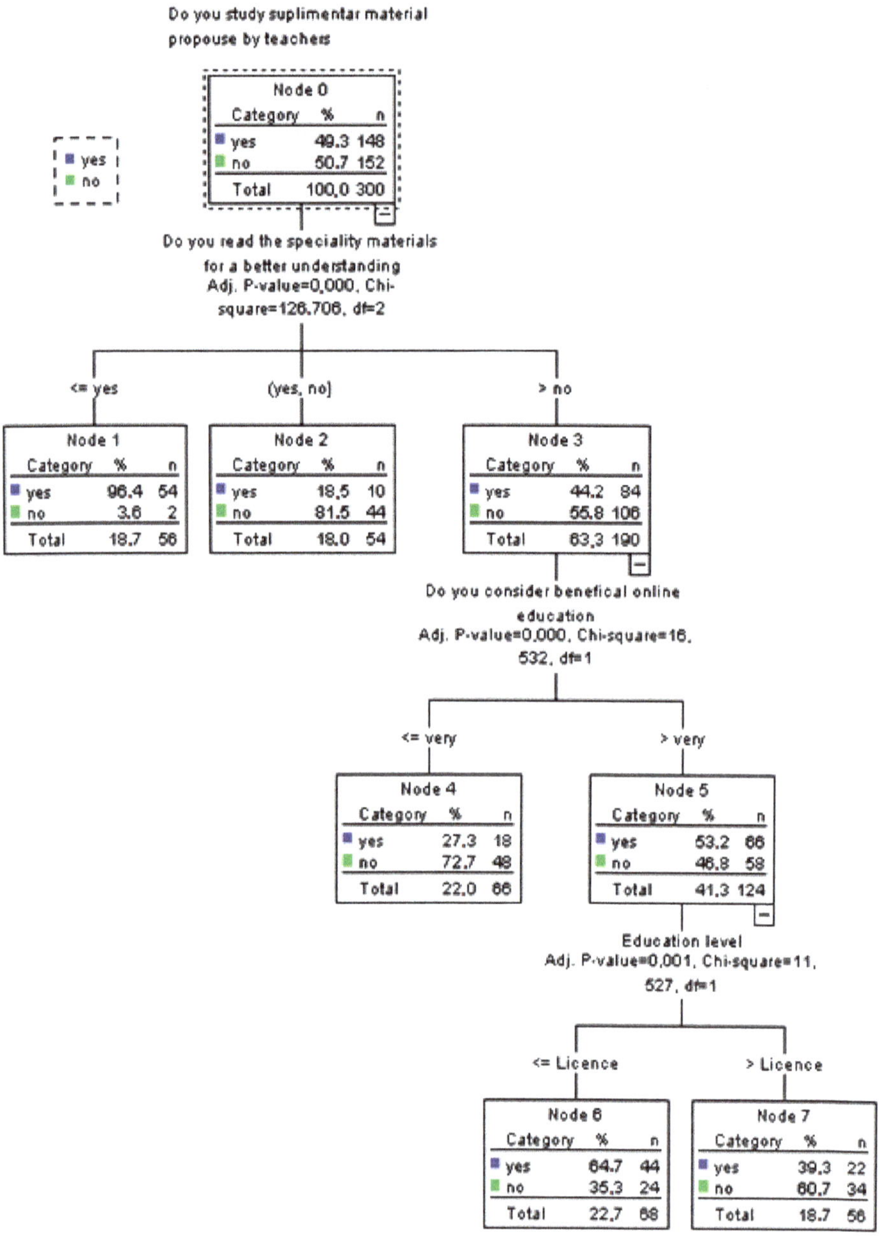

Figure 2. Classification for student's behavior using supplementary materials. Source: By author.

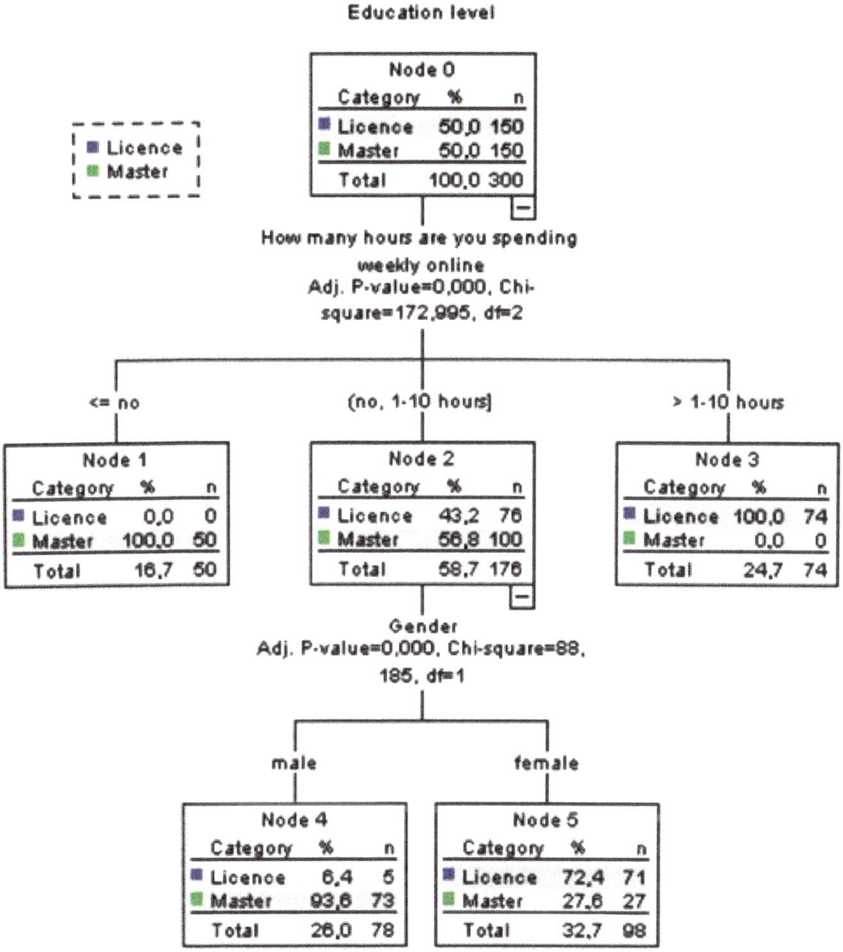

Figure 3. Classification for student's behavior using online education.

5.2. Correlations between Items

Using the database obtained from the 15 items of questionnaries from SPSS using the Lisrel 8.7 program, four groups of students with similar profiles regarding the online education were identified.

The model take into consideration the students characteristics (gender, age, education level), how the students are involved in diferent activities for a better asimilation of information and their knowledge from different activities in their culture (U, S, H, F), student needs for evaluation (E1, E2, E3, E4 and E5) and student quality satisfaction about online education (R, Q, QE1). The program solution for the students behavior model is presented in Figure 4.

Students behavior using online education during the pandemic period gives us the following correlations.

The highest positive value of 0.89 was obtained for the correlation between Individual characteristics and students Knowledge. The results represent students from the Baia Mare faculty, the future managers and economists or entrepreneurs, having strong knowledge of the online education and of how they are evaluated.

Another strong relation, of 0.07 value, was between students Needs and Quality of education, in our case online education, so the student involvement and virtual approach is necessary and beneficial by accessing the materials provided by teachers on platforms so that the educational process does not lose quality.

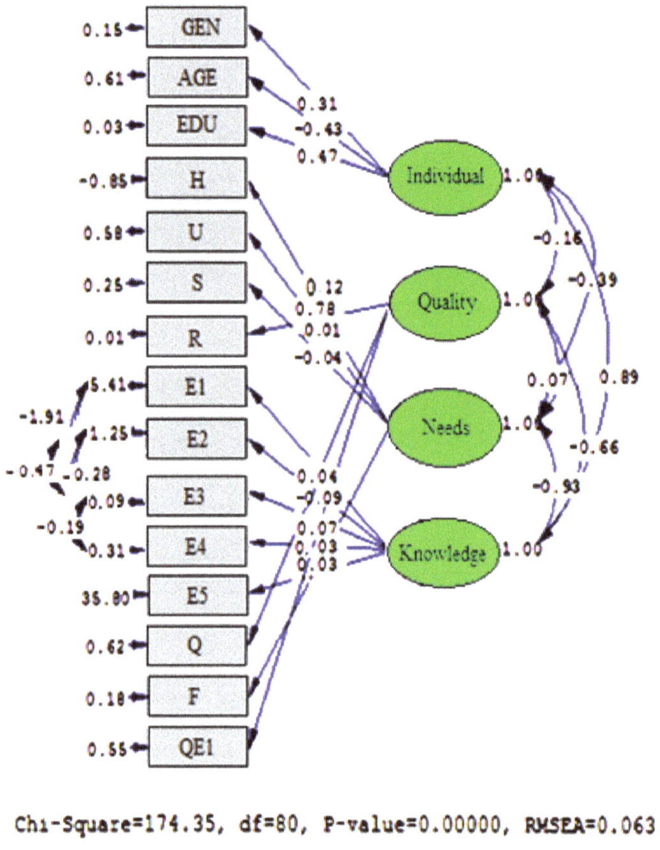

Figure 4. Factors influencing students' behavior using online education.

By using the online platform, from the Quality characteristic point of view, the reading (R), of supplementary and speciality materials obtained the strongest value of 0.78.

Additionally for Individual charactersitics of students the level of education EDU obtained the bigest value of 0.47,that means the students are informed and enjoy use of the platforms and being involved in online education, with the finalization on specialization.

For Needs, the highest value of 0.12 was obtained by the number of hours spent by students online (H), followed by the frequency (F) with which they access the platform, with a value of 0.03.

For Knowledge, the highest value of 0.07 was obtained by evauation using mulitple choice test and the lowest by written examination, with a value of −0.9.

In conclusion, between students individual characteristics and students knowledge's using digital platforms, there is a strongest correlation and the quality of educational process, it is not influenced by the individual characteristics. Student's attitudes towards online education is influenced by their needs and the platform quality improves student's knowledge and behavior.

5.3. Conventional Students Cluster

In order to be able to study students' behavior regarding the online education, a cluster analysis was also performed, this time taking into account the order of their preferences. The sample of 300 students was subjected to a k mean and hierarchical grouping, we proceeded to identify four groups of students with similar educational profiles. According to the features of the adult students (Table 10) four clusters were identified: Needs of students (characterized by online benefits, and individual study of specific materials), students Knowledge (characterized by the hours spent weekly online) Quality for students (characterized by quality of materials and speciality materials,), students Preference for evaluation (characterized by different types of online evaluation, multiple choice, portfolio and written version).

Table 10. Students conventional clusters.

Items	Clusters	Number of Students
Gender	Needs	73
Education level		
Do you consider beneficial online education		
Do you study supplementary material propose by teachers		
Age	Knowledge	73
How many hours are you spending weekly online		
Do you read the specialty materials for a better understanding	Quality	32
What kind of examination do you prefer	Preferences	122

Cluster 1: for 73 students Needs of online education and supplementary materials posted on the platform by the teachers are important, expressing their attitude to spend time and hours to collect the information.

Cluster 2: for 73 respondents Knowledge is important, the number of hours spent online or on platforms depended on the information and their connection with teachers and participate in different activities. Age confirms once again student's responsibilities for their professional preparation and skills impact in using new technology.

Cluster 3: Quality obtained the lowest value for using additional materials. The low value shows that teachers provide enough information through online courses and materials posted on platforms. The university's own platform and for each specialization allows students access at any time to have access to information.

Cluster 4: the highest value, for 122 students, was obtained in terms of the cluster for the student Preferences for the final evaluation, during examination or periodical evaluation. Among the verification options proposed: online, multiple choice, test and written verification, students have shown that assessment is very important for them in the time of pandemic also.

6. Conclusions

The article presents the results obtained following the application of questionnaires applied to identify student's behavior and attitudes towards online education during the COVID-19 pandemic.

Based on the literature, the results were able to create a student profile model and establish the factors which influence student's behavior and attitudes concerning online education. Online education has been a great challenge for both teachers and students. At present, education is still in a period of adaptation, of identifying the factors that influence the educational act for an as yet unexplained period. The COVID-19 pandemic brought for the first time the widespread adoption of online education around the world, making it a necessity in difficult times.

The four factors taken into consideration for the model were the individual characteristics specific to each student, the students' knowledge, the students' needs and the preferences for the quality of online education influencing the students' behavior and attitude.

The student's behavior is influenced by their attitude regarding their needs and quality digital education. The students' preferences for quality platforms and materials in the changes of this period confirm the hypothesis and the model.

Students' have represented that teachers are those who adapt and reformulate their habits, making them closer to students through the digital environments of the future, even if there is further physical distance, which contributes to a categorical evolution of university education.

The feedback of the questionnaire confirms that there is a strong connection between student's needs and the quality of education and the teaching process that influences the student's behavior, and not only in the pandemic period [3,4,11].

The results confirm that the students' obtained knowledge by using online education during this pandemic, which is a useful lesson for future demands, confirming other researcher's results [43,48,77].

The study identified students' needs, and this data suggests that it is not very realistic to start from the assumption that switching to teaching exclusively online can be done easily. The study identifies and confirms also that there are some inequalities regarding Internet access (no telephone signal, or do not have a computer/laptop/tablet/mobile phone, as well as a fairly low level of digital skills) [8,18]. Students have begun to notice that it is possible to put in more effort and have more time to attend courses through online digital tools, even if they are in isolation, at home.

Additionally, the individual characteristics presented similar preferences regarding the digital education and needs, as the correlation between individual characteristics and needs was confirmed [37,38,40].

The data obtained on the basis of statistical observations can highlight the attitude towards the use of a mixture of educational tools to ensure a new orientation towards a new vision for the future of Education 4.0, that changes their behavior towards online education not only in crises or pandemics (Table 11).

Table 11. The transfer to Education 4.0.

PAST	FUTURE
Traditional	Digital
Schedule	Flexible program
Staying in university	Staying at home
Input focus	Output focus
Face to Face	Communication tools
Focus on knowledge	Adaptive learning
Present information	Share information
Classroom	Virtual Classroom
Medium condition	Comfort environment
Rigidity	Flexibility
Presentation centered	Student centered
Socialization	Isolation

The study also confirmed that students prefer online education, but cannot replace the classic face-to-face method. However, students believe that there are opportunities for improvement and easing of the educational process.

The study can be a provocation for teachers to adapt teaching methods and techniques for online education using the new digital generation to use new technical methods such as inverted class, case studies and playing games which can be used for online technology, such as virtual platforms [39,41,42].

A SWOT Analyze of Online Education

As a final conclusion, a SWOT analyses has been made, with the following conclusion:

The study captured, as a last dimension, the students' opinions regarding the present educational context, a unique context for both teachers and students.

Thus, the mood that the students declare is a good one, they are connected to the information about the current situation, and they consider that the return to previous interactions will occur relatively quickly, but at the same time, they declare a low involvement in community volunteering activities.

It is natural, in conditions of uncertainty, for the decision to involve the community of whether to carry on the alternative, limiting the risk.

A strong element reported by students was that the use of digital resources in learning is perceived as a positive thing; also, media tools are preferred by the Facebook generation of students.

A signal is the new profile of the outlined student in the new educational context and the students' perception consists in the following weak elements:

- The topics and tasks proposed to the students following the courses and seminars are several;
- Teachers tend to monitor the student's progress;
- There is final and ongoing evaluation for the continuity of the learning process;
- The transfer of educational activities in the online environment rather negatively affects only the seminar activities;
- The teacher and student no longer interact enough;
- Total transfer to virtualization of activities.

The most important perceived disadvantages of transferring educational activities in the online environment are related to the long time spent in front of the computer to participate in teaching activities and to solve the tasks received, and the perceived excessive volume of homework and tasks. The s perceived as non-problematic can be listed as: the ease of use of digital tools and platforms, but also the accessibility of teaching and learning resources provided by teachers.

An opportunity for generations of students is that the Internet and online browsers, the Google search engine and social media platforms are the environment in which their generation was born and raised. The digital world is part of the lives of young students, and online education, through digital applications, is a language of learning that they have been using for a long time.

We can also mention that students have started to notice that it is possible to put more effort and have a longer time to participate in courses and applications through online digital tools, even if they are in isolation, at home.

This study was conducted in the Technical University Cluj Napoca, Romania, and the model and results can be used by other universities to identify students behavior and attitudes toward online learning and methods which can be used in future activities or other pandemic situations.

Funding: This research received no external funding.

Institutional Review Board Statement: Not applicable.

Informed Consent Statement: Not applicable.

Data Availability Statement: Data sharing not applicable. The data are not publicly available due to participants' privacy.

Conflicts of Interest: The author declares no conflict of interest.

References

1. Frecker, K.; Bieniarz, E. Why Online Education Is Here to Stay. Available online: https://www.lighthouselabs.ca/en/blog/why-education-is-moving-online-for-good (accessed on 29 March 2021).
2. Zhu, Z.-T.; Yu, M.-H.; Riezebos, P. A research framework of smart education. *Smart Learn. Environ.* **2016**, *3*, 1. [CrossRef]
3. Hartono, S.; Kosala, R.; Supangkat, S.H.; Ranti, B. Smart Hybrid Learning Framework Based on Three-Layer Architecture to Bolster Up Education 4.0. In Proceedings of the 2018 International Conference on ICT for Smart Society (ICISS), West Java, Indonesia, 10–12 October 2018; pp. 1–5.
4. Bond, M.A.; Lockee, B.B. Evaluating the effectiveness of faculty inquiry groups as communities of practice for faculty pro-fessional development. *J. Form. Design Learn.* **2018**, 1–7. [CrossRef]
5. Jackson, B.L. Online professional development courses in online teaching for higher education faculty. In Proceedings of the Society for Information Technology and Teacher Education (SITE 2018), San Diego, CA, USA, 26–30 March 2018; pp. 187–195.
6. Elzainy, A.; El Sadik, A.; Al Abdulmonem, W. Experience of e-learning and online assessment during the COVID-19 pandemic at the College of Medicine, Qassim University. *J. Taibah Univ. Med. Sci.* **2020**, *15*, 456–462. [CrossRef]
7. Intentional Futures, Instructional Design in Higher Education: A Report on the Role, Workflow, and Experience of Instructional Designers. 2016. Available online: https://intentionalfutures.com/wp-content/uploads/2017/08/Instructional-Design-in-Higher-Education-Report.pdf (accessed on 29 March 2021).
8. McGee, P.; Windes, D.; Torres, M. Experienced online instructors: Beliefs and preferred supports regarding online teaching. *J. Comput. High. Educ.* **2017**, *29*, 331–352. [CrossRef]
9. Rhode, J.; Krishnamurthi, M. Preparing Faculty to Teach Online: Recommendations for Developing. *Int. J. Inf. Educ. Technol.* **2016**, *6*, 376–382. [CrossRef]
10. Iwai, Y. Online Learning during the COVID-19 Pandemic: What do we gain and what do we lose when classrooms go virtual? *Sci. Am.* **2020**, *13*, 32–37.
11. Dandara, O. Platformele electronice—Mijloc de modernizare a tehnologiilor educaționale. *Studia Univ. Mold.* **2013**, *5*, 13.
12. Allen, I.E.; Seaman, J. *Digital Learning Compass: Distance Education Enrollment Report 2017*; Babson Survey Research Group, Babson College: Wellesley, MA, USA, 2017.
13. Espiritu, J.L.; Budhrani, K. Cultivating an e-learning culture. *Sci. Pedagog. Exp.* **2019**, *56*, 3–32.
14. Dhawan, S. Online Learning: A Panacea in the Time of COVID-19 Crisis. *J. Educ. Technol. Syst.* **2020**, *49*, 5–22. [CrossRef]
15. Jæger, M.M.; Blaabæk, E.H. Inequality in learning opportunities during Covid-19: Evidence from library takeout. *Res. Soc. Strat. Mobil.* **2020**, *68*, 100524. [CrossRef]
16. European Commission. Digital Education Action Plan (2021–2027) | Education and Training. Available online: https://ec.europa.eu/education/education-in-the-eu/digital-education-action-plan_en (accessed on 27 April 2021).
17. Schmidt, S.W.; Tschida, C.M.; Hodge, E.M. How faculty learns to teach online: What administrators need to know. *Online J. Distance Learn. Adm.* **2016**, *19*, 1–10.
18. Mishra, L.; Gupta, T.; Shree, A. Online teaching-learning in higher education during lockdown period of COVID-19 pandemic. *Int. J. Educ. Res. Open* **2020**, *1*, 100012. [CrossRef]
19. Bojović, Ž.; Bojović, P.D.; Vujošević, D.; Šuh, J. Education in times of crisis: Rapid transition to distance learning. *Comput. Appl. Eng. Educ.* **2020**, *28*, 1467–1489. [CrossRef]
20. Moskal, P.; Thompson, K.; Futch, L. Enrollment, Engagement and Satisfaction in the BlendKit Faculty Development Open, Online Course. *Online Learn.* **2015**, *19*, 1–12. [CrossRef]
21. Moore, R.L.; Fodrey, B.P.; Piña, A.A.; Lowell, V.L.; Harris, B.R. Distance Education and Technology Infrastructure: Strategies and Opportunities. In *Leading and Managing e-Learning*; Springer: Berlin/Heidelberg, Germany, 2017; pp. 87–100.
22. Mohr, S.C.; Shelton, K. Best Practices Framework for Online Faculty Professional Development: A Delphi Study. *Online Learn.* **2017**, *21*, 123–140. [CrossRef]
23. Baran, E.; Correia, A.-P. A professional development framework for online teaching. *TechTrends* **2014**, *58*, 95–101. [CrossRef]
24. Baran, E.; Correia, A.P.; Thompson, A. Tracing successful online teaching in high education: Voices of exemplary online teachers. *Teach. Coll. Rec.* **2013**, *115*, 1–41.
25. Brinthaupt, T.M.; Cruz, L.; Otto, S.; Pinter, M. A framework for the strategic leveraging of outside resources to enhance CTL effectiveness. *Improv. Acad.* **2019**, *38*, 82–94. [CrossRef]
26. Elliott, M.; Rhoades, N.; Jackson, C.; Mandernach, J. Professional Development: Designing Initiatives to Meet the Needs of Online Faculty. *J. Educ. Online* **2015**, *12*, 160–188. [CrossRef]
27. Martin, F.; Wang, C.; Sadaf, A. Student perception of helpfulness of facilitation strategies that enhance instructor presence, connectedness, engagement and learning in online courses. *Internet High. Educ.* **2018**, *37*, 52–65. [CrossRef]
28. Cazacu, D. Utilizarea Platformelor Educaționale în Predarea Geografiei. Available online: https://edict.ro/utilizarea-platformelor-educationale-in-predarea-geografiei/ (accessed on 29 March 2021).
29. Chatterjee, I.; Chakraborty, P. Use of Information Communication Technology by Medical Educators Amid COVID-19 Pandemic and Beyond. *J. Educ. Technol. Syst.* **2021**, *49*, 310–324. [CrossRef]
30. Grishchenko, N. The gap not only closes: Resistance and reverse shifts in the digital divide in Russia. *Telecommun. Policy* **2020**, *44*, 102004. [CrossRef]

31. Hasan, N.; Bao, Y. Impact of "e-Learning crack-up" perception on psychological distress among college students during COVID-19 pandemic: A mediating role of "fear of academic year loss". *Child. Youth Serv. Rev.* **2020**, *118*, 105355. [CrossRef]
32. Ionescu, C.; Paschia, L.; Nicolau, N.G.; Stanescu, S.; Stancescu, V.N.; Coman, M.; Uzlau, M. Sustainability Analysis of the E-Learning Education System during Pandemic Period—COVID-19 in Romania. *Sustainability* **2020**, *12*, 9030. [CrossRef]
33. Is Online Learning a Sustainable Model for Higher Education? Available online: https://u2b.com/2020/03/20/online-learning-sustainable/ (accessed on 16 June 2021).
34. Singh, E.P.; Adhikary, K.; Gupta, M.S.; Singh, S. User Interface Consideration for ScalableC. An Online Collaborative Platforms. In Proceedings of the International Conference on Enterprise Information Systems and Web Technologies, Madeira, Portugal, 8–12 June 2010; pp. 111–116.
35. Diaz, M.C.G.; Walsh, B.M. Telesimulation-based education during COVID-19. *Clin. Teach.* **2020**. [CrossRef] [PubMed]
36. Becker, S.A.; Cummins, M.; Davis, A.; Freeman, A.; Hall, C.G.; Ananthanarayanan, V. *NMC Horizon Report: 2017 Higher Education Edition*; The New Media Consortium: Austin, TX, USA, 2017.
37. Budhrani, K.; Ji, Y.; Lim, J.H. Unpacking conceptual elements of smart learning in the Korean scholarly discourse. *Smart Learn. Environ.* **2018**, *5*, 23. [CrossRef]
38. Martin, F.; Bolliger, D.U. Engagement Matters: Student Perceptions on the Importance of Engagement Strategies in the Online Learning Environment. *Online Learn.* **2018**, *22*, 205–222. [CrossRef]
39. Beaunoyer, E.; Dupéré, S.; Guitton, M.J. COVID-19 and digital inequalities: Reciprocal impacts and ditigation strategies. *Comput. Hum. Behav.* **2020**, *111*, 106424. [CrossRef]
40. Zhu, X.; Liu, J. Education in and After Covid-19: Immediate Responses and Long-Term Visions. *Post Digit. Sci. Educ.* **2020**, 2. Available online: https://www.researchgate.net/publication/340936542_Education_in_and_After_Covid-19_Immediate_Responses_and_Long-Term_Visions (accessed on 29 March 2021).
41. George-Andrei, U.; Olga, P.; Maria, U. Simulation of Electrical Circuits Using Data Acquisition Boards for Online Education. In Proceedings of the 2020 International Conference and Exposition on Electrical and Power Engineering (EPE), Iasi, Romania, 22–23 October 2020; pp. 484–487.
42. Windes, D.L.; Lesht, F.L. The effects of online teaching experience and institution type on faculty perceptions of teaching online. *Online J. Distance Learn. Adm.* **2014**, *17*, 12.
43. Chakraborty, P.; Mittal, P.; Gupca, M.P.; Yadav, S.; Arira, A. Opinion of students online education during the Covid-19 pandemic. *Hum. Behav. Emerg. Technol.* **2020**, 1–9. [CrossRef]
44. Aguilera-Hermida, A.P. College students' use and acceptance of emergency online learning due to COVID-19. *Int. J. Educ. Res. Open* **2020**, *1*, 100011. [CrossRef]
45. Hodges, C.; Moore, S.; Lockee, B.; Trust, T.; Bond, A. The difference between emergency remote teaching and online learning. *Educ. Rev.* **2020**, *27*, 1–12.
46. Si, J.; Kong, H.-H.; Lee, S.-H. Exploring medical educators' readiness and the priority of their educational needs for online teaching. *Korean J. Med. Educ.* **2021**, *33*, 37–44. [CrossRef] [PubMed]
47. Lunsford, L.; Baker, V.; Pifer, M. Faculty mentoring faculty: Career stages, relationship quality, and job satisfaction. *Int. J. Mentor. Coach. Educ.* **2018**, *7*, 139–154. [CrossRef]
48. Uden, L.; Liberona, D.; Sanchez, G.; Rodriguez Gonzalez, S. (Eds.) *Learning Technology for Education Challenges, Proceedings of the 8th International Workshop, LTEC 2019, Zamora, Spain, 15–18 July 2019*; Springer: Berlin/Heidelberg, Germany, 2019; Available online: https://www.springer.com/gp/book/9783030207977 (accessed on 9 June 2021).
49. Carroll, N.; Conboy, K. Normalising the "new normal": Changing tech-driven work practices under pandemic time pressure. *Int. J. Inf. Manag.* **2020**, *55*, 102186. [CrossRef]
50. Polly, D.; Martin, F.; Guilbaud, T.C. Examining barriers and desired supports to increase faculty members' use of digital technologies: Perspectives of faculty, staff and administrators. *J. Comput. High. Educ.* **2021**, *33*, 135–156. [CrossRef]
51. Coman, C.; Țîru, L.G.; Meseșan-Schmitz, L.; Stanciu, C.; Bularca, M.C. Online Teaching and Learning in Higher Education during the Coronavirus Pandemic: Students' Perspective. *Sustain. J. Rec.* **2020**, *12*, 10367. [CrossRef]
52. Arora, A.; Chakraborty, P.; Bhatia, M.P.S.; Mittal, P. Role of emotion in addictive use of twitter during COVID-19 imposed lockdown in India. *J. Technol. Behav. Sci.* **2020**, 1–8. [CrossRef]
53. Trammell, B.; LaForge, C. Common challenges for instructors in large online courses: Strategies to mitigate student and in-structor frustration. *J. Educ. Online* **2017**, *14*, 10–19.
54. Stark, A.M.; Smith, G.A. Communities of practice as agents of future faculty development. *J. Fac. Dev.* **2016**, *30*, 59–67.
55. Sheffield, S.L.-M.; McSweeney, J.M.; Panych, A. Exploring Future Teachers' Awareness, Competence, Confidence, and Attitudes Regarding Teaching Online: Incorporating Blended/Online Experience into the Teaching and Learning in Higher Education Course for Graduate Students. *Can. J. High. Educ.* **2015**, *45*, 1–14. [CrossRef]
56. Trust, T.; Krutka, D.G.; Carpenter, J.P. "Together we are better": Professional learning networks for teachers. *Comput. Educ.* **2016**, *102*, 15–34. [CrossRef]
57. Rhode, J.; Richter, S.; Miller, T. Designing Personalized Online Teaching Professional Development through Self-Assessment. *TechTrends* **2017**, *61*, 444–451. [CrossRef]

58. Adhikary, K.; Gupta, M.S.; Singh, E.P.; Singh, S. Collaborating Towards Learning: Using Web 2.0 for Educational Idea Development. In Proceedings of the Fifth Conference of Learning International Networks Consortium, Cambridge, MA, USA, 23–26 March 2010; pp. 465–474.
59. Skulmoski, A.; Rey, G.D. COVID-19 as an accelerator for digitalization at a German university. Establishing hybrid campuses in times of crisis. *Hum. Behav. Emerg. Technol.* **2020**, *2*, 212–216. [CrossRef]
60. Toquero, C.M.D.; Talidong, K.J.B. Socio-educational implications of technology use during COVID-19: A case study in general Santos City, Philippines. *Hum. Behav. Emerg. Technol.* **2020**. [CrossRef] [PubMed]
61. Zhang, W.; Wang, Y.; Yang, L.; Wang, C. Suspending classes without stopping learning: China's education emergency management policy in the COVID-19 outbreak. *J. Risk Financ. Manag.* **2020**, *13*, 55. [CrossRef]
62. Islam, M.A.; Barna, S.D.; Raihan, H.; Khan, M.N.A.; Hossain, M.T. Depression and anxiety among university students during the COVID-19 pandemic in Bangladesh: A web-based cross-sectional survey. *PLoS ONE* **2020**, *15*, e0238162. [CrossRef]
63. Davis, F.D. User acceptance of information technology: System characteristics, user perceptions and behavioral impacts. *Int. J. Man-Mach. Stud.* **1993**, *38*, 475–487. [CrossRef]
64. Liu, I.-F.; Chen, M.C.; Sun, Y.S.; Wible, D.; Kuo, C.-H. Extending the TAM model to explore the factors that affect Intention to Use an Online Learning Community. *Comput. Educ.* **2010**, *54*, 600–610. [CrossRef]
65. Al Kurdi, B.; Alshurideh, M.; Salloum, S.A.; Obeidat, Z.M.; Al-Dweeri, R.M. An Empirical Investigation into Examination of Factors Influencing University Students' Behavior towards Elearning Acceptance Using SEM Approach. *Int. J. Interact. Mob. Technol.* **2020**, *14*, 19–41. [CrossRef]
66. Al Kurdi, B.; Alshurideh, M.; Salloum, S.A. Investigating a theoretical framework for e-learning technology acceptance. *Int. J. Electr. Comput. Eng.* **2020**, *10*, 6484–6496. [CrossRef]
67. Mailizar, M.; Burg, D.; Maulina, S. Examining university students' behavioural intention to use e-learning during the COVID-19 pandemic: An extended TAM model. *Educ. Inf. Technol.* **2021**, 1–21. [CrossRef]
68. Bao, W. COVID-19 and online teaching in higher education: A case study of Peking University. *Hum. Behav. Emerg. Technol.* **2020**, *2*, 113–115. [CrossRef] [PubMed]
69. Lee, J.W. Online support service quality, online learning acceptance, and student satisfaction. *Internet High. Educ.* **2010**, *13*, 277–283. [CrossRef]
70. Hung, W.-C.; Jeng, I. Factors influencing future educational technologists' intentions to participate in online teaching. *Br. J. Educ. Technol.* **2012**, *44*, 255–272. [CrossRef]
71. Hatabu, A.; Mao, X.; Zhou, Y.; Kawashita, N.; Wen, Z.; Ueda, M.; Takagi, T.; Tian, Y.-S. Knowledge, attitudes, and practices toward COVID-19 among university students in Japan and associated factors: An online cross-sectional survey. *PLoS ONE* **2020**, *15*, e0244350. [CrossRef]
72. Zia, A. Exploring factors influencing online classes due to social distancing in COVID-19 pandemic: A business students perspective. *Int. J. Inf. Learn. Technol.* **2020**, *37*, 197–211. [CrossRef]
73. Alzahrani, L.; Seth, K.P. Factors influencing students' satisfaction with continuous use of learning management systems during the COVID-19 pandemic: An empirical study. *Educ. Inf. Technol.* **2021**, 1–19. [CrossRef]
74. Yunus, M.; Ang, W.S.; Hashim, H. Factors Affecting Teaching English as a Second Language (TESL) Postgraduate Stu dents' Behavioural Intention for Online Learning during the COVID-19 Pandemic. *Sustainability* **2021**, *13*, 3524. [CrossRef]
75. Essadek, A.; Rabevron, T. Mental health of French students during the Covid-19 pandemic. *J. Affect. Disord.* **2020**, *277*, 392–393. [CrossRef]
76. Paea, S.; Sharma, B.; Bulivou, G.; Paea, M.K. Card Sorting: A new pedagogy for understanding challenges in Mathematics during Emergencies and Crises. *Res. Sq.* 2021. [CrossRef]
77. Cao, W.; Fang, Z.; Hou, G.; Han, M.; Xu, X.; Dong, J.; Zheng, J. The psychological impact of the COVID-19 epidemic on college students in China. *Psychiatry Res.* **2020**, *287*, 112934. [CrossRef]
78. Hrastinski, S. A Study of Asynchronous and Synchronous e-Learning Methods Discovered that Each Supports Different Purposes. *Educ. Q.* **2008**, *38*, 51–55. Available online: https://www.researchgate.net/publication/238767486 (accessed on 8 March 2021).
79. Amiti, F. Synchronous and asynchronous e-learning. *Eur. J. Open Educ. E-Learn. Stud.* **2020**, *5*, 60–70.
80. Saraçli, S.; Yilmaz, V.; Arslan, T. The Effects of Mothers' Educational Levels on University Students' Environmental Protection Commitments and Environmental Behaviors. *Eurasian J. Educ. Res.* **2014**, *55*, 177–200. [CrossRef]
81. Gumus, H.; Saracli, S.; Yagmur, R.; Isik, O.; Ersoz, Y. The investigation of sportsmanship behaviors of university students. *J. Sport Psychol.* **2020**, *29*, 13–20.
82. What Does Cronbach's Alpha Mean? UCLA, Institute for Digital Research and Education, Statistical Consulting. Available online: https://www.statisticshowto.com/probability-and-statistics/statistics-definitions/cronbachs-alpha-spss/ (accessed on 15 June 2021).
83. Brown, J.D. The Cronbach alpha reliability estimate. *JALT Test. Eval. SIG Newsl.* **2002**, *6*, 17–18.
84. Schermelleh-Engel, K.; Moosbrugger, H.; Müller, H. Evaluating the Fit of Structural Equation Models: Tests of Significance and Descriptive Goodness-of-Fit Measures. *Methods Psychol. Res.* **2003**, *8*, 23–74.

sustainability

MDPI

Article

Residents' Training in COVID-19 Pandemic Times: An Integrated Survey of Educational Process, Institutional Support, Anxiety and Depression by the Saudi Commission for Health Specialties (SCFHS)

Basim Alsaywid [1,2,3,*], Abdulrahman Housawi [1,*], Miltiadis Lytras [4,*], Huda Halabi [5], Maha Abuzenada [1], Sami A. Alhaidar [6] and Wesam Abuznadah [6]

1 Planning and Organizational Excellence Administration, Saudi Commission for Health Specialties, Riyadh 11614, Saudi Arabia; m.abuzenada@scfhs.org
2 College of Medicine, King Saud Bin-Abdul-Aziz University for Health Sciences, Jeddah 14611, Saudi Arabia
3 Urology Section, Department of Surgery, King Abdulaziz Medical City, Ministry of National Guard, Jeddah 14815, Saudi Arabia
4 Effat College of Engineering, Effat University, P.O. Box 34689, Jeddah 22332, Saudi Arabia
5 College of Medicine, Ibn Sina National College, Jeddah 22421, Saudi Arabia; Huda.Halabi11@gmail.com
6 Executive Presidency of Academic Affairs, Saudi Commission for Health Specialties, Riyadh 11614, Saudi Arabia; s.alhaidar@scfhs.org (S.A.A.); w.abuznadah@scfhs.org (W.A.)
* Correspondence: b.alsaywid@scfhs.org (B.A.); a.housawi@scfhs.org (A.H.); mlytras@acg.edu (M.L.)

Received: 12 November 2020; Accepted: 3 December 2020; Published: 16 December 2020

check for updates

Abstract: In late December of 2019, the outbreak of coronavirus disease (COVID-19) was first reported in the city of Wuhan, the capital of Hubei province in China, and was declared a pandemic by the World Health Organization in March 2020. Globally, as of 8 July 2020, there have been 11,669,259 confirmed cases of COVID-19, including 539,906 deaths. In Saudi Arabia, the confirmed cases have already reached 223,327, with 161,096 patients confirmed to have recovered, and 2100 deaths. This study aims to determine the effect of the COVID-19 pandemic on the training programs of the Saudi Commission for Health Specialties (SCFHS) and assess trainees' mental health status (i.e., anxiety and depression). Trainee evaluations on training programs were also sought in order to obtain insights for strategic planning necessary for curricular modifications or improvements to address the clinical learning needs of trainees during this pandemic. The main contribution of our work is an investigation of the incidence of depression and anxiety regarding COVID-19 within the community of residents and fellows. Furthermore, we elaborate on key responsive actions towards the enhancement of the mental health of trainees. Last but not least, we propose the Saudi Commission for Health Specialties (SCFHS) Model for Residents' Mental Health Enhancement during the COVID-19 Pandemic, which consists of five integrative value layers for medical education and training, namely: the knowledge creation process and innovation; technological capabilities for personalized medicine and patient-centric healthcare with a social impact; innovative applications of technology-enhanced learning and web-based active learning approaches for medical training and education; residents' wellbeing and the impact of COVID-19 in strategic layers. In our future work, we intend to enhance the proposed framework with studies on trainee satisfaction and the efficiency of different technology-enhanced learning platforms for medical education.

Keywords: medical training; residents training; quality; COVID-19; medical education assurance; training; accreditation; satisfaction; governance; Saudi Commission for Health Specialties; framework; best practices; healthcare

1. Introduction

The spread of the COVID-19 pandemic has had a significant impact on diverse aspects of medical training worldwide. Several researchers worldwide claim the need to enhance medical training programs and comprehensive continuing medical education with an emphasis on skills and competencies development [1]. Furthermore, the suspension of medical education and clinical attachments of students [2] due to COVID-19, in several cases [2], poses critical questions about the delivery of medical training at a top-quality level. Additionally, various psychological factors, including stress, anxiety and burnout, are extensively discussed in research studies [3] emphasizing the need for integrated psychological intervention programs for the support of physicians and medical staff. We aim to move towards prevention, alleviation, and treatment of the increased psychological load.

Since the emergence and pandemic spread of highly contagious COVID-19, the pace of medical education and training programs has drastically changed. The Saudi government, with the collaboration of health agencies worldwide, imposed new policies and measures to contain and mitigate the spread of infection to individuals and communities. This has limited medical students' and trainees' roles in the clinical environment, and their opportunities to gain new experiences.

Within this new, demanding context, several complimentary aspects of medical education and training should be reconsidered. On one hand, aspects related to human factors and psychological pressure of residents and fellows as well as healthcare workers and senior medical staff have a direct impact on the training process. On the other hand, the need to support training with technology-enhanced learning methods and active learning strategies becomes a necessity that requires strategic support. Finally, training institutions for medical trainees have often served also as COVID-19 hospitals, and this adds to the complexity of the phenomenon.

In our research, we focus on the phenomenon of the medical training during COVID-19 times and we emphasize the analysis of the psychological factors that have an impact on the quality and efficiency of medical training. For the purpose of our research, we delivered an integrated survey in the Saudi Commission for the Health Specialties with a focus on training programs of fellows and residents.

The Saudi Commission for Health Specialties (SCFHS) is a government agency that regulates health care-related practices and accreditation at all levels in Saudi Arabia. This is home to over 160 residency programs with more than 15,000 residents. There are over 600 institutions accredited by SCFHS as training centers and all trainees, including residents and fellows, are designated in a clinical area corresponding to their chosen specialty. In accordance with the precautionary health measures recommended by the competent authorities to prevent the spread of the emerging COVID-19, SCFHS endorsed a bundle of initiatives which were directed to support the residents and fellows. Firstly, a 24/7 hotline was established to offer psychological support through Da'em to all trainees and practitioners.

The health academy's programs, which are set up in training centers in universities, hospitals, and medical centers, are suspended until further notice. Thus, SCFHS ensured that virtual education is fully activated for this category of trainees through the educational platforms in place to ensure the continuation of the educational process. Moreover, the first part the exam for all trainees at all levels, including the final written exam, was postponed. Accordingly, failure in those exams, whether the first part or the final exam, is not counted as an attempt among the limited number of possible attempts. Additionally, SCFHS extended the period of personal interviews for new candidates and applicants to all training programs.

Other helpful decisions included exempting residents and fellows from administrative fees to register in electronic exams for the current training year, converting the promotion test to become part of the continuous evaluation aimed at giving feedback and not attributing success or failure to the programs that were decided by the Scientific Council to have a promotion test. Furthermore, the possibility was introduced to carry over two weeks of leave to the next year for those trainees who did not have the opportunity to take parts of their annual leave this year in accordance with the regulations governing this, and the re-training of the training year within the maximum duration of the program was not counted.

The previously mentioned actions are timely responses to the COVID-19 pandemic. It is extremely important to investigate (with a scientific approach) the capacity of these actions to support the mental health of trainees and also to investigate which plans it is necessary to support.

Our research paper is organized as follows. In Section 2, we provide a summary of our critical literature review on the domain with an emphasis on studies undertaken during the COVID-19 pandemic related to residents training. In Section 3, we discuss our research methodology. The key results of our analysis are presented in Section 4; in Section 5 we constructively interpret the key findings and we proposed a methodological framework for well-being and training quality in residents training' programs. Finally, in Section 6, we provide our conclusions and future research directions.

2. Critical Literature Review

The COVID-19 pandemic, beyond its health determinants, has had a variety of indirect implications on society, sustainable health strategies, and medical training and policy making. It also poses critical challenges for the digital transformation of health and for the adoption of emerging and streamline technologies for new value-adding medical services. In the abundant recent literature on the COVID-19 pandemic, our study is linked to diverse research areas. Our special interest and our research problem are related to the impact of the COVID-19 pandemic on medical training and its subsequent psychological implications for trainees. In this context, the assessment of psychological pressure among health experts is a key priority for the justification of responsive actions and the updating of policies relates to the psychological support of doctors [4,5]. Mental health, well-being, and quality of life of residents and fellows in medical training programs is another significant component of sustainable health.

The analysis of the impact of COVID-19 on residents' health during the COVID-19 pandemic in a recent study [6] provides interesting insights. Twenty-nine residents that participated in the relevant study emphasized that social isolation due to a colleague's inability to engage in outdoor activities and social gatherings provide significant psychological pressure on their personalities. Furthermore, they expressed their concern about the impact of COVID-19 on the quality and mode of their didactic education as well as on their clinical rotations.

In another recent study in Saudi Arabia [7], researchers investigated the issue of mental health among healthcare providers during the COVID-19 pandemic. Some of their key findings are also related to the objectives of this research. The respondents in the survey from a variety of health specialties at a high level had depressive disorders which ranged from mild (24.9%), moderate (14.5%), and moderately severe (10%) to severe (5.8%). Half of the sample had generalized anxiety disorder (51.4%), which ranged from mild (25.1%) and moderate (11%) to severe (15.3%). In our study, we intend to investigate the anxiety and depression rates in residents and fellows of medical training programs in Saudi Arabia and also to understand which are the determinants of the COVID-19 pandemic that have a direct association with psychological factors of trainees.

In a closely related study [8], researchers analyzed the psychological status of healthcare workers during the COVID-19 pandemic. In this interesting research, authors tried to combine and to understand the association of depressive behavior and anxiety of 745 healthcare workers from 15 hospitals. The main finding of the research is that almost half of the respondents faced depression and anxiety symptoms (56.3% had depressive symptoms, while 46.7% had anxiety symptoms). The main factors investigated and confirmed for causal effect with depression are age, residency status, department, stigmatization, and living in a conflict zone. Anxiety symptoms were found to be linked with characteristics of healthcare workers like age, department, years of experience, working hours per week, internal displacement, stigmatization, living in a conflict zone, and verbal abuse. In our study, we focus also on determinant factors of depression and anxiety and we also want to investigate to what extent the training institutions of residents, the educational process, and the mode of education during the COVID-19 pandemic are related to psychological factors. One of the greatest concerns of the respondents (n = 73; response rate—73.7%) was the decline in operative exposure. Another significant finding of this research is that residents are more worried and concerned about the health of their loved ones than about their own risk and exposure of contracting COVID-19 [9].

In another study in Singapore [10] with 122 participants (frontline healthcare workers), the main finding was increased level of anxiety and depression during the COVID-19 pandemic. The key aspects of increased depression rate were associated with factors such as physical exhaustion, loss of appetite, poor sleep quality and the use of a sedative.

Similar findings also appear in a study among 906 healthcare staff from 5 hospitals in India and Singapore [11]. Increased anxiety and severe depression among participants were associated with various factors. Almost 17% of the healthcare workers showed moderate to very severe symptoms. Out of the 906 healthcare workers who participated in the survey, almost 14% showed moderate to very-severe depression and almost 11% moderate to extremely severe anxiety.

Studies in other countries like the USA [12] investigate the mental health toll on healthcare staff. The stark findings also reveal increased levels of anxiety and depression among the participants in a large sample (n = 657).

Numerous other studies [13–18] provide interesting insights for the multidimensional psychological pressure of COVID-19 to healthcare workers, residents, fellows and medical staff. In the complimentary scientific approaches that are deployed and beyond the various research tools that are deployed, there is a systematic argument for the association between COVID-19 and depression, anxiety, and stress of participants in the surveys. Cross-sectional studies and focused group studies in China [13,14] investigated and monitored the evolution of the COVID-19 pandemic and its association with psychological factors of health specialists. Other studies [15] focused on symptoms of posttraumatic stress, anxiety, depression, levels of resilience and burnout in Spain with significant findings that are closely related to the overall trend of the impact of the COVID-19 pandemic on the mental health of healthcare staff. Other interesting studies focus on issues like job satisfaction and the health conditions of medical staff [16–18] during the pandemic.

Medical education has faced many interruptions throughout history. As a result, the quality of training has been negatively affected. For example, some medical schools shortened their medical programs by several years to cover the acute need of doctors during the World War [19].

A heavy burden is placed on society and health care systems to provide treatment to a continuously increasing number of COVID-19 patients and to afford extensive measures to contain or limit the individual and community transmission of infection. It is almost 10 months since the cluster of patients confirmed with COVID-19 disease was reported in Wuhan, China [20]. Several countries have effectively managed the COVID-19 health crisis situation; however, others are still battling due to the increasing number of cases and limitations caused by lack of treatment resources and facilities. Currently, there are ongoing clinical trials for the treatment and prevention of COVID-19 that serve as the hope for people to overcome this pandemic crisis.

Health care workers are at greater risk of acquiring the infection because they are directly engaged in the diagnosis, treatment, and care for patients with COVID-19. In the United States, it was reported that redeployment of residents increased by 11% and almost 50% in states with higher incidences of COVID-19 and 70% exposure to patients who tested positive for COVID-19, which is similar to this study (i.e., 72.77%) [21]. In a single tertiary center study in Wuhan, China, the infection rates of first-line and non-first-line health care workers were shown to be 0.5% and 1.45%, respectively [22].

The current COVID-19 pandemic is similar to other historical events that posed major challenges to the global health system. Postgraduate medical trainees (residents and fellows) had to reduce their routine clinical and surgical practice due to the increasing number of COVID-19 patients [23]. Non-urgent surgeries are postponed, and outpatients' appointments have been cancelled or transformed into telehealth clinics. To limit the exposure to the COVID-19 virus, some trainees have been placed on standby. All these changes will affect their skills and reduce their learning opportunities to achieve competency in their fields of specialties [24].

This pandemic is also affecting the accreditation system. As the impact of COVID-19 differs from one region to another, the Accreditation Council for Graduate Medical Education (ACGME) allocated three stages and each training program will function in one of them during the period of the COVID-19 pandemic:

- Stage 1—the workflow will continue as usual with no changes in requirements.
- Stage 2 will be implemented according to the increase in clinical demands, and some residents may be shifted to COVID-19 patient care duties. In this stage, the impact on education will be acknowledged by suspending accreditation and self-study activities. Residents will be supervised by via telemedicine. Virtual lectures and journal clubs are allowed. Fellows will function by attending the same working hours as previously. The decision of graduation requirements will be left to the program director.
- Stage 3 is considered high Pandemic Emergency Status. All educational activities are suspended, and care for COVID-19 patients is prioritized [25].

On 9 March 2020, the ACGME also postponed all scheduled accreditation visits [26]. The American Board of Radiology has delayed the board certification exam to September. Such interruption in graduation and credentialing might delay the onboarding time for new residents [27]. Radiology and otolaryngology training programs in the USA suggested additional didactic sessions to fulfill the minimum requirement for graduation [28,29].

Many residency training programs have designed innovative online platforms to bridge the educational gap during the COVID-19 era [25,30–33]. Chick R.C. proposed innovative solutions for continuing the academic education for surgical residents that do not require large group meetings [25]. Solutions are summarized in Table 1.

Table 1. Innovative solutions for continuing the academic education (summary from other research contributions).

Solution	Aim
Flipped online classroom model	Provide learners with educational material in the form of a prerecorded videos that they can watch anytime remotely before the conference
Novel social-media-based platform	Discussion of surgical topics, daily exposure to practice questions
Weekly academic conferences/webinars via teleconference	Live online lectures
Tele-clinics	Keep trainees involved with outpatients' clinical activities
Surgical videos	Self-review of surgeries

Similarly, Schwartz et al. strongly [30] recommended the use of videoconferencing and tele-clinics to maintain residents' education. Stambough et al. [31] reported on orthopedic surgical education by using online meetings, journal clubs, and a surgical videos database such as cadaver, bone-substitute simulations, and the Orthopedic Video Theater. Similarly, in Singapore, cadaver is reported as an intelligent sourcing for self-directed learning in orthopedics. Another study reported that a virtual reality (VR) procedure improves the residents' surgical skills when transferred to the real world [32]. Vargo et al. reported on the experience of urology residents in virtual conference teaching of high-priority robotic cases [33].

Research activity has been disrupted too. To ensure the continuation of residency research programs, research mentors encourage residents to work from home and develop research projects during their off days by meeting remotely [28,34]. Crosby et al. reported on the Southern Otolaryngology Department at the Illinois University School of Medicine holding a weekly virtual research meeting for all residents to review the projects' updates in depth [28]. In-parallel, radiology programs in North America encourage research mentors and their residents to have continuing meetings to discuss their projects during the pandemic, including methodology, statistical analysis, funding, and reshaping the project into publishable material [34].

The COVID-19 pandemic caused an increased burden on the mental wellness of trainees [34]. Previous studies on Ebola and SARS show a severe emotional distress during the outbreaks [35]. Changes related to the COVID-19 pandemic have led to an unprecedented impact on residents' education and may contribute to resident burnout. One of the greatest challenges for programs at this time is to maintain the well-being of residents in the face of uncertainty, stress, and anxiety. Recently, ACGME emphasized the role of program directors to maintain a high-quality educational environment which directly affect residents' well-being [36]. The American College of Radiology recently conducted webinars, podcasts and released fitness apps as recourses for residents' self-care and well-being [36]. Khallafallah reported on a neurosurgical resident who had concerns about not achieving the surgical milestones and the future of their training [37]. In the USA, about 54.4% of residents and fellows have shown at least one symptom of burnout in the form of depersonalization, emotional exhaustion, or reduced senses of accomplishment [38].

The literature summarized in this section highlights a multifaceted research problem. In this context, several diverse variables are interconnected, including the mental health, the quality of educational programs, as well as the variety of technology-enhanced learning tools capable of enhancing the educational process. In the next section, we provide an overview of our research methodology.

3. Research Methodology

The key methodological approach for our research is based on an integrated strategy. The focus is on the detailed analysis of the impact of COVID-19 on psychological factors of residents and the impact of COVID-19 on the educational process and performance. In Figure 1, we summarize the key aspects of our methodology.

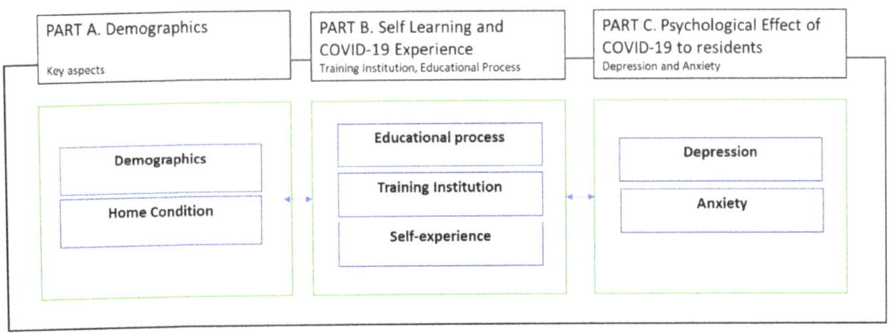

Figure 1. The research approaches.

This research study is part of a greater systematic research in the Saudi Health Commission for Health Specialties. Our main aim is to understand the self-experience of residents in training programs during the pandemic. For this reason, the training institution and the educational process are set as targets in our analysis and provide meaningful pillars in our research tool. The understanding of how COVID-19 was perceived from residents during their training and their service to hospitals is a key development for enhanced decision making. As explained later, we constructed a survey emphasizing complementary aspects of daily professional practice and training activities of residents. In the third component of our research approach, we intend to analyze the psychological effect of COVID-19 on residents with a key emphasis to be placed on two critical factors, namely depression and anxiety.

Our research utilizes a population-based cross-sectional study conducted among trainees from various training centers affiliated with the Saudi Health Commission for Health Specialties (SCFHS). The survey questionnaire was distributed online to 14,000 trainees (i.e., residents and fellows). Consequently, participants who completely filled out the questionnaire will be included in the analysis.

This study used a self-designed questionnaire composed of four parts:

1. Socio-demographic characteristics of the participants.
2. Participants' evaluation on the training program during the COVID-19 pandemic. The result of training program evaluation was used as supplementary data to support the discussion of study findings.
3. Assessment of participants' mental health status (i.e., anxiety and depression) utilizing the Generalized Anxiety Scale (GAD-7) [39] and Patient Health Questionnaire depression module (PHQ-9) [40].
4. Determining factors of anxiety and depression of participants.

GAD-7 is a seven-item anxiety scale use for screening and assessing severity of general anxiety disorder. This had good reliability with sensitivity of 89% and specificity of 82% [27]. Severity score is calculated by assigning scores of 0, 1, 2, and 3 to the response categories "not at all", "several days", "more than half the days", and "nearly every day", respectively. The total score for the seven items ranges from 0 to 21. The obtained score of 0–4 correspond to minimal anxiety, 5–9 mild anxiety, 10–14 moderate anxiety and 15–21 severe anxiety.

PHQ-9 on the other hand is a nine-item depression scale with good reliability (i.e., sensitivity of 88% and a specificity of 88%) and considered as a valid measure of depression severity [28]. PHQ-9 scoring is calculated by assigning scores of 0, 1, 2, 3 to the response categories "not at all," "several days," "more than half the days," and "nearly every day", respectively. The total score for the nine items ranges from 0 to 27. The obtained score of 0–4 corresponds to minimal depression, 5–9 to mild depression, 10–14 to moderate depression; 15–29 relates to moderately severe depression and 20–27 to severe depression.

In Figure 2, we provide a more detailed overview of the research design of our study. In Appendix A, we also provide the full questionnaire designed for the purpose of our research. The components of our research tool are summarized as follows:

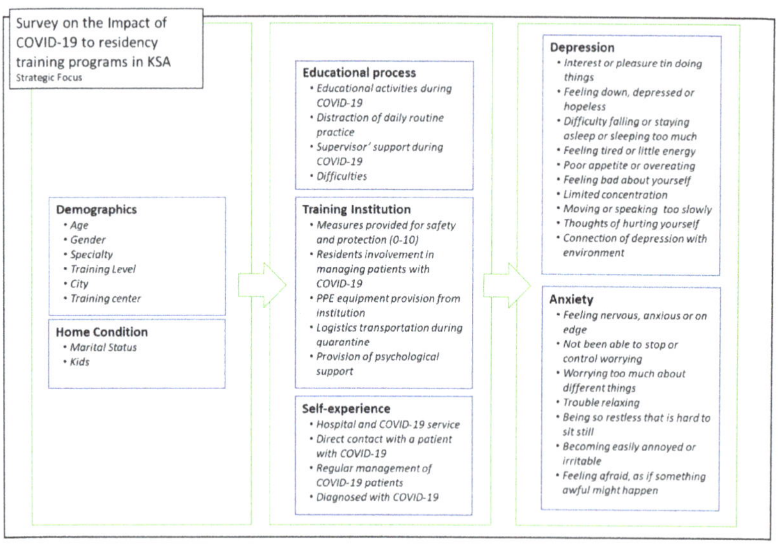

Figure 2. The research model of COVID-19 impact on training.

Educational process: The analysis of the COVID-19 impact on the medical educational process of residents is a key priority in our research. We are interested in understanding how the pandemic posed changes to the educational activities during the COVID-19 pandemic and to interpret them in the discussion and conclusions of our research. In addition, we aim to understand how the distraction of the daily educational routine practice posed psychological pressure to residents and fellows. Another key aspect of our research analysis is related to supervisors' contribution to residents' perceptions of the quality of education in the context of COVID-19. Finally, we are interested in understanding the difficulties of residents to deal with their training and professional activity during the COVID-19 pandemic. The analysis of the key findings in the discussion section informs the scientific dialogue for the required actions towards the enhancement of the educational process during the COVID-19 outbreak. Below we summarize the key focus of the educational process analysis of our research:

- *Educational activities during the COVID-19 pandemic.*
- *Distraction of daily routine practice.*
- *Supervisor' support during the COVID-19 pandemic.*
- *Difficulties.*

Training Institution: In this variable we emphasize on the significant value components of the training institution capability to support residents during the COVID-19 pandemic. One of the most important concerns and research targets is to analyze the perception of residents regarding the measures provided to them for personal safety and protection. We are very much interested in this aspect as we do believe there is also an association with the psychological pressure of residents in training institutions that also serve as hospitals for COVID-19. Another dimension is related to the realization of residents' involvement in managing patients with COVID-19 as well as practical aspects of the support provided to them by the institution. This includes PPE equipment and logistics transportation during quarantine. In addition, the provision of psychological support from the training institution to residents is the focus of our research, since we want to understand the ways it is implemented, and its efficiency as valued by the residents themselves.

- *Measures provided for safety and protection (0–10).*
- *Residents involvement in managing patients with COVID-19.*
- *PPE equipment provision from institution.*
- *Logistics transportation during quarantine.*
- *Provision of psychological support.*

Self-experience: We are very much interested in understanding residents' self-experience with COVID-19 in their training institutions. Several complementary aspects of the phenomenon include but are not limited to the following: the analysis of residents' involvement and service in hospitals that serve patients with COVID-19, and the measurement of the population of residents' that had direct contact with COVID-19 patients and their service in regular COVID-19 units. In addition, we investigate the number of residents that diagnosed with COVID-19. The self-experience component in our research tool also provides significant insights regarding the psychological impact of COVID-19 upon residents. In the discussion and conclusions of our research, we elaborate on the key findings and we make significant recommendations.

- *Hospital and COVID-19 service.*
- *Direct contact with a patient with COVID-19.*
- *Regular management of COVID-19 patients.*
- *Diagnosed with COVID-19.*

The third pillar of our research approach focuses on the depression and anxiety of residents as it is imposed by the various aspects of the pandemic and the special interconnections with the educational process and the training institution. These two aspects are set as key priorities and key research objectives of our research. We intend to analyze the impact of the pandemic on the anxiety and depression of residents and to recommend actions for an integrative support of residents in their training and professional conduct. We understand from the beginning that this is an ambitious objective given the limitations of our study that are discussed in the relevant sections.

Depression: For the analysis of Depression, as we already discussed, we used PHQ-9 which is based on a nine-item depression scale with good reliability. We believe that during the COVID-19 pandemic, such a scale provides a good measurement for the estimation of depression provided by the COVID-19 pandemic. It would be also significant through research to integrate into such scales specific items related to the interaction of residents or medical staff with COVID-19. In the discussion of the key findings of our research we provide our insights relating to this direction. The key components of this scale include the following factors.

- *Interest or pleasure in doing things.*
- *Feeling down, depressed, or hopeless.*
- *Difficulty falling or staying asleep or sleeping too much.*
- *Feeling tired or little energy.*
- *Poor appetite or overeating.*
- *Feeling bad about yourself.*
- *Limited concentration.*
- *Moving or speaking too slowly.*
- *Thoughts of hurting yourself.*
- *Connection of depression with environment.*

Anxiety: For the analysis of anxiety, we adopted the Generalized Anxiety Scale (GAD-7) which includes five measures. It is extremely important in our research to measure anxiety and to categorize our sample into the mild, moderate, and severe clusters. We also need to interpret, to the best possible degree, how these aspects are interconnected to COVID-19 or they are also related to the training and professional conduct of residents. We understand that this is an ambitious effort given the limitations, but we will try to provide our key interpretation in the discussion and conclusions section. The following are some of the key components of the GAD-7 scale for anxiety measurement:

- *Feeling nervous, anxious or on edge.*
- *Not being able to stop or control worrying.*
- *Worrying too much about different things.*
- *Trouble relaxing.*
- *Being so restless that is hard to sit still.*
- *Becoming easily annoyed or irritable.*
- *Feeling afraid, as if something awful might happen.*

The research objectives of our study are summarized as follows:

- ***Research Objective 1***: What are the key implications of the COVID-19 pandemic on the training and educational process of residents in KSA?
- ***Research Objective 2***: What are the key perceptions of residents regarding the support from training institutions during the COVID-19 pandemic?
- ***Research Objective 3***: What are the key aspects of psychological pressure including depression and anxiety of residents in KSA during the COVID-19 pandemic?
- ***Research Objective 4***: What are the key implications of this study for future recommendations towards best practices of training, psychological support, and technology-enhanced learning of residents' medical training?

Data analysis was conducted using a statistical analysis Statistical Package for the Social Sciences, version 21.0 (SPSS, version 21.0; Chicago, IL, USA). Descriptive statistics (frequencies and percentages) were used to describe the participants' socio-demographics, determining factors of anxiety and depression and trainees' evaluation on training programs. Forward stepwise multinomial regression analysis was then configured to use selection criterion obtained from demographic variables and determining factors to identify the odds of anxiety and depression among participants. The level of statistical significance was set to $p < 0.05$.

In the next section we summarize the key facts from our analysis.

4. Analysis and Main Findings

In the previous section, we presented the key aspects of our research methodology. In this section, we provide the basic aspects of our analysis. We organize our presentation according to the relevance of the three methodological pillars, namely, Demographics, Educational Process, and Training Institution and Psychological Factors, and we focused on the anxiety and depression of residents.

The survey questionnaire was distributed online to 14,000 trainees (i.e., residents and fellows). Of the 160 medical specialty training programs in 600 local and international training centers affiliated by Saudi Commission for Health Specialties, 1985 trainees participated in the survey and 1528 completed the GAD-7 anxiety and PHQ-9 depression questionnaires and were included in final data analysis. In Tables 2 and 3 below, we present the key demographic data for our participants.

Table 2. Trainee evaluation of services provided during the COVID-19 pandemic.

	Items	Mean	Standard Deviation
1.	How happy are you with the measures provided by your institution for your own safety and protection during the COVID-19 pandemic?	4.39	3226
2.	How supportive is your supervisor during the COVID-19 pandemic?	4.26	3484
3.	How would you rate the institution's provided psychological support?	3.16	2756
4.	How would you rate the Da'em program provided by SCFHS?	1.46	2369
5.	How would you rate the curriculum educational experience during the COVID-19 pandemic?	3.59	3184

4.1. Demographics

The trainees' mean age was 29.4, they were mostly male (see Figure 3, below), married (see Figure 4, below), and had no children (see Figure 5, below).

Concerning the number of children, most of the respondents had no children (58.2%) while 33.8% had 1 to 2 children, 7.3% three to four children and 0.7% had more than four children.

Some more facts about the marital status of participants in our survey are summarized as follows:

- Single (living with your parents): 32.9%.
- Single (living alone): 12.2%.
- Married (living with your family): 40.6%.
- Married (living alone): 12.3%.
- Unspecified: 2.0%.

Table 3. A summary of key findings of the statistical analysis

Items	Anxiety	Depression
Residents Level 1	less likely to have mild and moderate anxiety	
Residents Level 2	less likely to have mild anxiety	less likely to have mild (and moderate depression
Residents Level 3	less likely to have moderate anxiety	
Residents Level 4	more likely to have moderate and severe anxiety	
Residents Level 5	-	are more likely to have severe depression
Residents Level 6	-	
Fellows Level 1	-	
Fellows Level 2	-	less likely to have moderate depression
Fellows Level 3	-	
Trainees provided with enough PPE	less likely to have mild, moderate, and severe anxiety	
Those who regularly managing COVID-19 patients	more likely to have moderate and severe anxiety	more likely to have severe depression
Those who had direct contact with COVID-19 patients		more likely to have moderately severe depression
Those who were provided with transportations logistics to training center		less likely to have moderate depression and severe depression
Males		less likely to have mild depression, moderate severe and severe depression

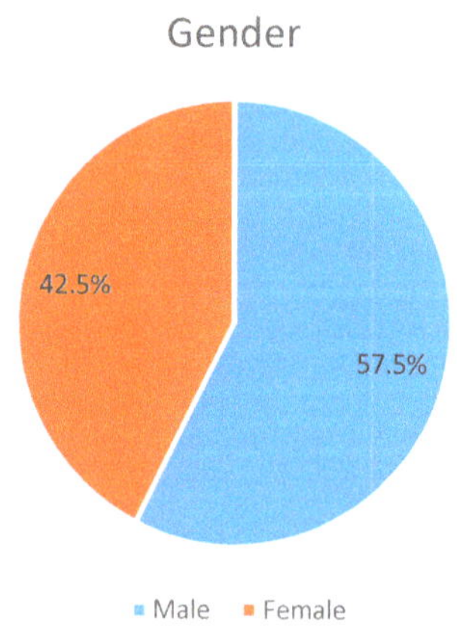

Figure 3. Demographics: Gender.

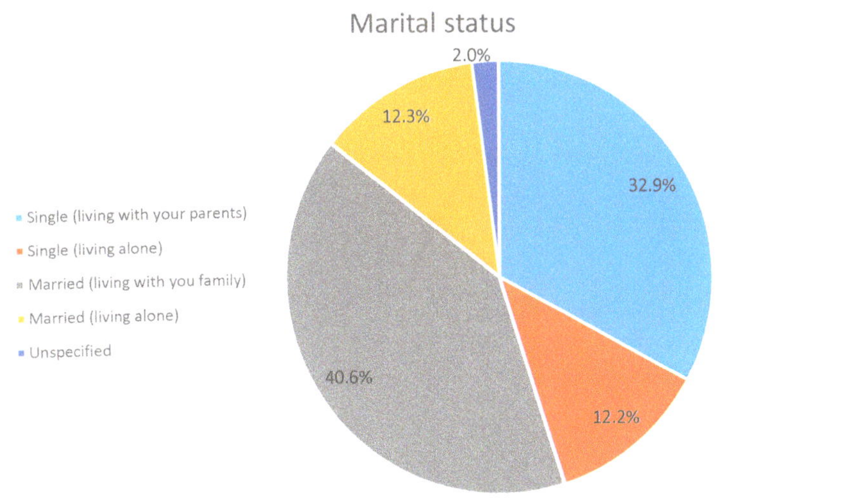

Figure 4. Demographics: Marital status.

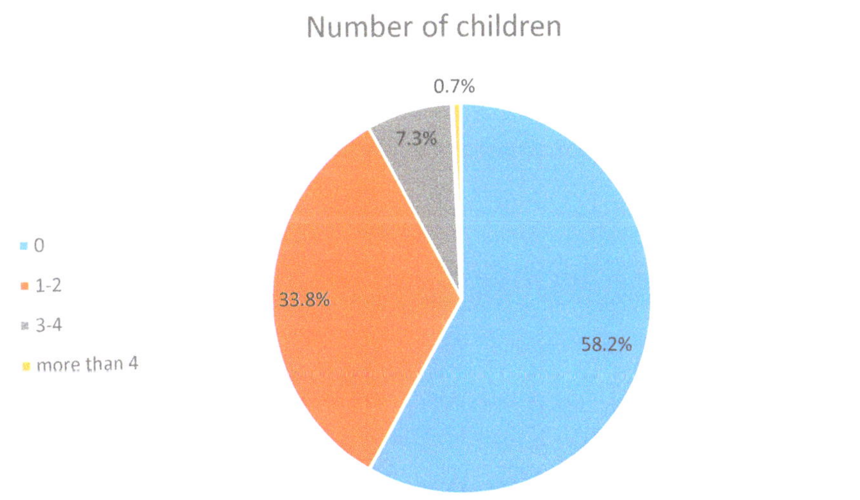

Figure 5. Demographics: Number of children.

The number of trainees in the residency training program is higher than those in the fellowship training program (see Figure 6 below).

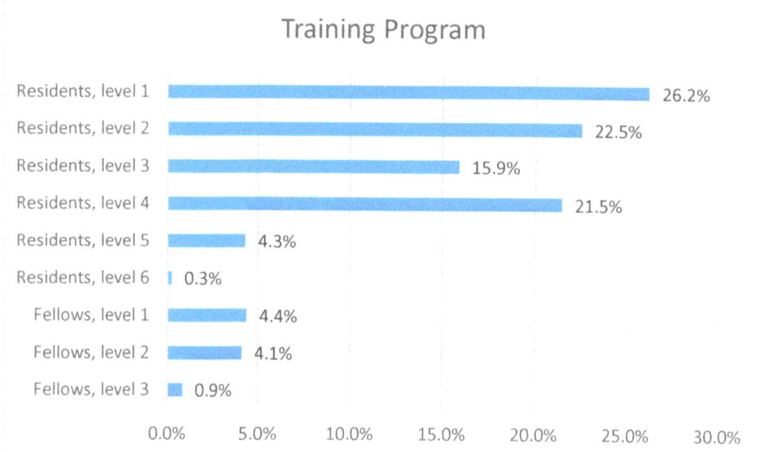

Figure 6. Demographics: Training program type (Resident/Fellow).

The number of trainees in the residency training program is higher (1385, 90.6%) than that in the fellowship training program (143, 9.4%). In Figure 6, more data on the allocation of respondents to different residency or fellow programs are provided. In the next section, we focus on the second pillar of our methodological approach, which is related to the educational process and training institution components.

4.2. Educational Process and Training Institution

Two of the most significant research objectives of the research, as presented in Section 3 above, are related to the understanding of the key implications of the COVID-19 pandemic on the training and educational process of residents and fellows in KSA as well as an analysis of the residents' key perceptions of the training institution's support during the COVID-19 pandemic.

In Figure 7, we provide a basic analysis related to the trainee's evaluation of the services provided to them during the COVID-19 outbreak from the training institution. The overall attitude is not satisfactory. In the scale of 0 to 10 (0 = not supportive at all 10 = completely supportive), the average score and the standard deviation points to several possible enhancements, as we will discuss in the conclusion. As it is summarized in Table 2, there is a limited satisfaction in terms of the measures provided by training institutions' regard for safety and protection during the COVID-19 pandemic. The low mean of 4.39 out of 10 is indicative of residents and fellows feeling somehow unprotected, or at least, expecting higher protection. In the same overview, the psychological support service during the COVID-19 outbreak launched for this purpose (Da'em in Saudi) was poorly rated (mean 1.46) by trainees. This is a very significant finding that poses many questions for its future development and its redesign. We will comment on this key finding in the discussion section. Furthermore, the support provided by supervisors was also rated neutrally (mean 4.26). Another critical aspect for the mode of training during the COVID-19 pandemic is the perception of trainees of the rating of the curriculum educational experience, which received a mean of 3.59.

The synthesis of the various findings of our analysis is that residents and fellows felt a multidimensional pressure in their educational process during the COVID-19 outbreak. Given the fact that many of them also served hospitals with COVID-19 patients, this adds to the complexity of the phenomenon, especially in the context of psychological pressure. It is extremely important for our study to further investigate the psychological pressure of trainees in the form of anxiety and depression.

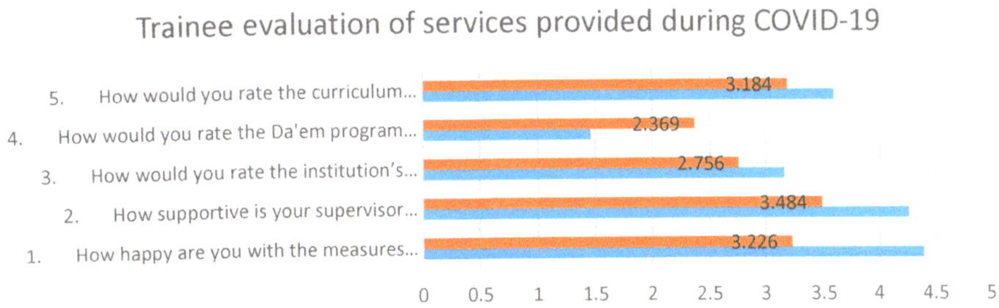

	1. How happy are you with the measures provided by your institution for your own safety and protection during COVID-19 pandemic?	2. How supportive is your supervisor during COVID-19 pandemic?	3. How would you rate the institution's provided psychological support?	4. How would you rate the Da'em program provided by SCFHS?	5. How would you rate the curriculum educational experience during COVID-19?
Standard deviation	3.226	3.484	2.756	2.369	3.184
Mean	4.39	4.26	3.16	1.46	3.59

Standard deviation   Mean

Figure 7. Trainee evaluation of services provided during the COVID-19 pandemic.

In the next section, we emphasize the study of the psychological pressure on trainees during the COVID-19 pandemic.

4.3. Psychological Factors: Depression and Anxiety

Some of the data presented in the previous section highlight the psychological pressure on residents and fellows during the COVID-19 pandemic. It is important before we present the key findings related to the assessment of participants' mental health status (i.e., anxiety and depression) utilizing the Generalized Anxiety Scale (GAD-7) and Patient Health Questionnaire depression module (PHQ-9) to investigate more facts about the psychological pressure on trainees.

In Figure 8 below, we provide a summary of the opinions of trainees with key aspects of their interaction and self-experience during the COVID-19 pandemic. A first interpretation of the key findings is provided below.

Most trainees have been in direct contact with a patient with COVID-19 at their training hospital (73%). This is a significant finding which also highlights the key contribution of residents and fellows to the management of the pandemic crisis. It is extremely important to investigate, through our methodological tools (GAD-7 and PHQ-9 scales), the impact of this fact on the anxiety and depression of trainees. Additionally, almost half of the trainees (53%) regularly managed patients with COVID-19. This exposure can also contribute to the psychological pressure of trainees. Given the fact that the educational process and the educational activities still take place with variations it is interesting to investigate the situation further. Six percent (6%) of the trainees have been diagnosed with COVID-19 due to their service to the training institution.

On the other hand, respondents provided useful insights into the institution's support during the COVID-19 pandemic. Special training for a pandemic crisis was offered to 36% of the respondents. It is quite surprising that the remaining sixty four percent (64%) received no special training in such a demanding situation. Almost half of the respondents (49%) stated that they received enough Personal Protective Equipment (PPE). Unfortunately, the other half of the respondents claim that they did not receive enough protective equipment, and this is one more key finding. We will comment on these findings in our discussion section.

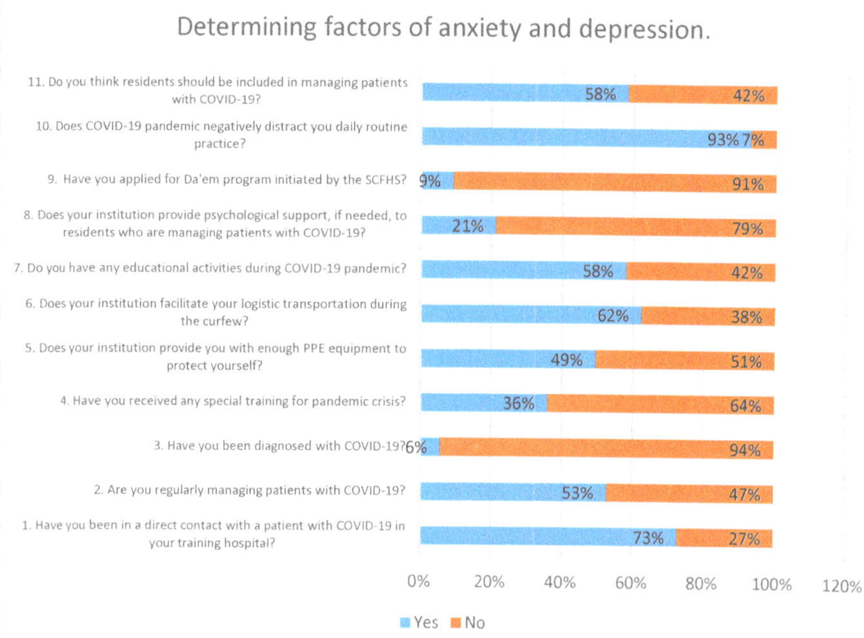

Figure 8. Determining factors for trainee's anxiety and depression during the COVID-19 pandemic.

The training institution facilitates the logistic transportation of 62% of the trainees during the COVID-19 pandemic and also, at a much lower rate (21%), provides psychological support, if needed, to residents who are managing patients with COVID-19. Surprisingly, only 9% of trainees applied for psychological support through the Da'em program for psychological support initiated by the SCFHS. This is also unsatisfactory and in the discussion page we elaborate further with some propositions. Most of the residents (58%) still have educational activities during the COVID-19 pandemic and almost all of them (93%) agree that the COVID-19 pandemic negatively distracts their daily routine practice. While all the previous aspects of increased psychological pressure are evident, the majority of trainees (58%) believe that residents should be included in managing patients with COVID-19.

The median scores of trainees' anxiety and depression were 3.00 (1.00, 4.00) and 3.00 (1, 5.00). Of the 1528 trainees, 201 (13.2%) have minimal anxiety, and 408 (26.7%), 375 (24.5%), and 544 (35.6%) have mild, moderate, and severe anxiety, respectively (see Figure 9, below). This is a stark finding of our research. It appears that the COVID-19 pandemic has increased the levels of anxiety over time.

From a decision-making point of view, it is extremely important to investigate ways to support the residents and fellows in this context, since anxiety has an impact on their efficiency and decision-making capability. In our discussion section, we elaborate further on this issue.

The findings related to the depression level of trainees are summarized in Figure 10. Given the fact that we do not have comparative studies about the depression rates that were recorded before COVID-19, the findings show a rather increased depression rate among trainees. The fact that almost one tenth of the full sample have minimal depression is an indication that the COVID-19 pandemic has put more pressure on trainees.

Some more data on the analysis of depression with the use of PHQ-9 scale are provided below. In our sample, 152 respondents (9.9%) have minimal depression and 358 (23.4%), 373 (24.4%), 341 (22.3%) and 304 (19.9%) have mild, moderate, moderate severe and severe depression, respectively.

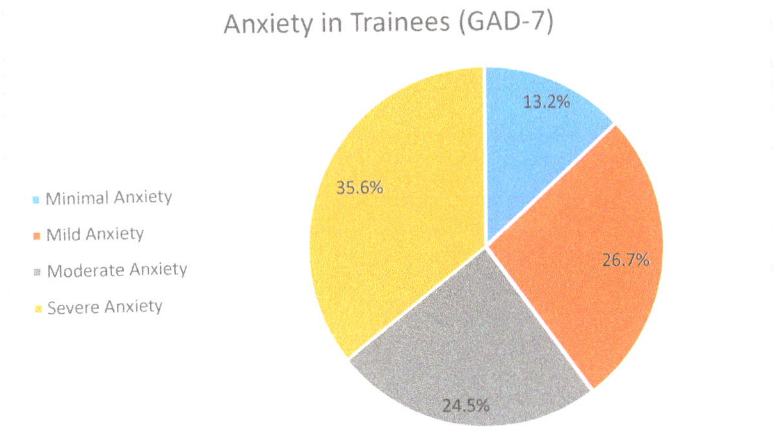

Figure 9. Anxiety in trainees based on Generalized Anxiety Scale (GAD-7) scale during the COVID-19 pandemic.

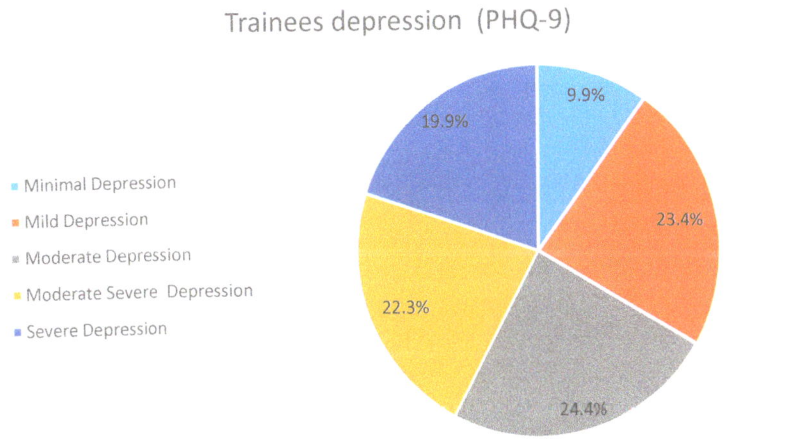

Figure 10. Depression in trainees based on Patient Health Questionnaire depression module (PHQ-9) scale during the COVID-19 pandmeic.

To further investigate the correlation between trainee's anxiety and trainee's depression, we deployed relevant statistics tests. This is directly related to the intention of our research objective to understand how the COVID-19 pandemic had an impact on trainee's mental health, as stated in our research methodology section.

Using Pearson correlation analysis, the correlation coefficient between trainees' anxiety score and depression score was 0.791 ($p < 0.00$), which indicates that ***trainees' anxiety and depression are highly positively correlated with the COVID-19 pandemic***.

This is a significant finding of our research. While someone can associate COVID-19 with psychological pressure on trainees and medical staff, it is a bold finding in such a big sample (1528 respondents) to approximate the depression and anxiety rates.

We intend soon to measure the same variables again with an enhanced research instrument in order also to understand the causal effect of COVID-19 depression and anxiety of residents and fellows.

In order to further investigate the connections of anxiety and depression, we deployed further statistics, including the Odds Ratio (OR) that quantifies the strength of the association between two events. Additionally, we deployed multinomial regression analysis utilizing a forward stepwise approach to automatically select criterion or variables of trainees' anxiety and depression. Two models were established (anxiety and depression, Tables 3, A1 and A2 in Appendix A). Below are some of the key findings.

Using minimal anxiety as a baseline data:

- Level 2 trainees of a residency training program are less likely to have mild anxiety *(OR 0.547; CI 95% 0.338–0.887, p = 0.014)*.
- Level 1 trainees of a fellowship training program are less likely to have mild (*OR 0.451; CI 95% 0.212–0.957, p = 0.038*) and moderate anxiety (*OR 0.323; CI 95% 0.137–0.764, p = 0.010*).
- Level 3 trainees of a fellowship training program are less likely to have moderate anxiety *(OR 0.094; CI 95% 0.010–0.853, p = 0.036)*.
- Males are less likely to have mild *(OR 0.520; CI 95% 0.352–0.770, p = 0.001)*, moderate *(OR 0.445; CI 95% 0.299–0.662, p = 0.000)* and severe anxiety *(OR 0.325; CI 95% 0.222–0.477, p = 0.000)*.
- Trainees provided with enough PPE are less likely to have mild *(OR 0.511; CI 95% 0.353–0.742, p = 0.000)*, moderate *(OR 0.481; CI 95% 0.329–0.702, p = 0.000)* and severe anxiety *(OR 0.353; CI 95% 0.245–0.509, p = 0.000)*.
- Level 4 trainees of a residency training program are more likely to have moderate *(OR 2.101; CI 95% 1.142–3.865, p = 0.017)* and severe anxiety *(OR 2.484; CI 95% 1.365–4.519, p = 0.003)*.
- Those who regularly managing COVID-19 patients are more likely to have moderate *(OR 1.771; CI 95% 1.218–2.576, p = 0.003)* and severe anxiety *(OR 2.626; CI 95% 1.829–3.772, p = 0.000)*. Complete results are shown in Appendix A.

Using minimal depression as a baseline data:

- Males are less likely to have mild depression (OR 0.631; CI 95% 0.402–0.990, *p* = 0.045), moderate severe *(OR 0.398; CI 95% 0.252–0.629, p = 0.000)* and severe depression *(OR 0.379; CI 95% 0.237–0.607, p = 0.000)*.
- Level 2 trainees of a residency training program are less likely to have mild (OR 0.465; CI 95% 0.264–0.819, *p* = 0.008) and moderate depression (OR 0.460; CI 95% 0.261–0.809, *p* = 0.007).
- Level 2 trainees of a fellowship training program are less likely to have moderate depression (OR 0.314; CI 95% 0.118–0.833, *p* = 0.020).
- Level 5 trainees of a residency training program are more likely to have severe depression *(OR 10.746; CI 95% 1.313–87.968, p = 0.027)*.
- Trainees that have been in direct contact with COVID-19 patient are more likely to have moderately severe depression *(OR 2.441; CI 95% 1.418–4.201, p = 0.001)*.
- Those who are regularly managing COVID-19 patients are more likely to have severe depression *(OR 2.253; CI 95% 1.346–3.771, p = 0.002)*.
- However, those trainees that are provided with transportation are less like to have moderate (OR 0.575; CI 95% 0.361–0.914, *p* = 0.019) depression and severe depression *(OR 0.464; CI 95% 0.287–0.749, p = 0.002)*. Complete results are shown in Tables A1 and A2 in Appendix A with reference also to statistically significant results.

In Figures 11–13 below, for reference purposes we summarize some observations related to the anxiety in various groups without all of them to be statistically significantly. These findings require further investigation, and further discussion is provided in the section below.

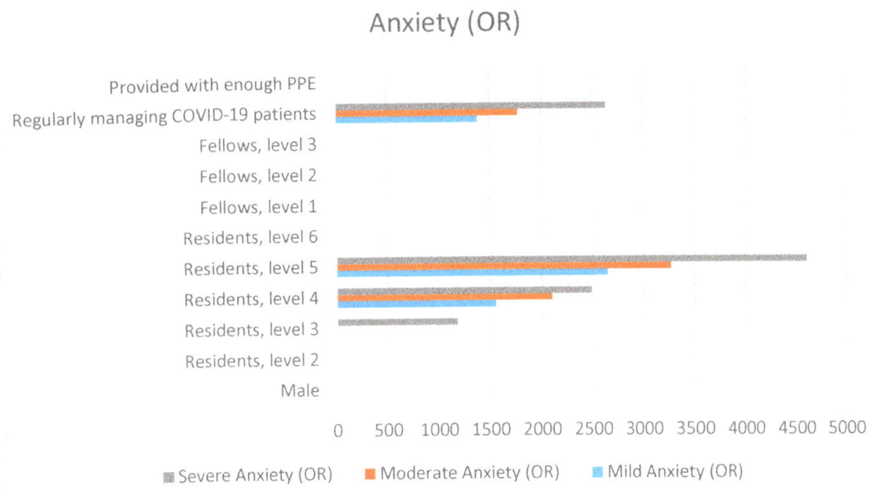

Figure 11. Anxiety of trainees based on GAD-7 scale during the COVID-19 pandemic.

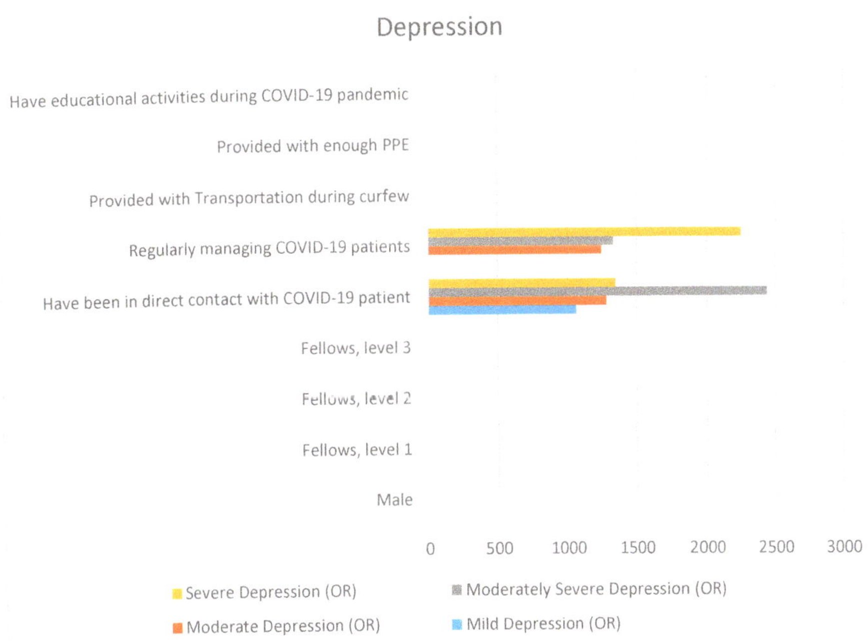

Figure 12. Depression in trainees based on PHQ-9 scale during the COVID-19 pandemic.

The analysis in this section is not exhaustive. It refers mostly to the three research objectives of our study. Various limitations of our study that will be explained in the relevant section pose some methodological questions. For example, no previous study has measured the degrees of anxiety and depression. However, the scientific approach followed in this study permits us to discuss in the next section the significant findings and also the key interpretations of our research for decision making.

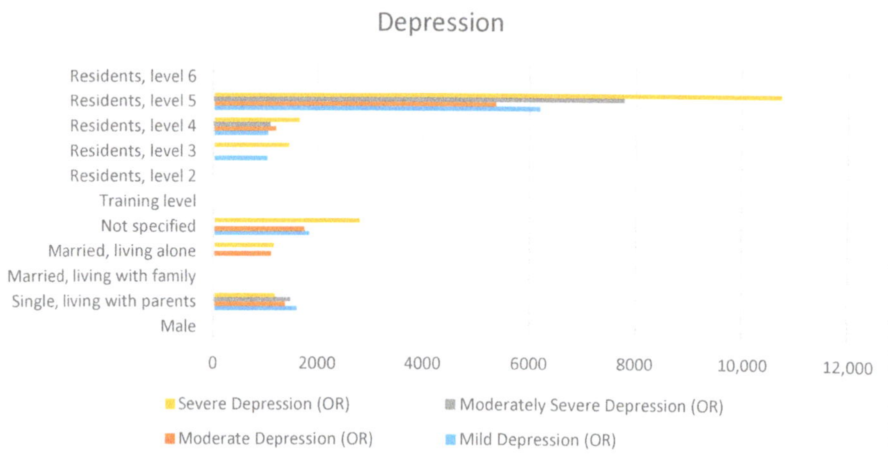

Figure 13. Depression in Trainees based on PHQ-9 scale during the COVID-19 pandemic.

5. Discussion, Limitations, and Interpretation of Our Research

The analysis of the previous section provided useful insights into the impact of the COVID-19 pandemic in the training programs and in the professional service of residents and fellows. In this section, we will elaborate upon the key findings and their implications. We will also try to go beyond the numbers and the statistics to interpret the key findings.

It is important before providing the key discussion to refer to some significant limitations of our research.

5.1. Limitations

In this research study, we investigate diverse and complicated research variables that also provide a challenging research context for COVID-19. Our focus on the training programs and the institutional support to residents and fellows of medical training programs in times of COVID-19 is by itself a very demanding context. Furthermore, we face an inability to compare our findings with the times before COVID-19. There is no relevant or existing research for residents and fellows in Saudi Arabia. Thus, we have a limited capability to compare the anxiety or depression rates with the situation before COVID-19. However, our research collected significant, trusted, and accurate data from a large sample of residents and fellows in KSA.

One more limitation of our research study is that we decided in this study not to include facts or detailed data on learning interventions during the COVID-19 pandemic in residents' training programs. We did this on purpose since we plan another integrated study on the matter of how technology-enhanced learning supports revised educational strategies during the COVID-19 pandemic.

One more significant limitation of our study is also that, since we wanted to secure a great participation from respondents, we did not include many cause and effect questions. Thus, for some facts revealed by our study, we cannot provide concrete interpretations without making significant assumptions. However, we prefer to communicate facts, and our interpretations will be based solely on these facts. For example, the various levels of depression and anxiety are presented as facts, without an effort to explain why the X or the Y subgroup of residents or fellows in medical training programs communicates this attitude or feeling. This will be our objective in future research focusing on the causal effect of psychological pressure. We do believe though that the accurate, trusted data that respondents provided allowed us to make significant propositions and interpretations. For example, it is significant to reveal that the non-provision of PPE equipment is associated with increased anxiety and depression and, thus, the institutions should take care of this in a transparent way as a responsive action to the impact of COVID-19.

The last limitation of our study is the comparability of findings with other similar studies abroad. We do believe though that the findings and their interpretations that are discussed in the next sections are a contribution that can be valued by researchers and institutions worldwide.

5.2. *Discussion, Key Findings and Interpretation*

We will organize the discussion in this section regarding our four research objectives

- Combined Discussion on *Research Objective 1*: What are the key implications of the COVID-19 pandemic on the training and educational process of residents in KSA and *Research Objective 2*: What are the key perceptions of residents regarding the training institution support during the COVID-19 pandemic?

In Figure 7, we provided a basic analysis related to the trainee's evaluation of the services provided to them during the COVID-19 pandemic from the training institution. The overall attitude is not satisfactory. On a scale of 0 to 10 (0 = not supportive at all 10 = completely supportive), the average score and the standard deviation show a rather unsatisfactory attitude of trainees regarding support by their training institution.

These are the key findings:

- Residents claim limited satisfaction for the measures provided by their training institution for their own safety and protection during the COVID-19 pandemic. The low mean of 4.39 out of 10 is indicative that residents and fellows felt somehow unprotected or at least they were expecting higher protection.
- The psychological support service during the COVID-19 pandemic launched for this purpose (Da'em in Saudi) was poorly rated (mean 1.46) by trainees. This is a very significant finding that poses many questions for its future development and its redesign. We will comment on this key finding in the discussion section.
- The support provided by supervisors was also rated neutrally (mean 4.26).
- The perception of trainees for the rating of the curriculum educational experience received a mean of 3.59.
- Special training for a pandemic crisis was offered to 36% of the respondents. It is quite surprising that the remaining 64% received no special training in such a demanding situation.
- Almost half of the respondents (49%) stated that they received enough Personal Protective Equipment (PPE). Unfortunately, the other half of the respondents claim that they did not receive enough protective equipment.
- The training institution facilitated the logistic transportation of 62% of the trainees during the COVID-19 pandemic.
- At a much lower rate (21%), the institution provided psychological support, if needed, to residents who were managing patients with COVID-19.
- Surprisingly, only 9% of trainees applied for psychological support through the Da'em program for psychological support initiated by the SCFHS.
- Most of the residents (58%) still have educational activities during the COVID-19 pandemic.
- Almost all of them (93%) agreed that the COVID-19 pandemic negatively distracted them from their daily routine practice.

This detailed picture of the perception of residents towards support by their training institution should initiate a dialogue and reflective actions. One of the key measurements should be the enhancement of PPE provision to residents and fellows. Furthermore, the low evaluation of the curriculum educational experience should set new directions for technology-enhanced learning and development of active learning engagements of residents. In this direction, new forms of virtual and augmented reality and simulation labs can also be exploited. In a future research study, we intend to discuss the experimentation with various technology-enhanced learning tools. The limited psychological support to trainees as well as the limited use of the Da'em service, a platform for psychological support initiated by the SCFHS, proves that (in this context) a lot more work and effort is needed. The platform for Da'em provided integrated services for support, but it seems that residents do not use it. Thus, a new awareness and communication campaign is needed to promote to trainees the value and the capacity of the new digital channel for psychological support. Moreover, the rather limited registration of trainees to Da'em proves that a new campaign for this is required. This is a key finding of our study. In order for the sophisticated digital platform to be efficient, the development of a learning curve is required from users, and, sometimes, psychological barriers do not allow targeted users to deploy the services. Thus, it is necessary to investigate the main reasons for the limited use of the Da'em platform soon.

Supervisors also must put more effort into supporting residents and trainees, especially in the context of COVID-19. We assume that supervisors also have increased workload and psychological pressure, so a new future survey must also target the population of supervisors in residents' programs. Technology can also support the supervisors with sophisticated services. Technology-enhanced supervision must be the focus of recommendations.

Finally, the distraction of the educational process must be seen as a key challenge for SCFCS.

- ***Research Objective 3***: What are the key aspects of psychological pressure including depression and anxiety of residents in KSA during the COVID-19 pandemic?

The key aspects of the psychological pressure on residents and fellows are summarized as follows. There is a continuous distraction from the educational process and limited support by supervisors, or, at least, support is not provided at a preferable level. Trainees offer their services to hospitals with COVID-19 patients and have direct contact with them while the provision of PPE equipment is, according to them, not satisfactory. Our research also revealed a direct association between COVID-19 times and significant rates of anxiety and depression. The relevant rates of anxiety and depression are significant and can initiate an integrated discussion for the relief of trainees. Towards this direction, several actions can be implemented, including:

- Hiring of more medical staff during the pandemic.
- Release of working time for research and rehabilitation.
- Access to technology-enhanced services for psychological support and recreation.
- Enhancement of social community engagement of trainees and building of optimistic faith for the future and their role in the pandemic.
- Increase in rewards and appreciation for their services during the pandemic.
- Access to sophisticated active learning engagement resources.
- Provision of virtual and augmented reality labs and simulation platforms.

The previous list is not exhaustive but contributes to a strategic plan for the enhancement of the psychology of trainees. One of the most important findings of our research is that ***trainees' anxiety and depression are highly positively correlated with the COVID-19 pandemic***.

Some additional key findings of our research based on significant statistics are summarized in Table 3 below.

Some key implications from this sophisticated analysis of depression and anxiety rates, though quite simplistic, are as follows.

- Women should be provided with additional support.
- The provision of transportation logistics to trainees seems to be a catalyst for depression. This may have a psychological connection with the feeling of caring. Maybe residents feel that the institution is caring for them and this has a good impact on their psychology.

- The provision of PPE equipment to trainees seems to be a key catalyst for the elimination of anxiety. This is a bold recommendation of our research that all the medical staff should be provided with sufficient and efficient PPE equipment during COVID-19 times and there must be a transparent procedure for the management of this equipment.
- Special and continuous psychological support must be provided to all the residents and fellows that regularly manage COVID-19 patients or have direct contact with COVID-19 patients. From our analysis, it is evident that these two categories show moderate severe and severe rates of depression and anxiety. Special programs for their support and recreation and reward must be designed and this is a high-level recommendation for policy making that has emerged from our research.

In this study, however, for almost one-quarter of residents and fellows that had a direct contact with patients who tested positive for COVID-19, 7% of those acquired the disease, which is approximately 4 to 14-fold higher as compared to a latter study [41]. This infection rate among the trainees must be investigated to provide insights for strategic planning on improving COVID-19 transmission management practices among trainees. In addition, 41.8% of resident physicians and fellows in this study argued that they should not be included in managing patients with COVID-19. This indicates the fear and anxiety of trainees associated with managing patients with COVID-19.

Nonetheless, frontline healthcare workers were indeed at a higher risk of symptoms of depression and anxiety. As this study shows, 98% and 90% of the trainees have anxiety and depression, ranging from mild to severe cases, respectively. Level 4 and 5 residents, those with direct contact and continuous management of patients with COVID-19, and females were found in this study as the groups that have a greater risk for severe anxiety and depression. One of the variables identified in this study is similar to the study done in Italy that found that being exposed to patients with COVID-19 was associated with symptoms of depression. These identified vulnerable groups need prompt attention, psychological guidance, and management necessary to help them overcome difficulties during this COVID-19 crisis and to prevent significant functionality problems that might affect their training path. The Saudi Commission for Health Specialties are very keen to provide support for trainees, and launched the Da'em program. This program particularly aims to improve the productivity and reduce the risk of psychological and professional burnout among trainees. Despite the high percentages of trainees with anxiety and depression in this study, only 141 or 9.22% applied for the program and rated the program at 3.60/10 (i.e., 0 = very poor to 10 = excellent). Although training centers also provided psychological support to trainees, trainees rated the service at 6.15/10 (i.e., 0 = very poor to 10 = excellent). Supervisor support to trainees during the COVID-19 pandemic was also evaluated and rated at 4.26/10 (i.e., 0 = very poor to 10 = excellent). Finally, trainees evaluated the curriculum educational experience during the COVID-19 pandemic at 3.59/10 (i.e., 0 = very poor to 10 = excellent). This indicates that the training program, curriculum delivery, psychological support, and guidance among trainees are unsatisfactory, need to be addressed, and require further improvement.

Providing trainees with necessary support such as transportation during curfews and lockdowns, continuous educational activities, and most importantly an adequate supply of protective personal equipment (PPE) were found in this study to be the significant factors that reduce risk of anxiety and depression among trainees. However, in the evaluation questionnaires about safety and protection measures provided to trainees by the training centers, trainees' evaluations were based on their happiness level, which was rated at 4.39/10 (0 = not happy at all to 10 = very happy). This indicates that the measures provided by the training centers are still unsatisfactory.

5.3. Implications of Our Research Recommendations

In this section, we are elaborating upon the last research objective of our research which is summarized as follows: *Research Objective 4: What are the key implications of this study for future recommendations towards best practices of training, psychological support, and technology-enhanced learning for medical training residents?*

The highly demanding COVID-19 pandemic challenges not only the way of delivering education, but also the integrated strategies for the management of the human capital in times of emergency and crisis. In the recent literature, various aspects of the phenomenon have been addressed, including psychological factors, social dynamics, and technological capability. The overload in the health systems worldwide is also flagging the capacity of modern management theories and applied systems for monitoring and control of resources to deliver the promised social value. The analysis of our survey provided interesting insights into the impact of COVID-19 on the training of residents. In this section, we communicate our interpretation of this survey in terms of an integrated approach for residents' training during the pandemic crisis.

In Figure 14, we communicate a high-level abstraction of our strategic approach at SCFHS and the intended integrative residents' training approach for the near future while COVID-19 is still spreading. In fact, five distinct pillars are collaborating towards the efficiency of the health services, medical training, and quality of life and wellbeing for residents and supervisors during the COVID-19 pandemic.

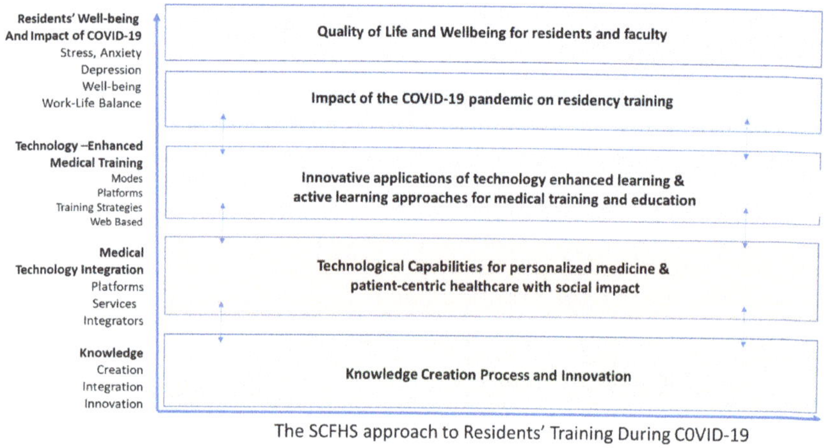

Figure 14. The SCFHS Model for Residents' Mental Health Enhancement during the COVID-19 Pandemic.

At the lower level, knowledge creation process and innovation serves as the scientific and applied medicine domain enabler for knowledge dissemination and applications. Key issues like knowledge creation, integration, and innovation can be integrated into training programs and educational curricula.

Technological capabilities for personalized medicine and patient-centric healthcare with social impact, in layer 2, refer to medical technology integration, including platforms, services and value integrators in the health domain. Various consideration and strategic decisions at this level have an impact on the readiness of the health system to manage the current time of crisis with COVID-19.

In the context of the quality of medical training and innovation, the third layer is related to innovative applications of technology-enhanced learning and web-based active learning approaches for medical training and education. In the strategic agenda of this layer, we are dealing with strategies and services for Technology-Enhanced Medical Training (TEME), modes and platforms for TEME, with training strategies for online learning and integrators for learning capabilities within daily medical practices.

Finally, residents' well-being and the impact of COVID-19 is the upper strategic layer with various health and psychological issues involved including stress, anxiety, and depression management, strategies and actions for well-being and work-life balance. We intend to present in a future research paper some more detailed findings on factors integrated into this model with an emphasis placed on satisfaction of trainees and effectiveness of distance and online learning for clinical and surgical training.

Within such a demanding context, there are various complementary facets of the research phenomenon which need further investigation.

- ***Impact of the COVID-19 pandemic on residency training***. For example, Amparore et al. [42] focus on urology residency training in Italy. Their key findings are quite interesting. For example, almost 97% of the participants reported that their surgical exposure was reduced due to the pandemic. They also recommend that continuing to revise steadily using webinars, podcasts, prerecorded sessions, and social media is a key way to maintain high-quality medical education. This key proposition leads directly to the new significant aspect of the literature. Finally, they also propose that routine activities such as journal clubs and departmental teaching should be facilitated through webinars if possible. In a similar study, Vargo et al. [33] studied the impact of COVID-19 in resident urology training in Cleveland. The findings are quite similar since the COVID-19 pandemic has resulted in reduced surgical volume and has disrupted established urology residency program curriculums nationwide. The authors propose a new framework based on a revised curriculum with emphasis on daily virtual learning as a key means of collaboration between faculty and residents. They also conclude that tele-medicine will be a key area for the near future.
- **Innovative applications of technology-enhanced learning and active learning approaches** for the support of medical education during COVID-19 times. For example, Tomlinson et al. [43] discuss in a recent research study the innovations in neurosurgical education during the COVID-19 pandemic and they question if it is time to reexamine the current neurosurgical training models. One of the bold findings of their research is the rapid increase in web traffic in relation to medical training resources and the adoption of emerging technologies, including virtual reality. The adoption of medical educational content on the internet by residency programs has supported a fair continuation of the curriculum despite the variety and complexity of barriers. Their key message is that the medical education community is currently eagerly awaiting significant enhancement in medical simulation software and virtual and augmented reality industry in order to have access to sophisticated tools that provide readiness to face similar crises in the future.
- **Quality of Life and Wellbeing for residents and faculty**. The aspect of well-being for residents and medical staff is a very delicate matter. The increased anxiety, technostress and human factors require significant research and support. Chong et al. [44] in their recent research study, focus on this dimension. In our research, we have a special interest in understanding the degrees of anxiety and depression of residents. We also intend to interpret the findings and to present an integrated framework for supporting residents.
- **Knowledge Creation Processes and Learning Excellence**. The knowledge creation in COVID-19 times as a responsive action to the pandemic requires enhancement and integration of the training and learning initiatives. Naeve et al. [45] describe efficient knowledge creation modes based on Nonaka's model of knowledge creation, and set a basis for parametrizing the technology-enhanced learning support of residents' education.
- **Capabilities of Emerging Technologies**. In recent years, the evolution of streamline and emerging technologies including artificial intelligence, virtual and augmented reality, medical big data and analytics, data warehouses, image processing, and cloud computing, has resulted in a brand-new era of capabilities of exploitation in the health specialties. Spruit and Lytras [46] integrated various aspects of data science for the healthcare domain towards patient-centric healthcare and personalized medicine. These also challenge the domain of medical education and there is an extremely significant need to study and understand how any of these technologies offer added value for residents during COVID-19 times. This is another aspect of our ongoing research in the Saudi Commission for the Health Specialties (SCFHS). The Quality Initiative, as described by Housawi et al. [47] in a recent study, targets the understanding of key performance indicators in residents' programs in Saudi Arabia and proposes an integrated framework for sustainable medical

education and trainees satisfaction [48]. Also, issues related to the use of ICTs, mobile learning and social media to enhance learning for special needs should be considered in the future [49].

As a summary of strategic implications of our research, here are some key suggestions (see also Figure 15 below):

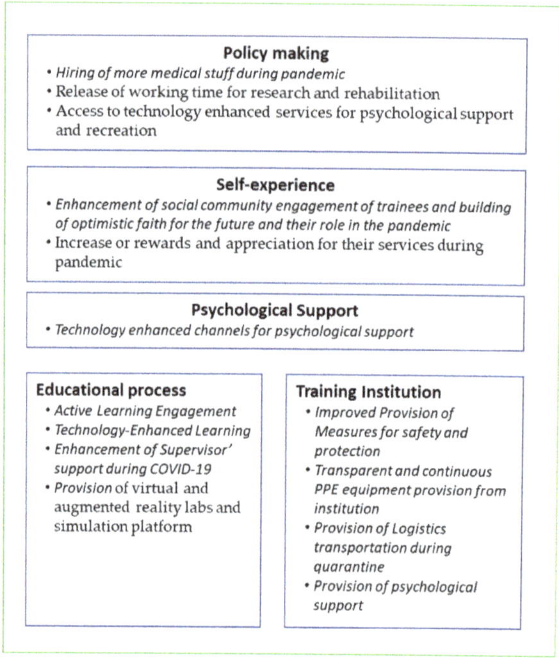

Figure 15. Strategic implications and recommendations.

Policymaking

- Hiring of more medical staff during the pandemic;
- Release of working time for research and rehabilitation;
- Access to technology-enhanced services for psychological support and recreation.

Self-experience

- Enhancement of social community engagement of trainees and building of optimistic faith for the future and their role in the pandemic;
- Increase in rewards and appreciation for their services during pandemic.

Psychological Support

- Technology-enhanced channels for psychological support.

Educational process

- Active learning engagement;
- Technology-enhanced learning;
- Enhancement of supervisor' support during the COVID-19 pandemic;

- Provision of virtual and augmented reality labs and simulation platform.

Training Institution

- Improved provision of measures for safety and protection;
- Transparent and continuous PPE equipment provision from institution;
- Provision of logistics transportation during quarantine;
- Provision of psychological support.

6. Conclusions

In conclusion, residents and fellows are a group that is vulnerable to COVID-19 infection, with a high proportion also experiencing significant moderate to severe anxiety and depression, merely due to dissatisfaction with training programs and curriculum delivery that was undeniably affected by the COVID-19 pandemic. Dissatisfaction regarding the provided safety and protection measures has a great impact on their mental status. The individual trainees experiencing moderate to severe anxiety must be identified and provided with prompt attention, support and psychological management. This study also suggests a need for improvement in the psychological support provided by training centers to trainees, promote trainees' awareness of the available psychological services, such as the Da'em program and its benefits, and provide additional service initiatives to improve trainee's safety and protection regarding COVID-19. Most importantly, training centers should identify and manage the factors associated with the high infection rate of COVID-19 among trainees.

The key findings of our research can be a starting point for the strategic consultation and crafting of new strategies. In the future, we will communicate the key findings of two complimentary research studies. One is related to the job satisfaction of trainees and the other one explores the efficiency of various technology-enhanced learning methods during the COVID-19 pandemic.

Author Contributions: The authors contributed equally to this research work and they involved in an integrated way in the stages of this research including Conceptualization, methodology, software, validation, formal analysis, investigation, resources, data curation, writing—original draft preparation, writing—review and editing, visualization, supervision, project administration, funding acquisition. All authors have read and agreed to the published version of the manuscript.

Funding: This research received no external funding.

Conflicts of Interest: The authors declare no conflict of interest.

Appendix A

Table A1. Multinomial-logistic regression analysis of the factors of anxiety among trainees.

Variables	Mild Anxiety				Moderate Anxiety				Severe Anxiety			
	OR	(CI 95%)		p	OR	(CI 95%)		p	OR	(CI 95%)		p
Male	0.520	0.352	0.770	0.001	0.445	0.299	0.662	0.000	0.325	0.222	0.477	0.000
Training Level												
Residents, level 1 **												
Residents, level 2	0.547	0.338	0.887	0.014	0.653	0.400	1.065	0.088	0.779	0.484	1.253	0.302
Residents, level 3	0.994	0.558	1.768	0.982	0.990	0.549	1.788	0.974	1.181	0.666	2.094	0.568
Residents, level 4	1.558	0.847	2.866	0.154	2.101	1.142	3.865	0.017	2.484	1.365	4.519	0.003
Residents, level 5	2.645	0.736	9.507	0.136	3.269	0.904	11.819	0.071	4.602	1.303	16.253	0.018
Residents, level 6	0.387	0.022	6.783	0.516	1.862×10^{-9}	1.862×10^{-9}	1.862×10^{-9}	.	0.706	0.052	9.590	0.794
Fellows, level 1	0.451	0.212	0.957	0.038	0.323	0.137	0.764	0.010	0.566	0.265	1.207	0.141
Fellows, level 2	0.878	0.397	1.943	0.748	0.564	0.229	1.388	0.212	0.670	0.281	1.597	0.366
Fellows, level 3	0.247	0.055	1.104	0.067	0.094	0.010	0.853	0.036	0.295	0.069	1.262	0.100
Regularly managing COVID-19 patients	1.385	0.958	2.001	0.083	1.771	1.218	2.576	0.003	2.626	1.829	3.772	0.000
Provided with enough PPE	0.511	0.353	0.742	0.000	0.481	0.329	0.702	0.000	0.353	0.245	0.509	0.000

Odds ratio (OR) > 1.0; High-risk for anxiety, Significant at $p < 0.05$ level. ** Baseline data. Note for the info in the tables: The blue colored boxes represent key findings. Values for OR > 1 are associated with significant findings. The red number for p test represents significant statistical findings.

Table A2. Multinomial-logistic regression analysis of the factors of depression among trainees.

Variables	Mild Depression				Moderate Depression				Moderately Severe Depression				Severe Depression			
	OR	(CI 95%)		p	OR	(CI 95%)		p	OR	(CI 95%)		p	OR	(CI 95%)		p
Male	0.631	0.402	0.990	0.045	0.644	0.409	1.012	0.057	0.398	0.252	0.629	0.000	0.379	0.237	0.607	0.000
Marital Status																
Single, living alone **																
Single, living with parents	1.602	0.798	3.217	0.185	1.374	0.691	2.732	0.365	1.497	0.743	3.016	0.259	1.174	0.570	2.418	0.663
Married, living with family	0.975	0.509	1.868	0.940	0.696	0.366	1.324	0.270	0.774	0.401	1.491	0.443	0.580	0.294	1.144	0.116
Married, living alone	0.988	0.433	2.258	0.978	1.104	0.496	2.456	0.808	0.858	0.373	1.975	0.719	1.155	0.502	2.656	0.734
Not specified	1.835	0.330	10.200	0.488	1.744	0.318	9.570	0.522	0.442	0.054	3.599	0.445	2.778	0.508	15.202	0.239
Training level																
Residents, level 1 **																
Residents, level 2	0.465	0.264	0.819	0.008	0.460	0.261	0.809	0.007	0.577	0.327	1.021	0.059	0.661	0.361	1.210	0.180
Residents, level 3	1.030	0.519	2.045	0.933	0.954	0.480	1.896	0.893	0.673	0.328	1.380	0.280	1.455	0.710	2.985	0.306
Residents, level 4	1.044	0.536	2.034	0.899	1.203	0.622	2.325	0.583	1.094	0.559	2.142	0.793	1.652	0.826	3.307	0.156
Residents, level 5	6.179	0.779	49.010	0.085	5.358	0.666	43.126	0.115	7.779	0.971	62.344	0.053	10.746	1.313	87.968	0.027
Residents, level 6	2.312×10^{-9}	2.312×10^{-9}	2.312×10^{-9}		0.241	0.011	5.086	0.360	0.392	0.019	8.005	0.543	0.309	0.013	7.483	0.470

Variables	Mild Depression				Moderate Depression				Moderately Severe Depression				Severe Depression			
	OR	(CI 95%)		p	OR	(CI 95%)		p	OR	(CI 95%)		p	OR	(CI 95%)		p
Male	0.631	0.402	0.990	0.045	0.644	0.409	1.012	0.057	0.398	0.252	0.629	0.000	0.379	0.237	0.607	0.000
Fellows, level 1	0.807	0.332	1.963	0.637	0.638	0.252	1.617	0.344	0.689	0.261	1.819	0.452	0.542	0.177	1.656	0.282
Fellows, level 2	0.572	0.243	1.348	0.202	0.314	0.118	0.833	0.020	0.405	0.154	1.067	0.067	0.582	0.206	1.642	0.306
Fellows, level 3	0.181	0.030	1.091	0.062	0.277	0.055	1.383	0.118	0.214	0.035	1.299	0.094	0.306	0.047	1.985	0.214
Have been in direct contact with COVID-19 patients	1.064	0.643	1.763	0.809	1.281	0.765	2.142	0.346	2.441	1.418	4.201	0.001	1.346	0.765	2.368	0.303
Regularly managing COVID-19 patients	0.938	0.576	1.526	0.796	1.245	0.767	2.021	0.375	1.332	0.817	2.171	0.250	2.253	1.346	3.771	0.002
Provided with Transportation during curfew	0.672	0.422	1.070	0.094	0.575	0.361	0.914	0.019	0.731	0.455	1.176	0.196	0.464	0.287	0.749	0.002
Provided with enough PPE	0.460	0.295	0.717	0.001	0.435	0.280	0.678	0.000	0.423	0.270	0.662	0.000	0.297	0.187	0.471	0.000
Have educational activities during COVID-19 pandemic	0.827	0.536	1.276	0.391	0.706	0.459	1.087	0.114	0.544	0.352	0.841	0.006	0.598	0.381	0.937	0.025

Odds ratio (OR) > 1.0; High-risk for anxiety, Significant at $p < 0.05$ level. ** Baseline data. Note for the info in the tables: The blue colored boxes represent key findings. Values for OR > 1 are associated with significant findings. The red number for p test represents significant statistical findings.

References

1. Li, L.; Xv, Q.; Yan, J. COVID-19: The Need for Continuous Medical Education and Training. *Lancet Respir. Med.* **2020**, *8*, e23. [CrossRef]
2. Ahmed, H.; Allaf, M.; Elghazaly, H. COVID-19 and Medical Education. *Lancet Infect. Dis.* **2020**, *20*, 777–778. [CrossRef]
3. Dubey, S.; Biswas, P.; Ghosh, R.; Chatterjee, S.; Dubey, M.J.; Chatterjee, S.; Lahiri, D.; Lavie, C.J. Psychosocial Impact of COVID-19. *Diabetes Metab. Syndr.* **2020**, *14*, 779–788. [CrossRef]
4. Liu, C.Y.; Yang, Y.Z.; Zhang, X.M.; Xu, X.; Dou, Q.-L.; Zhang, W.-W.; Cheng, A.S.K. The Prevalence and Influencing Factors in Anxiety in Medical Workers Fighting COVID-19 in China: A Cross-Sectional Survey. *Epidemiol. Infect.* **2020**, *148*, e98. [CrossRef]
5. Farrukh, S.; Hussain, W.; Siddiqui, Z.S. Assessment of Anxiety among Healthcare Professionals Working on Frontline against Covid-19. *Biomedica* **2020**, *36*, 270–274.
6. Sanghavi, P.B.; Yeung, K.A.; Sosa, C.E.; Veesenmeyer, A.F.; Limon, J.A.; Vijayan, V. Effect of the Coronavirus Disease 2019 (Covid-19) Pandemic on Pediatric Resident Well-Being. *J. Med. Educ. Curric. Dev.* **2020**, *7*, 2382120520947062. [CrossRef]
7. AlAteeq, D.A.; Aljhani, S.; Althiyabi, I.; Majzoub, S. Mental Health among Healthcare Providers during Coronavirus Disease (covid-19) Outbreak in Saudi Arabia. *J. Infect. Public Health* **2020**, *13*, 1432–1437. [CrossRef]
8. Elhadi, M.; Msherghi, A.; Elgzairi, M.; Alhashimi, A.; Bouhuwaish, A.; Biala, M.; Abuelmeda, S.; Khel, S.; Khaled, A.; Alsoufi, A.; et al. Psychological Status of Healthcare Workers during the Civil War and Covid-19 Pandemic: A Cross-Sectional Study. *J. Psychosom. Res.* **2020**, *137*. [CrossRef]
9. Collins, C.; Mahuron, K.; Bongiovanni, T.; Lancaster, E.; Sosa, J.A.; Wick, E. Stress and the Surgical Resident in the Covid-19 Pandemic. *J. Surg. Educ.* **2020**. [CrossRef]
10. Teo, W.Z.Y.; Soo, Y.E.; Yip, C.; Lizhen, O.; Chun-Tsu, L. The Psychological Impact of Covid-19 on 'hidden' Frontline Healthcare Workers. *Int. J. Soc. Psychiatry* **2020**. [CrossRef]
11. Chew, N.W.S.; Lee, G.K.H.; Tan, B.Y.Q.; Jing, M.; Goh, Y.; Ngiam, N.J.H.; Yeo, L.L.L.; Ahmad, A.; Khan, F.A.; Shanmugam, G.N.; et al. A Multinational, Multicentre Study on the Psychological Outcomes and Associated Physical Symptoms Amongst Healthcare Workers during COVID-19 Outbreak. *Brain Behav. Immun.* **2020**, *88*, 559–565. [CrossRef] [PubMed]
12. Shechter, A.; Diaz, F.; Moise, N.; Anstey, D.E.; Ye, S.; Agarwal, S.; Birk, J.L.; Brodie, D.; Cannone, D.E.; Chang, B.; et al. Psychological Distress, Coping Behaviors, and Preferences for Support among New York Healthcare Workers during the COVID-19 Pandemic. *Gen. Hosp. Psychiatry* **2020**, *66*, 1–8. [CrossRef] [PubMed]
13. Que, J.; Le Shi, J.D.; Liu, J.; Zhang, L.; Wu, S.; Gong, Y.; Huang, W.; Yuan, K.; Yan, W.; Sun, Y. Psychological Impact of the covid-19 Pandemic on Healthcare Workers: A Cross-Sectional Study in China. *Gen. Psychiatry* **2020**, *33*, e100259. [CrossRef] [PubMed]
14. Du, J.; Dong, L.; Wang, T.; Yuan, C.; Fu, R.; Zhang, L.; Liu, B.; Zhang, M.; Yin, Y.; Qin, J. Psychological Symptoms among Frontline Healthcare Workers during COVID-19 Outbreak in Wuhan. *Gen. Hosp. Psychiatry* **2020**. [CrossRef] [PubMed]
15. Luceño-Moreno, L.; Talavera-Velasco, B.; García-Albuerne, Y.; Martín-García, J. Symptoms of Posttraumatic Stress, Anxiety, Depression, Levels of Resilience and Burnout in Spanish Health Personnel during the COVID-19 Pandemic. *Int. J. Environ. Res. Public Health* **2020**, *17*, 5514. [CrossRef] [PubMed]
16. Zhang, S.X.; Liu, J.; Jahanshahi, A.A.; Nawaser, K.; Yousefi, A.; Li, J.; Sun, S. At the Height of the Storm: Healthcare Staff's Health Conditions and Job Satisfaction and Their Associated Predictors during the Epidemic Peak of COVID-19. *Brain Behav. Immun.* **2020**. [CrossRef]
17. Ornell, F.; Halpern, S.C.; Kessler, F.H.P.; Narvaez, J.C.d.M. The Impact of the COVID-19 Pandemic on the Mental Health of Healthcare Professionals. *Cad. Saúde Pública* **2020**, *36*, e00063520. [CrossRef]
18. Gold, J.A. Covid-19: Adverse Mental Health Outcomes for Healthcare Workers. *Br. Med. J. Publ. Group* **2020**. [CrossRef]
19. O'Flynn, K. Medical Education in London during 1939–1941, with Special Reference to the Blitz. *Med. Educ.* **2006**, *40*, 235–242. [CrossRef]

20. Huang, C.; Wang, Y.; Li, X.; Ren, L.; Zhao, J.; Hu, Y.; Zhang, L.; Fan, G.; Xu, J.; Gu, X.; et al. Clinical Features of Patients Infected with 2019 Novel Coronavirus in Wuhan, China. *Lancet* **2020**, *395*, 497–506. [CrossRef]
21. Rosen, G.H.; Murray, K.S.; Greene, K.L.; Pruthi, R.S.; Richstone, L.; Mirza, M. Effect of COVID-19 on Urology Residency Training: A Nationwide Survey of Program Directors by the Society of Academic Urologists. *J. Urol.* **2020**, *204*, 1039–1045. [CrossRef]
22. Lai, X.; Wang, M.; Qin, C.; Tan, L.; Ran, L.; Chen, D.; Zhang, H.; Shang, K.; Xia, C.; Wang, S.; et al. Coronavirus Disease 2019 (COVID-2019) Infection Among Health Care Workers and Implications for Prevention Measures in a Tertiary Hospital in Wuhan, China. *JAMA Netw. Open.* **2020**, *3*, e209666. [CrossRef] [PubMed]
23. Yuen, J.; Xie, F. Medical Education during the COVID-19 Pandemic: Perspectives from UK Trainees. *Postgrad. Med. J.* **2020**. [CrossRef] [PubMed]
24. Edigin, E.; Eseaton, P.O.; Shaka, H.; Ojemolon, P.E.; Asemota, I.R.; Akuna, E. Impact of COVID-19 Pandemic on Medical Postgraduate Training in the United States. *Med Educ. Online* **2020**, *25*. [CrossRef] [PubMed]
25. Chick, R.C.; Clifton, G.T.; Peace, K.M.; Propper, B.W.; Hale, D.F.; Alseidi, A.A.; Vreeland, T.J. Using Technology to Maintain the Education of Residents during the COVID-19 Pandemic. *J. Surg. Educ.* **2020**, *77*, 729–732. [CrossRef] [PubMed]
26. Li, Y.; Zhao, R.; Zheng, S.; Chen, X.; Wang, J.; Sheng, X.; Zhou, J.; Cai, H.; Fang, Q.; Yu, F.; et al. Lack of Vertical Transmission of Severe Acute Respiratory Syndrome Coronavirus 2, China. *Emerg. Infect. Dis.* **2020**, *26*, 1335–1336. [CrossRef]
27. American Board of Radiology. Coronavirus Information. 2020. Available online: http://www.theabr.org/announcements/coronavirus-updates (accessed on 5 August 2020).
28. Crosby, D.L.; Sharma, A. Insights on Otolaryngology Residency Training during the covid-19 Pandemic. *Otolaryngol. Head Neck Surg.* **2020**. [CrossRef]
29. Alvin, M.D.; George, E.; Deng, F.; Warhadpande, S.; Lee, S.I. *The Impact of COVID-19 on Radiology Trainees*; Radiological Society of North America: Oak Brook, IL, USA, 2020. [CrossRef]
30. Andrew, S.; Jacob, W.; Scott, B.; Thomas, M.; Thomas, B.; Nicholas, F. Managing Resident Workforce and Education during the COVID-19 Pandemic. *JBJS Open Access* **2020**, *5*, e0045. [CrossRef]
31. Stambough, J.B.; Curtin, B.M.; Gililland, J.M.; Guild, G.N., III; Kain, M.S.; Karas, V.; Keeney, J.A.; Plancher, K.D.; Moskal, J.T. The past, present, and future of orthopaedic education: Lessons learned from the COVID-19 pandemic. *J. Arthroplast.* **2020**, *35*, S60–S64. [CrossRef]
32. Wong, C.S.; Tay, W.C.; Hap, X.F.; Chia, F.L.A. Love in the time of coronavirus: Training and service during COVID-19. *Singap. Med. J.* **2020**, *1*. [CrossRef]
33. Vargo, E.; Ali, M.; Henry, F.; Kmetz, D.; Drevna, D.; Krishnan, J.; Bologna, R. Cleveland clinic akron general urology residency program's COVID-19 experience. *Urology* **2020**, *140*, 1–3. [CrossRef]
34. Alvin, M.D.; Horton, K.; Johnson, P. Training Radiology Residents to be Stewards in Healthcare. *Acad. Radiol.* **2017**. [CrossRef]
35. Shah, K.; Kamrai, D.; Mekala, H.; Mann, B.; Desai, K.; Patel, R.S. Focus on Mental Health During the Coronavirus (COVID-19) Pandemic: Applying Learnings from the Past Outbreaks. *Cureus* **2020**, *12*, e7405. [CrossRef]
36. England, E.; Patel, M.D.; Jordan, S.; Kalia, V.; Ali, K.; DeBenedectis, C.M.; Gaviola, G.C.; Ho, C.P.; Milburn, J.M.; Ong, S.; et al. Promoting Well-Being in Radiology Residency: A Primer for Program Directors. *Acad. Radiol.* **2020**, *27*, 720–723. [CrossRef]
37. Khalafallah, A.M.; Lam, S.; Gami, A.; Dornbos, D.L.; Sivakumar, W.; Johnson, J.N.; Mukherjee, D. A national survey on the impact of the COVID-19 pandemic upon burnout and career satisfaction among neurosurgery residents. *J. Clin. Neurosci.* **2020**, *80*, 137–142. [CrossRef]
38. Shah, K.; Chaudhari, G.; Kamrai, D.; Lail, A.; Patel, R.S. How Essential Is to Focus on Physician's Health and Burnout in Coronavirus (COVID-19) Pandemic? *Cureus* **2020**, *12*, e7538. [CrossRef]
39. Spitzer, R.L.; Kroenke, K.; Williams, J.B.; Löwe, B. A brief measure for assessing generalized anxiety disorder: The GAD-7. *Arch Intern Med.* **2006**, *166*, 1092–1097. [CrossRef]
40. Kroenke, K.; Spitzer, R.L.; Williams, J.B. The PHQ-9: Validity of a brief depression severity measure. *J. Gen. Intern. Med.* **2001**, *16*, 606–613. [CrossRef]
41. Lai, J.; Ma, S.; Wang, Y.; Cai, Z.; Hu, J.; Wei, N.; Wu, J.; Du, H.; Chen, T.; Li, R.; et al. Factors Associated with Mental Health Outcomes Among Health Care Workers Exposed to Coronavirus Disease 2019. *JAMA Netw. Open.* **2020**, *3*, e203976. [CrossRef]

42. Amparore, D.; Claps, F.; Cacciamani, G.E.; Esperto, F.; Fiori, C.; Liguori, G.; Serni, S.; Trombetta, C.; Carini, M.; Porpiglia, F.; et al. Impact of the COVID-19 pandemic on urology residency training in Italy. *Minerva Urol. Nefrol.* **2020**, *72*, 505–509, Epub 2020 Apr 7. [CrossRef] [PubMed]
43. Tomlinson, S.B.; Hendricks, B.K.; Cohen-Gadol, A.A. Innovations in neurosurgical education during the COVID-19 pandemic: Is it time to reexamine our neurosurgical training models? *J. Neurosurg.* **2020**, *1*, 1–2. [CrossRef]
44. Chong, A.; Kagetsu, N.J.; Yen, A.; Cooke, E.A. Radiology residency preparedness and response to the COVID-19 pandemic. *Acad Radiol.* **2020**, *27*, 856–861. [CrossRef]
45. Naeve, A.; Yli-Luoma, P.; Kravcik, M.; Lytras, M.D. A modelling approach to study learning processes with a focus on knowledge creation. *Int. J. Technol. Enhanc. Learn.* **2018**, *1*, 1–34. [CrossRef]
46. Spruit, M.; Lytras, M. Applied Data Science in Patient-centric Healthcare. *Telemat. Inform.* **2018**, *35*, 2018. [CrossRef]
47. Housawi, A.; Al Amoudi, A.; Alsaywid, B.; Lytras, M.; bin Moreba, Y.H.; Abuznadah, W.; Alhaidar, S.A. Evaluation of Key Performance Indicators (KPIs) for Sustainable Postgraduate Medical Training: An Opportunity for Implementing an Innovative Approach to Advance the Quality of Training Programs at the Saudi Commission for Health Specialties (SCFHS). *Sustainability* **2020**, *12*, 8030. [CrossRef]
48. Housawi, A.; Al Amoudi, A.; Alsaywid, B.; Lytras, M.; bin Moreba, Y.H.; Abuznadah, W.; Munshi, F.; Al Haider, S.; Tolah, A.W. A Progressive Model for Quality Benchmarks of Trainees' Satisfaction in Medical Education: Towards Strategic Enhancement of Residency Training Programs at Saudi Commission for Health Specialties (SCFHS). *Sustainability* **2020**, *12*, 10186. [CrossRef]
49. Drigas, A.; Ioannidou, R.E.; Kokkalia, G.; Lytras, M.D. ICTs, mobile learning and social media to enhance learning for attention difficulties. *J. UCS* **2014**, *20*, 1499–1510. [CrossRef]

sustainability

MDPI

Article

A Progressive Model for Quality Benchmarks of Trainees' Satisfaction in Medical Education: Towards Strategic Enhancement of Residency Training Programs at Saudi Commission for Health Specialties (SCFHS)

Abdulrahman Housawi [1], Amal Al Amoudi [1], Basim Alsaywid [1,2,3,*], Miltiadis Lytras [1,4,*], Yara H. bin Moreba [1], Wesam Abuznadah [5], Fadi Munshi [1], Sami Al Haider [5] and Abrar W. Tolah [1]

1 Planning and Organizational Excellence Administration, Saudi Commission for Health Specialties, Riyadh 11614, Saudi Arabia; a.housawi@scfhs.org (A.H.); A.AlAmoudi@scfhs.org (A.A.A.); y.binmoreba@scfhs.org (Y.H.b.M.); f.munshi@scfhs.org (F.M.); a.tolah@scfhs.org (A.W.T.)

2 College of Medicine, King Saud Bin-Abdul-Aziz University for Health Sciences, Jeddah 14611, Saudi Arabia

3 Urology Section, Department of Surgery, King Abdulaziz Medical City, Ministry of National Guard, Jeddah 11173, Saudi Arabia

4 Effat College of Engineering, Effat University, P.O. Box 34689, Jeddah 11426, Saudi Arabia

5 Executive Presidency of Academic Affairs, Saudi Commission for Health Specialties, Riyadh 11614, Saudi Arabia; w.abuznadah@scfhs.org (W.A.); s.alhaidar@scfhs.org (S.A.H.)

* Correspondence: b.alsaywid@scfhs.org (B.A.); mlytras@acg.edu (M.L.)

Received: 24 November 2020; Accepted: 4 December 2020; Published: 6 December 2020

Abstract: The latest developments in Sustainable Health focus on the provision of high quality medical training to health specialists, with a special focus on human factors. The need to promote effective Training Programs also reflects the job satisfaction needs of trainees. The objective of this study is to evaluate the trainees' satisfaction with the quality of Training Programs and assess the degree of achievement based on the defined parameters to provide baseline data based on which strategies for improvement can be formulated. Our study was conducted in Saudi Arabia and our targeted population was residents in medical programs supervised by the Saudi Commission for the Health Specialties (SCFHS). The trainees' response rate to the online survey was 27% (3696/13,688) and the key aspects of job satisfaction investigated include: Satisfaction with Academic Activities in the Center; Satisfaction with the Residents and Colleagues in the Center; Satisfaction with the Administrative Components in the Center; Satisfaction with the Training Programs; Satisfaction with the Specialty; Satisfaction with the Training Center; Satisfaction with the SCFHS. The main contribution of our work is a benchmark model for job satisfaction that can be used as a managerial tool for the enhancement of medical education with reference to the satisfaction of trainees. We analyze the key aspects and components of training satisfaction and we introduce our progressive model for Trainees' Satisfaction in Medical Training. In future work, we intend to enhance the proposed framework with a set of key performance indicators as well as with a focused cause and effect focused survey on factors related to the key benchmark of this study.

Keywords: job satisfaction; sustainable health; medical training; accreditation; satisfaction; health governance; Saudi Commission for Health Specialties; smart healthcare

1. Introduction

Medical education contributes significantly to the vision for Sustainability in Health. In our times, with the COVID-19 pandemic wide-spreading worldwide, the necessity to empower health experts with knowledge, skills, and competencies is a key priority for the management of healthcare. Furthermore, the increase in the positive motivation of health specialists and their willingness to promote their social role and the impact of their professional conduct is associated with various psychological factors like burnout or job satisfaction. The development of high-skilled experts with mental health and increased satisfaction from their work must be set as a key objective towards achieving Sustainable Health.

From a managerial point of view, Sustainable Health also requires the design, implementation, and continuous monitoring of benchmarks and Key Performance Indicators (KPIs) capable of offering a trusted, sustainable, and reliable set of measurements for performance monitoring.

Within this context, the assessment of medical Training Programs and the evaluation of their added value must be a strategic initiative for empowerment and sustainability. The development of medical competence through training and education is a bold action. Our research study is focusing on Residency Training Programs in the Kingdom of the Saudi Arabia supervised by the Saudi Commission for the Health Specialties.

Residency or postgraduate training is a stage of graduate medical education. Residents are trainees who are enrolled in programs recognized by the Saudi Commission for Health Sciences (SCFHS) and are fulfilling the requirements for primary discipline certification or a certification of special competence. A resident refers to a qualified physician, podiatrist, dentist, optometrist, veterinarian, or pharmacist who practices medicine, usually in a hospital or clinic, under the direct or indirect supervision of a senior clinician registered in that specialty such as an attending physician or consultant.

In Section 2, we provide a thorough critical literature review on the concept of Job Satisfaction and we present various research studies that contribute to our research problem specification. The SCFHS has a main responsibility to deliver high quality Training Programs to approximately 18,000 health specialists in the Kingdom of Saudi Arabia. It also supervises numerous training centers and academic programs with a critical objective to deliver value to the Health System and to the Society of the Kingdom. This value-driven strategy of SCFHS has an impact on all the activities undertaken by the institution.

In total, 1200 programs are conducted across Saudi Arabia and other countries in the Gulf region covering 79 health specialties [1]. One of the key strategic aims of the SCFHS is also to design, implement, and monitor a quality assurance system for the execution of all its Training Programs and also to access the efficiency of this system to enhance the update of strategies and policies. In this context, the SCFHS also supervises the quality of the implementation of residency medical Training Programs.

Trainees' satisfaction is a multidimensional concept that has been associated with environmental factors and is predicted by intrinsic (personal growth and perceived ability to work) and extrinsic (perceived social support) forms of motivation. The objective of this study is to evaluate the trainees' satisfaction level and to investigate its association with various features, including the quality of Training Programs.

The study of the trainees' satisfaction is part of a multi-dimensional research strategy in the Saudi Commission of Health Specialties. The SCFHS is deploying a value-driven strategy for the integration of research-based evidence to the quality of its Training Programs. It also collects and processes significant data related to the various Training Programs and their evaluations towards enhanced decision making. Various data are related to key aspects of the residents' perceptions related to the quality of the Training Program and institutions, as well as the self-assessments by the trainees of various psychological aspects of their educational and professional activities.

A continuous quality assurance strategy is in place focusing on different complementary aspects of medical training. In a recent study, we introduced the SCFHS Framework for Sustainable Medical Education and we introduced 23 integrated Key Performance Indicators for Implementing an Innovative

Approach to Advance the Quality of Training Programs at the Saudi Commission for Health Specialties. Other strategic goals include the measurement of the effectiveness of online-training strategies, as well as the deep understanding and interpretation of human factors involved in medical education. For examples, recently we delivered research on the impact of the COVID-19 pandemic on the anxiety and depression of residents in medical Training Programs.

In Figure 1 below, we summarize some of the key aspects of our research problem. The main motivation of our research is the direct association of four integrative pillars in training and professional life of residents. According to the literature that is presented below, the image of physicians has two critical components: the continuous development, and the seeking of life-work balance of the trainee/resident. Thus, the training satisfaction remains a core aspect of efficiency of Training Programs. The intention of our research is to analyze how specific aspects of trainee satisfaction lead to inefficiencies in medical Training Programs and their impact based on international benchmarks and our research approach.

Furthermore, we also focused on psychological factors, including the burnout syndrome of residents, as a key variable for the understanding of trainees' satisfaction. Our ultimate objective is to analyze and to propose enhancements on organizational good practices, and to lead innovative plans and strategies for the beneficiaries of Training Programs of the SCFHS.

As the image of physicians has always been associated with high levels of professionalism, it is notable that values, behavior, and relationships are the main factors that sustain this image in the eyes of people. These factors can be maintained by continuous development and the seeking of a life-work balance of the practitioner. Maintenance of this balance can lead to satisfaction [2].

Studies reported that trainee satisfaction has a major impact on their knowledge enhancement and outcomes of care. Also, this directly affects their education and maintenance of good organizational practices [3].

In a study conducted to measure the satisfaction of UK junior doctors (medical post graduate students), factors of "appropriate workload (sufficient to learn but not oppressive to educational opportunities and to wellbeing), good supervision of practice (clinical supervision) and the receipt of timely, good quality feedback" were found to be correlated with trainee satisfaction. Moreover, assignment of an appropriate workload was found to the most crucial factor affecting the satisfaction of trainees, followed by the remaining factors prospectively. A high level of dissatisfaction was reported by junior doctors assigned to work in environments with a high workload and this led to the appearance of burnout symptoms in many cases [3,4]. According to [5], burnout syndrome (BS) is a set of psychological symptoms resulting from the interaction between chronic occupational stress and individual factors. The syndrome is characterized by emotional exhaustion, depersonalization, and reduced personal achievement [5]. Emotional exhaustion (EE) indicates the feelings of burden and the weakening of emotional resources; depersonalization (DP) entails responding to others, such as associates and patients, in a cynical and isolated way; reduced personal accomplishment (PA) occurs when the subject feels less competent in his/her role. These factors validate the importance of addressing the problem of burnout as fast as it occurs at all levels (training, education, and practice). As discussed in relevant literature, addressing of the problem can be defined into four main actions: distinguish its existence, address its prevalence and strength, and then perform the preventive and treatment strategy, this should then be followed by periodic measurement of all implemented strategies' effectiveness [6].

This article is organized as follows: In Section 2, we provide an overview of our critical constructive review on the key aspects of our research phenomenon.

Then in Section 3, we summarize our research methodology and we communicate our research objectives. In Section 4, we provide analysis of the results and the key findings of our research. In Section 5, we discuss the contribution and the implications of our work towards Sustainable Health. We also in parallel build our key theoretical contribution, the progressive model for Trainees' Satisfaction in Medical Education for Sustainable Health. Finally, in Section 6, we provide they key conclusions and future research directions.

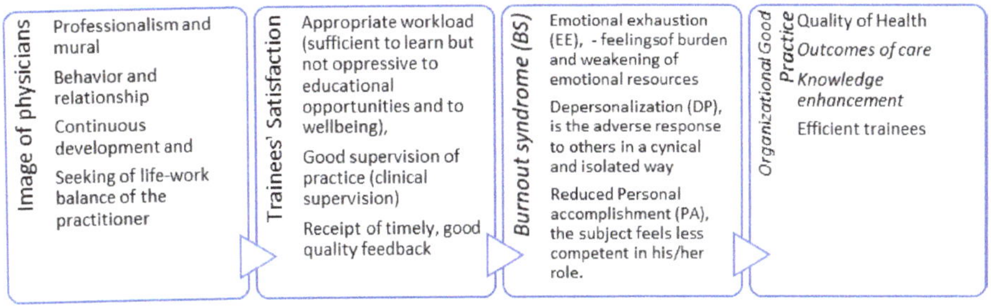

Figure 1. Research problem aspects.

2. Critical Literature Review on Trainee Satisfaction in Training Programs

The recent literature on medical training and healthcare workers research deals extensively with human and psychological factors of trainees, residents, and healthcare workers overall. Various recent studies explain the phenomenon and provide useful insights for our research. Trainee satisfaction appears to be a well discussed human soft factor in diverse studies on medical training. It is also a key variable in broader studies in other research and professional domains. According to literature, several integrated studies focus on the analysis of job satisfaction, professional burnout, the work environment, and health issues. Also, the recent pandemic of COVID-19 has led to new research on the further impact of the pandemic on the traditional human factors of healthcare workers, doctors, and residents. In Table 1, below, we provide an overview of 13 recent studies on the phenomenon.

Table 1. An overview of recent literature on job satisfaction and human factors in health specialties research.

Reference	Authorship	Key Emphasis					
		Job Satisfaction	*Professional Burnout*	*Healthcare Workers*	*COVID-19*	*Work Environment*	*Health*
[7]	Alsubaie et al.	x	Not reported	X	Not reported	x	Not reported
[8]	Chen et al.	x	Not reported	X	Not reported	x	x
[9]	Bawakid et al.	x	x	X	Not reported	Not reported	x
[10]	Zhang SX et al.			X	x	Not reported	Not reported
[11]	Leskovic et al.	x	x	X	x	Not reported	Not reported
[12]	Soto-Rubio et al.	x	x	X	x	Not reported	x
[13]	Aoyagi et al.	x (willingness to work)	Not reported	X	x (pandemic)	Not reported	Not reported
[14]	Liu et al.	x	Not reported	X (doctors)	Not reported	X (hospital)	Not reported
[15]	Platis et al.	x	Not reported	X(nurses)	Not reported	X (job performance)	x
[16]	Chao et al.	x	x	Not reported	Not reported	x	x
[17]	Maissiat et al.	x		x	Not reported	x	x
[18]	Pandey et al.	x	x	x			
[19]	Labrague et al.	x	x	x	x		

In Table 2 below, we provide the overview of various research studies related to our research problem. A brief discussion and elaboration on the key aspects of our research problem that is anchored in the selected studies is provided in the next few paragraphs.

Alsubaiea and Isouard [7] provide a meta-research analysis of four studies related to job satisfaction, and retention of Saudi nursing staff. The previous four studies involved 2362 nurses in Saudi Arabia, with a rather good rate of job satisfaction but limited job retention. In our research, we need to investigate the value adding components of residents in Training Programs supervised by the SCFHS. We also intend to develop a set of benchmarks for the relevant components of job satisfaction.

Table 2. Connection of our research to recent research on job satisfaction and relevant human factors.

Literature Review			
Author(s)	**Title of Article**	**Key Contribution**	**Impact on Our Research Model**
[7]	Job Satisfaction and Retention of Nursing Staff in Saudi Hospitals.	This paper reviews the research conducted on job satisfaction, and the retention of Saudi nursing staff.	We are interested in setting up benchmarks for the added value components of job satisfaction of residents in Training Programs supervised by the SCFHS.
[8]	Job Satisfaction Analysis in Rural China: A Qualitative Study of Doctors in a Township Hospital.	The goal was to understand the level of job satisfaction of doctors and to make recommendations for improvements.	In our research study, we want to understand the main picture of the job satisfaction phenomenon in the large population of residents in Saudi Arabia and to reveal some key positive and negative relationships of job satisfaction with specific factors. As a responsive action, our research also aims to provide key recommendations for improvement. We are also very much interested in developing a reliable mechanism to monitor the evolution of job satisfaction on residency medical programs.
[9]	Professional Satisfaction of Family Physicians Working in Primary Healthcare Centers: A Comparison of Two Saudi Regions.	The objectives of this study are to assess the level of professional satisfaction and to compare and identify the factors potentially associated with professional satisfaction/dissatisfaction among FPs in two regions.	In our research study, we need to understand the basic "cause and effect" relationships between job satisfaction and aspects of the psychological load of residents including aspects of the Training Program and educational activities, as well as the impact of burnout on satisfaction.
[10]	At the Height of the Storm: Healthcare Staff's Health Conditions and Job Satisfaction and Their Associated Predictors during the Epidemic Peak of Covid-19.	This study helps to identify the healthcare staff in need to enable more targeted help, as healthcare staff in many countries are facing peaks in their COVID-19 cases.	In our research study, we are also interested in analyzing the psychological factors that affect job satisfaction. We also in parallel run a complimentary study that focuses on the analysis of depression and anxiety in the context of the professional and educational activities of the residents in training centers.
[11]	Burnout and job satisfaction of healthcare workers in Slovenian nursing homes in rural areas during the COVID-19 pandemic.	This study aims to analyze job satisfaction and burnout levels of healthcare professionals working in Slovenian nursing homes in rural areas during the COVID-19 pandemic, and make a comparison with the results of the same services in 2013.	These findings also anchor our research in the relevant literature. As we explained, a key priority in our research to reveal hermeneutic factors for the job satisfaction of residents.
[12]	Effect of Emotional Intelligence and Psychosocial Risks on Burnout, Job Satisfaction, and Nurses' Health during the COVID-19 Pandemic.	The present study aimed to analyze the effect of psychosocial risks and emotional intelligence on nurses' health, well-being, burnout level, and job satisfaction during the rise and main peak of the COVID-19 pandemic in Spain.	This is one more interesting anchor for our study. We are interested in understanding which aspects of mental capabilities and skills of residents can work against psychological pressure and dissatisfaction based on the professional and academic conduct of residents.

Table 2. *Cont.*

Literature Review			
Author(s)	**Title of Article**	**Key Contribution**	**Impact on Our Research Model**
[13]	Healthcare workers' willingness to work during an influenza pandemic: a systematic review and meta-analysis	Meta-analyses of specific factors showed that for male Healthcare workers (HCWs), physicians, and nurses, full-time employment, perceived personal safety, awareness of pandemic risk and clinical knowledge of influenza pandemics, role-specific knowledge, pandemic response training, and confidence in personal skills were statistically significantly associated with increased willingness to work.	In our research, we need to understand the motivation of residents to perform effectively as trainees and also to understand how job satisfaction is associated with specific motivational factors.
[14]	Cross-sectional survey on job satisfaction and its associated factors among doctors in tertiary public hospitals in Shanghai, China	The results of the logistic regression analysis suggested that doctors' job satisfaction was related to their professional title, types of patients that doctors treated or expected to treat, as well as their work stress.	In our research, the focus is on the study of job satisfaction of residents and this differentiates our work from many other related works. Our survey is one of the first covering such a major population of residents.
[15]	Relation between job satisfaction and job performance in healthcare services, Procedia-Social and Behavioral Sciences	In this work, authors try to analyze the relationship between job satisfaction and job performance.	We do not intend to study the possible association of job satisfaction of trainees with job performance. This could be a direction for future research.
[16]	Workplace stress (WS), job satisfaction, job performance, and turnover intention (TI) of health care workers in rural Taiwan.	The results showed that WS had a positive effect on both TI and job performance (JP) but had a negative effect on satisfaction. JS did improve performance.	In future research, we would like to further study the factor of workplace stress, as it is related to job satisfaction of residents.
[17]	Work context, job satisfaction, and suffering in primary health care	To evaluate the work context, job satisfaction, and suffering from the perspective of workers in primary health care.	We are interested in understanding how special aspects of the training institutions of trainees and special aspects of the Training Program affect the job satisfaction of residents.
[18]	Donning the mask: effects of emotional labour strategies on burnout and job satisfaction in community healthcare.	This study brings forth the neglected issues of emotions and their implications for these healthcare workers in low and middle-income countries who are a vital link that delivers healthcare to weaker sections of the society.	In future research, we would also like to investigate the income and rewards parameter's impact on job satisfaction.
[19]	Fear of COVID-19, psychological distress, work satisfaction and turnover intention among frontline nurses.	To examine the relative influence of fear of COVID-19 on nurses' psychological distress, work satisfaction, and intent to leave their organisation and the profession.	We are interested in revealing key psychological barriers affecting the job satisfaction of residents.

Chen et al. [8] conducted a study on 39 doctors from five township hospitals in Guangxi Zhuang, China. The goal was to understand the level of job satisfaction of doctors and to make recommendations for improvements. The key findings of their research highlight that job satisfaction is associated with numerous factors including working conditions, financial rewards, and the doctor's relationships with patients. As a key policy making proposition, the study suggests that the increasing of income and fringe benefits of healthcare workers will have a positive impact on job satisfaction. Additionally, enhanced training and more opportunities for continuous improvement are set as key reflective actions against unsatisfactory conditions. In our research study, we want to obtain a rich picture of the phenomenon in the large population of residents in Saudi Arabia and to reveal some key positive and negative relationships of job satisfaction with specific factors. As a responsive action, our research also aims to provide key recommendation for improvement. We are also very much interested in developing a trusted way and mechanism to monitor the evolution of job satisfaction on residency medical programs.

In another recent study in KSA (Kingdom of Saudi Arabia), Bawakid et al. [9] conducted a study with 237 Family Physicians (FPs) working in primary healthcare centers under the Ministry of Health in two regions (Jeddah and the Eastern region). In the key findings of their research, it was evident that more than half of the FPs were satisfied in terms of their professional conduct. The majority though had a perception of being inferior to other specialties. As a bold recommendation, the relevant study proposed that the enhancement of self-esteem as well as the continuous support and improvement of the working environment, would limit stress and improve the health and psychological loads of physicians. In our research study, we need to understand the basic "cause and effect" relationships between job satisfaction and aspects of the psychological load of residents, including aspects of the Training Program and educational activities as well as the impact of burnout on satisfaction.

Zhang [10] conducted a recent research study with a sample of 304 healthcare staff (doctors, nurses, radiologists, technicians, etc.) in China during COVID-19 with an emphasis on the revealing of psychological pressure on their professional conduct and job satisfaction including through stress, anxiety, and depression. In the key findings of this research, it was found that a significant portion of this staff faced high levels of anxiety (28.0%), depression (30.6%), and distress (20.1%). Several demographics and factors related to the provision of personal protection equipment were associated negatively with job satisfaction. In our research study, we are also interested in analyzing the psychological factors that affect job satisfaction. We also in parallel run a complimentary study that focuses on the analysis of depression and anxiety in the context of the professional and educational activities of the residents in training centers.

In their study that conducted in Spring 2013 (n = 556) and Spring 2020 at the peak of the pandemic in Eastern Europe (n = 781) in Slovenia, Leskovic et al. [11] concluded that the COVID-19 pandemic significantly increased the burnout syndrome faced by nursing homes healthcare workers in Slovenian rural areas. In their study, they also dealt with job satisfaction and they proved that there is a direct association a predicting capability for the burnout syndrome. Additionally, they observed a negative correlation between job satisfaction in 2020 and feelings of emotional exhaustion and personal accomplishment in both 2013 and 2020. These findings also anchor our research in the relevant literature. As we explained, it is a key priority in our research to reveal hermeneutic factors for the job satisfaction of residents.

Soto–Rubio et al. [12] delivers integrated research in Spain that involved 125 nurses. In their key findings, they emphasize that emotional intelligence serves as a positive favorable effect on job satisfaction. This is one more interesting anchor for our study. We are also interested in understanding which aspects of the mental capabilities and skills of residents can work against psychological pressure and dissatisfaction stemming from the professional and academic conduct of residents.

Aoyagi et al. [13] performed a meta-analysis and narrative synthesis of available research and found out that respondents' willingness to work ranged from 23.1% to 95.8%, depending on their work context. A variety of factors determine an increased willingness to work and these include full-time employment, perceived personal safety, awareness of pandemic risk and clinical knowledge of influenza pandemics, role-specific knowledge, pandemic response training, and confidence in personal skills. In our research, we need to understand the motivation of residents to perform effectively as trainees and to understand how job satisfaction is associated with specific motivational factors.

Liu et al. [14]'s study of job satisfaction through a survey of 897 doctors from 11 tertiary public hospitals in Shanghai, China revealed that 64.8% of participants were dissatisfied with their jobs. A variety of factors were determined to have a direct relationship with dissatisfaction including professional title, the types of patients that doctors treated or expected to treat, as well as their work stress. In our research, the main interest is on the study of job satisfaction of residents and this differentiates our work from many other related works. Our survey is one of the first covering such a major population of residents.

Various other research studies contribute also to the domain of knowledge of job satisfaction and healthcare workers, nurses, physicians, licensed professionals, and social health activists of frontline nurses [15–19]. For example, in [15] a list of factors for the job satisfaction of nurses includes the following components: (satisfaction from the manager; satisfaction from the management administration; satisfaction with the ways of working; satisfaction of recognition; satisfactory working hours; satisfactory working security; self-satisfaction towards productivity; self-satisfaction of initiatives; self-satisfaction of working; self-satisfaction of quality improvements). Last but not least, a recent study during the COVID-19 pandemic [19] proved that COVID-19 has an increased impact on the level of fear that is also associated with decreased job satisfaction and increased psychological distress, among other implications.

Some further aspects for the anchors of our study to the literature are discussed in the next paragraphs. These provide some more directions for the design of our research tools and the key objectives of our research.

In a global systematic review study focused on the correlation between residents' burnout rate and their specialty, the highest prevalence of burnout was found to be among residents of radiology, neurology, and general surgery. In contrast, residents with the specialties of psychiatry, oncology, and family medicine were found to have the lowest level of burnout. Also, the same study reported the highest levels of burnout being experienced among males and older residents [20].

The presence of life-work balance and meaning in work were found to be associated with lower levels or the absence of burnout and was associated with career satisfaction among residents [21]. In a study of the prevalence of burnout among pediatric residents, burnout found to have a negative impact on the medical knowledge quality of care and professional conduct [22].

Studies reported higher levels of depression among medical students when compared to the normal population; recommendations to focus on students' psychological well-being and faculty support were raised. In the same vein, the advantages of including well-being as a curriculum area in addition to skills of stress management was highlighted [23]. A systematic review focused on identifying factors affecting psychological well-being listed sleep, resilience, residents' independence, building competence, and enhanced social relatedness as the main factors influencing residents' well-being and requiring further focus and research [24].

Psychological well-being featured in a study focused on measuring the burnout among 2nd year U.S. residents and regret towards career choices; the study found that a higher prevalence of burnout existed among the population associated with regret towards their career choices. Furthermore, the study recommended further research on this topic [25].

The effect of burnout is not limited to the medical trainee only; it also affects patients under treatment and the community. It plays a role in medical error as well as patient dissatisfaction. Studies urge prompt intervention; such interventions might be as simple as the limitation of working hours and mindfulness training [26].

In a national level study, physicians' burnout, exhaustion, and work unit safety were found to be associated with major medical errors, and prompt intervention was requested to minimize the medical error through addressing physicians' well-being and unit safety, as well by the assessment and modification of workload and alleviating the exhaustion of medical trainees and physicians [27]. Some studies found that self-blame is higher in female residents, which leads to resident distress [28].

A systematic review analysis conducted on the effect on different interventions on the level of trainees burnout and emotional exhaustion found that the limitation of working hours to the Accreditation Council for Graduate Medical Education (ACGME) work hour limits were associated with improvements in avoiding emotional exhaustion and burnout [29].

Physician wellness has a great significant impact on patient satisfaction, long-term physician satisfaction, and increased physician productivity. Therefore, medical educators, academic leaders, and researchers are focusing on the initiative of 'improving trainee well-being' or on satisfaction and analyzing burnout [26].

Management and dealing with the health professional practice requires stamina, good health, appropriate knowledge and skills, and the ability to respond positively to challenging experiences. Doctors needs to have good psychological and physical health to perform according to good practices and not to burn out. A recent study introduced the psychological concept of Resilience, which relates to the long-term ability of individuals to survive in and thrive on adversity; this concept is rarely used in medical practice but proved to increase the psychological stamina and interaction of individuals, which in turn showed significant impacts on the reduction of dissatisfaction and burnout. More research is highly recommended to be performed on this topic and the implementation of resilience assistance for medical trainees and practitioners [2].

The performance of prospective observational studies is highly recommended by the literature; the conducting of such studies using validated international scales will provide a clear assessment of the affected population over time (trainees), which will measure the prevalence of the burnout syndrome and associated factors at each level (education, intern, resident, or practice). This will help in building both prevention and treatment strategies and increase their effectiveness [30]. Within this context, our research study is timely and responds to a key requirement for resilient healthcare [31,32]. It is critical to monitor variations in trainees' satisfaction and burnout rates in the short and long term in order to be capable of recognizing the effect of novel didactical approaches in Training Programs and in best practices [33–36].

Lastly, such investments in the wellness of physicians is an ethical responsibility that needs to be upheld by the medical community to assure good practices are conducted by future generations of practitioners [37–49].

From this point of view, our research delivers this ethical responsibility and provides significant findings that can be exploited towards the enhancement of well-being, quality of life, and images of the residents in Training Programs. It is also a bold initiative for promoting the health specialties community's involvement in the recognition of performance in all the aspects of health practices and training.

In summary, the contribution of the critical literature review to the specification of our research problem can be summarized as follows:

- We are interested in setting up benchmarks to indicate the added value components of job satisfaction of residents in Training Programs supervised by the SCFHS.
- We want to obtain a comprehensive picture of the burnout phenomenon in a large population of residents in Saudi Arabia and to reveal some key positive and negative relationships of job satisfaction with specific factors.
- As a responsive action, our research also aims to provide key recommendations for improvement.
- We intend to develop a trusted mechanism to monitor the evolution of job satisfaction during residency medical programs.
- We want to understand the basic "cause and effect" relationships between job satisfaction and aspects of the psychological load of residents, including aspects of the Training Program and educational activities as well as the impact of burnout on satisfaction.
- We focus on the analysis of the psychological factors that affect job satisfaction. We also in parallel run a complimentary study that studies the rates and impact of depression and anxiety in the context of the professional and educational activities of the residents in training centers.

- We are also interested in understanding which aspects of mental capabilities and skills of residents can work against psychological pressure and dissatisfaction based on the professional and academic conduct of residents.
- We do not intend to study the possible association of the job satisfaction of trainees with job performance. This can be a direction for future research.
- We are interested in understanding how special aspects of the training institutions of trainees and special aspects of Training Programs affect the job satisfaction of residents.
- In future research we would also like to investigate the income and rewards parameter and how it affects job satisfaction.

In the next section, we provide our research methodology strategy and we outline our research objectives.

3. Methods

The previous critical literature review revealed several complementary aspects of the research area in the wider literature. The focus of our research is on factors that enhance trainees' satisfaction in post graduate medical Training Programs. We are very interested in analyzing, understanding, and interpreting perceptions from trainees of the quality of their residency programs. Our methodological approach is integrated into a greater research context. The SCFHS has launched the Quality Initiative, within which a multidimensional framework for quality assurance has been introduced. The core part of this framework is a set of 23 Key Performance Indicators (KPIs) that can be exploited as a significant managerial tool for enhanced decision making. In this research paper, we emphasize the perceptions of residents in postgraduate medical Training Programs as they refer to their job satisfaction. We also intend to add additional aspects for our Quality Assessment framework with an emphasis on KPIs related to job satisfaction.

Trainees' satisfaction in residents Training Programs is a key factor for the measurement of efficiency. In our approach, we integrate several value components of the trainees' satisfaction and we investigate the attitude of residents and fellows in the SCFHS's medical Training Programs.

The seven components of trainees' satisfaction under investigation in our research are summarized below (see Table 3, below):

Q1: ***Satisfaction with Academic Activities in the Training Center.*** This is a critical aspect of our study on residents' satisfaction. We are interested in analyzing and interpreting the degree of satisfaction of trainees related to the academic activities and learning strategies of their training center. The ultimate objective of our integrated research involving the SCFHS is to develop reflective strategies and actions towards improvement of the quality of Training Programs offered by the SCFHS.

Q2: ***Satisfaction with the Residents and Colleagues in the Training Center.*** This component is related to the collegiality and the professional social interaction of trainees with other parties in their training centers. It is important to investigate how trainees perceive the quality of their interaction with other residents and colleagues in the training center. We understand that several psychological and behavioral factors determine this relationship, but we intend to study its overall measure.

Q3: ***Satisfaction with the Administrative Components in the Training Center.*** We focus on the key aspects of interactions of residents with their administration and the execution of administrative tasks integrated in their training. We are interested in understanding the overall satisfaction and how to interpret it to provide significant and meaningful recommendations.

Q4: ***Satisfaction with the Training Program.*** The overall perception and evaluation of the Training Program by the trainees provides a holistic, subjective measure for the quality of education. We are interested in monitoring this rate over time and this study serves as a benchmark for future studies and initiatives related to the enhancement of the quality of Training Programs. It is also associated with our effort to update, and to enhance a compact set of Key Performance

Indicators (KPIs) associated to the Quality Initiative in SCFHS. A full set of these KPIs can be found in our recent previous publication.

Q5: ***Satisfaction with the Specialty***. The perception of residents and fellows towards their specialty and their internal association of different levels of satisfaction is another factor under investigation in our research study.

Q6: ***Satisfaction with the Training Center.*** The perception of trainees towards their training center is also an important area for understanding possible barriers, difficulties, and quality evaluations for residents. This quantifiable measure can also be used for strategic actions towards improvements.

Q7: ***Satisfaction with the SCFHS***. Trainees in their daily interaction with the Training Program in the training center also indirectly interact with the SCFHS as a supervising and accreditation body. It is extremely important to reveal how this relationship is valued by the residents and also how it informs reflective actions for the enhancement of the overall quality.

In Figure 2, below we summarize the core components of the job satisfaction as we approached them with our methodological tool. Namely, seven core components of job satisfaction were set in the center of our focus.

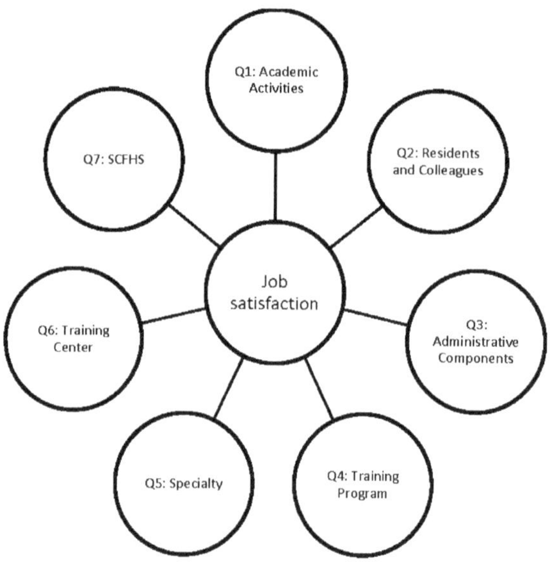

Figure 2. The research focus on job satisfaction.

Table 3. Core components of Job Satisfaction of Trainees in Residency Programs.

Q1	Satisfaction with Academic Activities in the Center
Q2	Satisfaction with the Residents and Colleagues in the Center
Q3	Satisfaction with the Administrative Components in the Center
Q4	Satisfaction with the Training Program
Q5	Satisfaction with the Speciality
Q6	Satisfaction with the Training Center
Q7	Satisfaction with the SCFHS

Our research has four critical key objectives:

Research Question 1: Which are the core components of job satisfaction in residency medical Training Programs as perceived by the trainees?

Research Question 2: Which are the current benchmarks of job satisfaction in the residency program supervised by the Saudi Commission for the Health Specialties?

Research Question 3: Which are the key recommendations for the enhancement of residency Training Programs for enhanced trainee satisfaction?

Research Question 4: How can we interpret the main perceptions of trainees' job satisfaction in terms of measurable, trusted, and reliable Key Performance Indicators for promoting the quality of post graduate medical Training Programs?

The key aspects of our research approach are summarized as follows:

This is an analytical, prospective, cross-sectional study design that represents the trainees' job satisfaction towards Training Programs supervised by the SCFHS in 2018. A self-administered, semi-structured questionnaire survey with both open and close ended questions was distributed to trainees through an online link.

The validation process of the questionnaire included assessing its content validity, which was performed by content experts, after its face validity was assessed by a medical educationist who found that the survey fulfilled the objectives of the study and that the flow of questions followed a logical sequence. To test the reliability of the questionnaire, a pilot study was conducted on 40 participants.

The survey questionnaires for residents were developed by the PGMT Quality Indicator Committee (QIC) to produce an error-free measure of the quality of care, based on characteristics of best practice such as validity, reliability, and transparency. The questionnaire comprised items and included domains pertaining to the training center evaluation, research and simulation training, personal issues, residency program evaluation, and recommendations.

In total, there were 13,688 residents working in different specialties throughout Saudi Arabia, of which only 3442 (25.14%) agreed to participate in the online survey. The trainers were excluded from the survey owing to time constraints. A total of 41 questions represented the indicators of the quality of Training Programs, which were validated by experts and QIC panels for clarity and content relevance.

The questionnaire comprised six sections (see Figure 3):

- The first section comprised eight questions on demographic characteristics
- The second section comprised 14 questions on trainees' educational activities
- The third section comprised three questions on satisfaction with the Training Program
- The fourth section comprised eight questions on perceptions and personal experience
- The fifth and sixth sections comprised three and five questions on research participation and satisfaction with SCFHS, respectively.

The Key Performance Indicators Working Group is responsible for collecting and analyzing data annually and ensuring that the indicators remain precise and appropriate. This goes against Toussaint et al.'s suggestion of releasing quarterly reports [22]. After respondents' data were collected and entered into Microsoft Excel, they were rechecked to ensure there were no typos or blank/empty data cells. The statistical program SPSS version 24 was used for online data analysis. A test was considered significant if the p-value < 0.05.

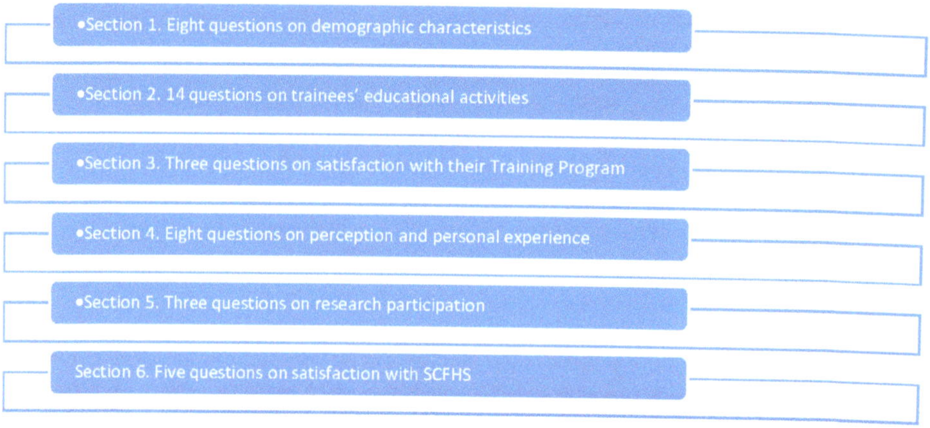

Figure 3. The structure of our research questionnaire.

In the next section, we provide a systematic overview of key findings organized into five distinct areas of interest:

- In Section 4.1, we present the key demographics of our study
- In Section 4.2, we present the key findings related to the Overall Job Satisfaction of trainees in residency programs
- In Section 4.3, we summarize the key aspects of core components of trainees' satisfaction per their discipline and gender
- In Section 4.4, we discuss the findings related to job satisfaction associated to the Training Center of the residents
- In Section 4.5, we elaborate on the key perceptions of residents for their job satisfaction associated to the Training Program they participated in
- Last but not least, in Section 4.6 we present some key aspects of the impact of burnout and sexual harassment on trainees' satisfaction.

In Section 5, we provide a thorough discussion of key findings and their implications. We also discuss the limitations of our study, as well as key recommendations related to sustainable health.

4. Results

The trainees' response rate to the online survey was 27% (3696/13,688) which is satisfactory and represents a significant sample for our interpretations and key findings. This section is organized as follows. First, we present the key demographics of our research. Then we emphasize the presentation of the overall job satisfaction of residents per their discipline, residency type, and gender. After this overall analysis, we emphasize key aspects affecting the perceptions of trainees towards the Training Program and the training institution. Finally, we discuss the important factors affecting burnout and other psychological aspects of trainees' satisfaction.

4.1. Demographics of Our Survey

There were 1932 (52.3%) male respondents (see Figure 4, below), and the majority (91.9%) were Saudi nationals.

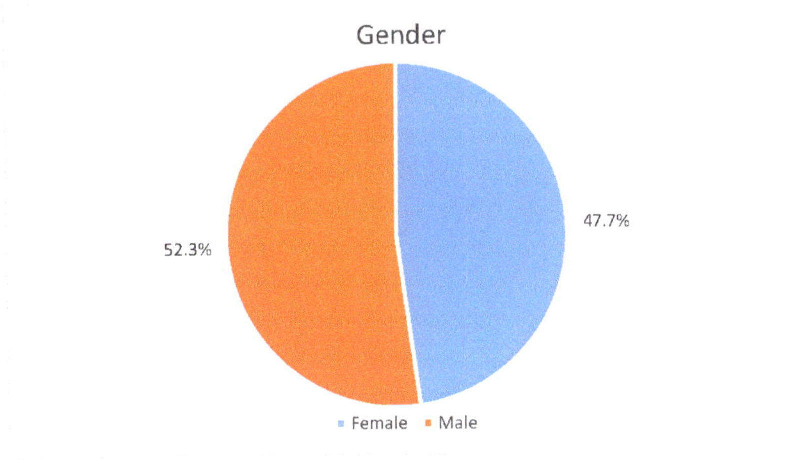

Figure 4. Demographics: Gender.

Concerning the discipline, as seen in Figure 5 below, the vast majority, almost 9 out of 10 respondents belong to the Medicine discipline (90.9%) while the second largest representative discipline is Dentistry with 5.9% (see Tables 4 and 5 below).

Table 4. Respondents per Discipline.

Discipline	Count	Percentage
Pharmacy	35	0.9%
Applied Health Sciences	21	0.6%
Dentistry	217	5.9%
Medicine	3361	90.9%
Nursing	62	1.7%
	3696	100.0%

Table 5. Respondents per Residency Type.

Type	Count	Percentage
F1	26	0.7%
F2	145	3.9%
F3	34	0.9%
R1	459	12.4%
R2	1270	34.4%
R3	860	23.3%
R4	709	19.2%
R5	182	4.9%
R6	11	0.3%
	3696	100.0%

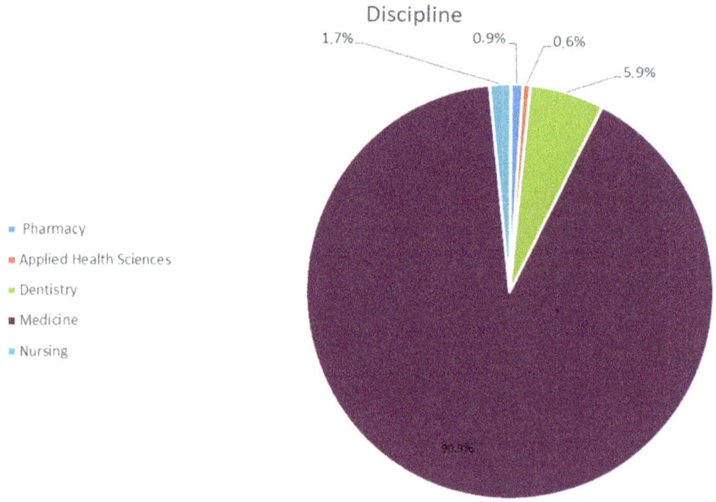

Figure 5. Demographics: Disciplines of respondents.

Almost third of the residents (1270 or 34.4%) were working in the R2 level, while 860 (23.3%) were working in the R3 level (see Figure 6 below).

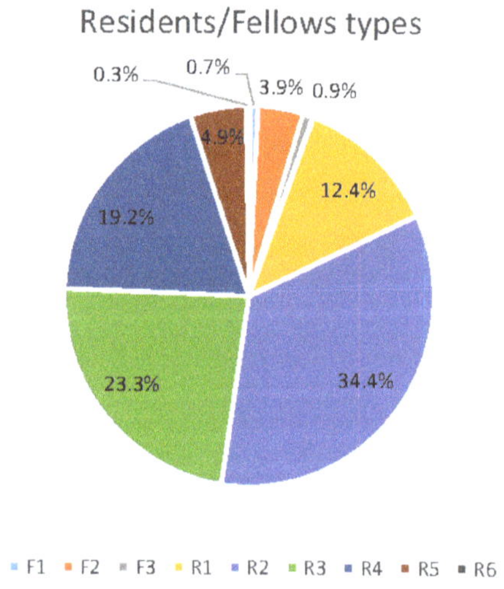

Figure 6. Demographics: Type of residency of respondents.

More than one-third (40.4%) of the trainees were in the Central Region of Saudi Arabia and approximately one-third (34.5%) were in the Western Region. The highest proportion of trainees (39.6%) were working in Riyadh City and 20.1% were working in Jeddah.

4.2. Analysis of Overall Job Satisfaction

One of the key objectives of our research was to develop a benchmark for the measurement of the job satisfaction of trainees in residency programs in the Kingdom of the Saudi Arabia. From this point of view, our work is novel and provides a snapshot for the measurement of trainees' satisfaction in this given period. This metric is significant and allows further developments related to its constitutional factors.

In Figure 7 below, we provide the overall trainees' satisfaction per discipline. The absolute rate for Medicine is 58.6%, a number which allows for significant improvement and sets an initial benchmark for the Training Programs of the SCFHS. From a decision-making point of view, an update of the integrate strategy for the integration of quality in residency programs is required. We intend to discuss this first significant finding of our research further in the Discussion and Implications of our research study section. The overall satisfaction for Pharmacy, is quite like the one related to Medicine and it equals 58.8% (see Table 6, below).

Table 6. Overall Trainees' Satisfaction per discipline.

Discipline	Overall Trainees' Satisfaction
Pharmacy	58.8
Applied Health Sciences	69.6
Dentistry	56.6
Medicine	58.6
Nursing	61.7
Average	58.6

The highest satisfaction rate is for the Applied Health Science trainees, at almost 70%, while the lowest is related to Dentistry at 56.6%. The first interpretation for this finding is that it can be used as a benchmark for the ongoing and future monitoring of trainees' satisfaction in the near future and also that the especially for the Medicine specialty, there is significant space for improvement. In the Discussion section, we provide our key recommendations.

Figure 7. Overall Trainees' Satisfaction per discipline.

Another interesting finding of our research is summarized in Figure 8 below, which is related to the overall satisfaction per residency type. It seems that fellows in residency programs have 8% greater satisfaction than junior and senior residents.

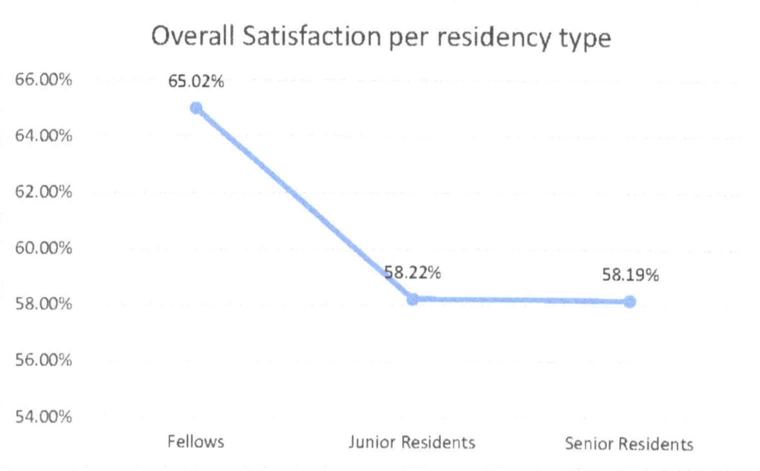

Figure 8. Overall Trainees' Satisfaction per residency type.

4.3. Analysis Trainees' Satisfaction Core Components

As discussed in the Research methodology section, one the key objectives of our survey is to understand the perception of trainees in residency medical programs related to seven core aspects of satisfaction including:

Q1: Academic Activities
Q2: Residents and Colleagues
Q3: Administrative Components
Q4: Training Program
Q5: Specialty
Q6: Training Center
Q7: SCFHS.

In Figures 9 and 10 below, we provide the key findings per Gender. The female participants rate showed high satisfaction gained from their interaction and co-existence with other residents and colleagues in the center (76.6%). They also appreciate the administrative components in the center at a fairly high rate (64.5%) and they are satisfied with their specialty at a good rate (62.1%). Their perceptions of their Training Center and the Training Program at the center are also rather fair (53.3% and 56.8%, respectively). It seems that there is lots of room for improvement for their academic activities in the center, as well as their interactions and integration with the SCFHS (41% and 47.8%) ((see Table 7, below).

Male participants in our survey have similar attitudes to the seven components with some minor differentiations. They rate their overall satisfaction (+1.5%), their satisfaction related to administration in the Training Center (+1.7%), and to academic activities in the center (+1.2%) a bit higher. They are happier with their specialty (+4.3%) and also they value their relation with the SCFHS (+2.4%) more highly. Similarly to females, males seem to need further developments and enhancement in their academic activities in the center.

Table 7. Job Satisfaction overview by core components (Q1–Q7).

	Q1: Academic Activities in the Center	Q2: Residents and Colleagues in the Center	Q3: Administrative Components in the Center	Q4: Training Program in the Center	Q5: Specialty	Q6: Training Center	Q7: SCFHS	Overall
Female	41.0	76.6	64.5	56.8	62.1	53.3	47.8	57.8
Male	42.2	76.0	66.2	59.1	66.4	51.7	50.2	59.3
Difference	1.2	−0.6	1.7	2.3	4.3	−1.6	2.4	1.5

The rather low degree of satisfaction of residents related to the academic activities in the training institution must initiate a debate on a strategic plan for quality enhancement. Various ideas and approaches can be integrated. Some ideas are related to:

- Access to trusted online academic content
- Teamwork and active learning strategies
- Extensive use of robotics, Artificial intelligence tools, virtual and augmented reality labs, and simulation labs
- Extensive research enhancement initiatives for the support of residents in the research domain
- Applied research-based training and learning
- Technology enhanced learning interventions
- Integration of academic activities with the healthcare industry
- Enhancement of the relation of residents with academic supervisors
- Psychological support and relief of burnout rates. A more balanced work-life-study approach

In the discussion section, we elaborate further on this key finding of our research. We also comment on our future research for a more sophisticated analysis of these core components of satisfaction.

In Figure 11 below, we add one more level of analysis concerning the core components of trainees' satisfaction. We summarize the values for the seven core components of residents' satisfaction per different discipline. In the attached table, we also highlight some significant findings for further discussion.

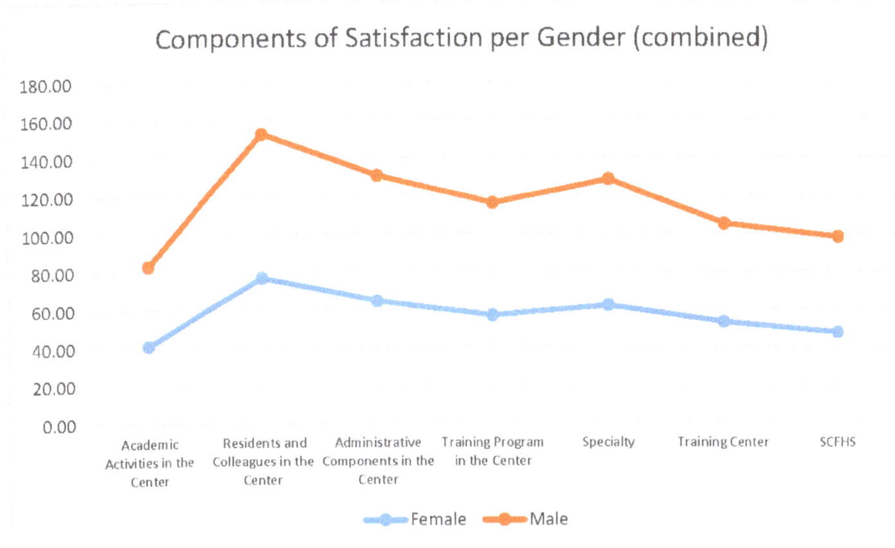

Figure 9. Analysis of Trainees' Satisfaction per Gender (combined).

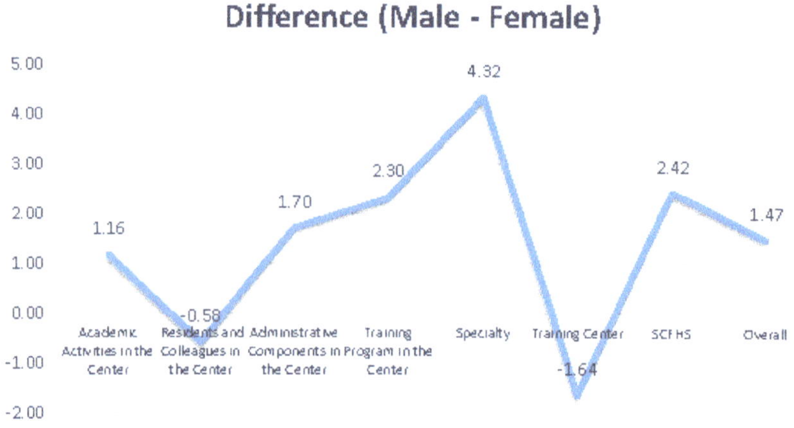

Figure 10. Analysis of Trainees' Satisfaction variations per Gender.

In the Pharmacy and Dentistry disciplines, there is a rather low satisfaction of residents towards their Training Center. The absolute rates (38.6% and 43.2%, respectively) require further study and investigation. Obviously, some aspects of the training centers do not satisfy residents.

As the key finding for most of our sample, the Medicine discipline residents are not satisfied with their academic activities in the Training Center. Their rate of 41.6% is a bit higher from the relevant rate for the Dentistry specialty, which is 38.9%. One of the key propositions of our study is that the Academic activities should be reconsidered and we should also initiate this satisfaction component as an absolute KPI for the quality of the Training Programs offered by the SCFHS.

Residents in Applied Health Sciences seem to be enthusiastic towards their Training Program in the center, with an extremely high rate of 94.1%. They also seem the happiest in terms of their specialty (94.1%) (see Table 8, below).

All the other findings follow the same trends and patterns of male and female participant averages. It seems from all the different disciplines that there is a need for focused strategic actions targeted towards:

- The multidimensional enhancement of the academic activities in the training center. For this purpose, we intend to initiate a new survey of the required actions and the suggestions of administrators and trainees.
- The cultivation of a trusted relationship between the residents and the SCFHS. It is necessary to update communication channels and to increase the awareness of residents to provide added value and increase its contribution to their careers.
- The enhancement of the Training Centers and a continuous improvement process and strategy over time. It is a recommendation to use the benchmarks of this study as a managerial tool for enhancing satisfaction rates over time for the next 5 years by 5–10% per year. This is a bold requirement that needs further investigation, planning, and implementation.
- The development of an institution-wide initiative for the enhancement of the Training Centers in all their aspects. In the next section of our survey, we provide numerous additional qualitative features of the perceptions of the residents towards their Training Center. We do believe that one of the bold findings of our research is related to the average perceptions of respondents towards the Training Center and the academic activities. These two aspects jointly summarize the core components for the residents' experiences and behavior.

Table 8. Core components of Job Satisfaction Per Discipline.

Discipline	Average Q1: Academic Activities in the Center	Average Q2: Residents and Colleagues in the Center	Average Q3: Administrative Components in the Center	Average Q4: Training Program in the Center	Average Q5: Specialty	Average Q6: Training Center	Average Q7: SCFHS	Average Overall Satisfaction
Pharmacy	45.3	78.4	69.0	63.6	62.5	38.6	45.5	58.8
Applied Health Sciences	51.0	85.4	77.8	94.1	82.4	**45.9**	39.4	69.6
Dentistry	38.9	78.7	66.5	54.6	58.7	43.2	48.0	56.6
Medicine	41.6	76.0	65.1	57.7	64.6	53.1	49.3	58.6
Nursing	48.2	79.3	67.1	72.2	66.0	54.0	44.5	61.7
Average	41.0	76.3	65.4	58.0	64.4	52.5	49.1	58.6

Note: Colored numbers refer to significant findings (for improvement or interpretation).

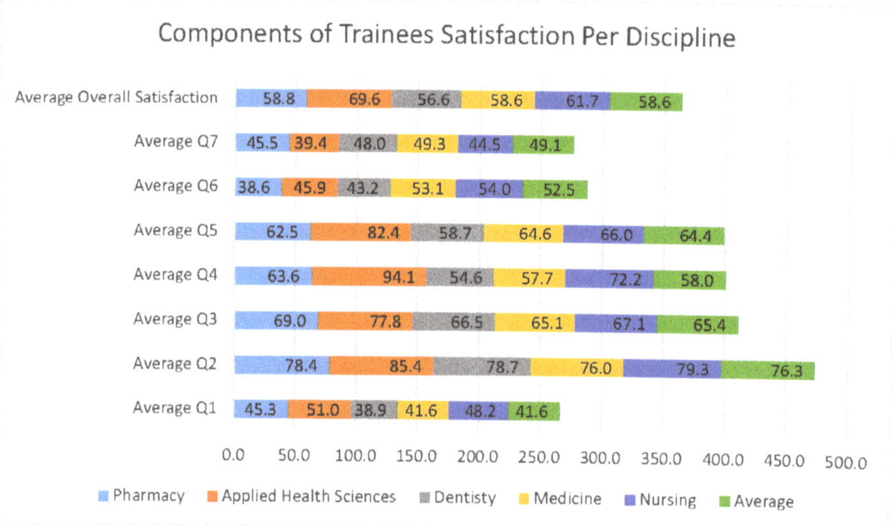

Figure 11. Analysis of Trainees' Satisfaction per Discipline (combined).

4.4. *Special Analysis of Trainees' Satisfaction Related to Training Center*

Our research study is one of the first efforts worldwide to identify, measure, and interpret the satisfaction of trainees in residency medical programs for the enhancement in their quality and their impacts. It is linked to the Sustainability and Health debate because it promotes a scientific debate for the constitutional value adding components in medical training. In the discussion section, we provide thorough case for the connection and the implications of our research for the Sustainable Health domain.

In this section, we elaborate further in the perceptions of our respondents for their satisfaction towards the Training Program and the Training Center. In Figure 12, we provide a high-level abstraction for the satisfaction rate of residents for 120 different Training Centers without uncovering the identity of each center. The picture provided in this figure is interesting. There are Training Centers that are perceived highly by residents, which provide high satisfaction and happiness. From the other side, residents also provide rather low satisfaction rates for numerous other Training Centers. The overall rate of satisfaction of trainees for all the Training Centers is approximately 55%. This is another interesting benchmark that our research study reveals.

It is necessary to study the findings of Figure 12 further. The SCFHS must have a continuous improvement process to understand in depth the causal factors for the low evaluation of specific training centers indicated by the respondents of our survey. We do understand that the current survey provides a snapshot of a given moment, but the overall findings have validity and require interpretation and revision of policies, as well as actions for enhanced decision making. In the next section, we try to provide more data and facts indicating the perceptions and attitudes of trainees towards their Training Center and program. We recommend a new run of a similar surveys in due time to understand how some reflective actions and recommendations had an impact on overall quality and satisfaction levels.

In Figure 13, we also provide the top Training Centers in terms of residents' satisfaction. We understand the limitations of our study, which will also be discussed in Section 5 of this paper, but we also consider that the ratings of such a large sample in our study provide trustworthy approximations of quality and satisfaction.

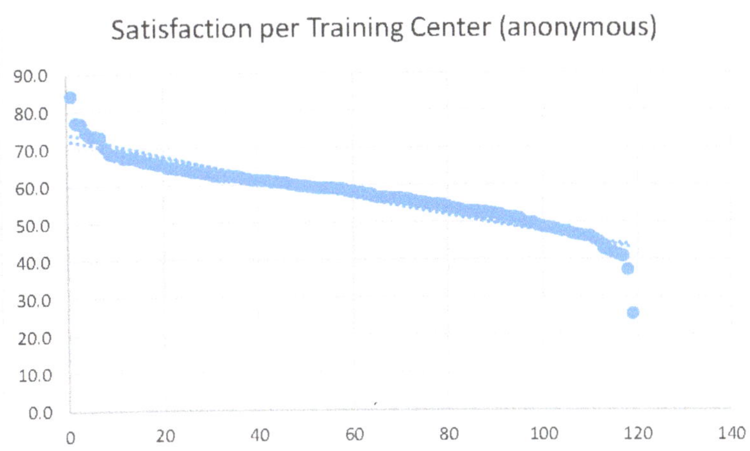

Figure 12. Overall Trainees' Satisfaction per Training Center.

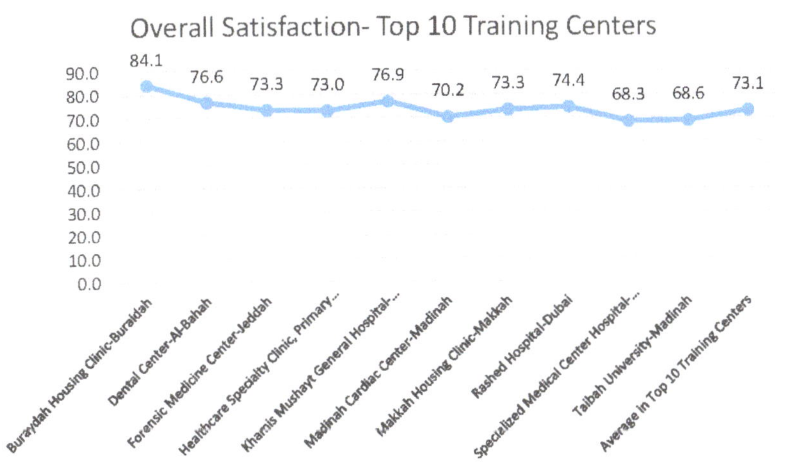

Figure 13. Perception of residents towards the quality of the Training Centers.

The concluding comment for the attitudes of residents related to the Training Centers is that overall residents feel that more value can be delivered in the academic program, as well as in the integration with the services and supervision of the SCFHS. In the next section, we provide additional key data and facts related to the disposition of trainees with the Training Program. Last but not least, we also provide some sensitive data related to burnout rates and sexual harassment figures.

4.5. Special Analysis of Trainees' Satisfaction Related to the Training Program

The fourth value component and interpretive factor for the overall job satisfaction of trainees in our methodological approach is related to the training program. The analysis of the key findings is presented in this section. Overall, there is some appreciation from the residents for their Training Programs. According to our collected data, 16.5% of trainees are satisfied with their current Training Programs, while almost 50% request improvements or have neutral attitudes. This is a significant finding.

We do believe that a thorough strategic initiative for the enhancement of Training Programs must be undertaken by the SCFHS and the other bodies that are involved in these programs. Some additional facts that are presented below provide directions for this enhancement.

Regarding the satisfaction rate of the trainees, 34.7% believe the programs are good, 16.5% believe that they can be improved, and 2.7% think they are useless (see Figure 14 below). One fourth of respondents also provided neutral perceptions, which highlights that currently there is a lot of space for enhancements in the Training Programs for residents. From this short overview of the satisfaction with Training Programs, it is highly recommended to use the current benchmarks the Useful as it is and the Useful but needs improvement rates as initial points for improving the satisfaction rate over the next two years by 20%.

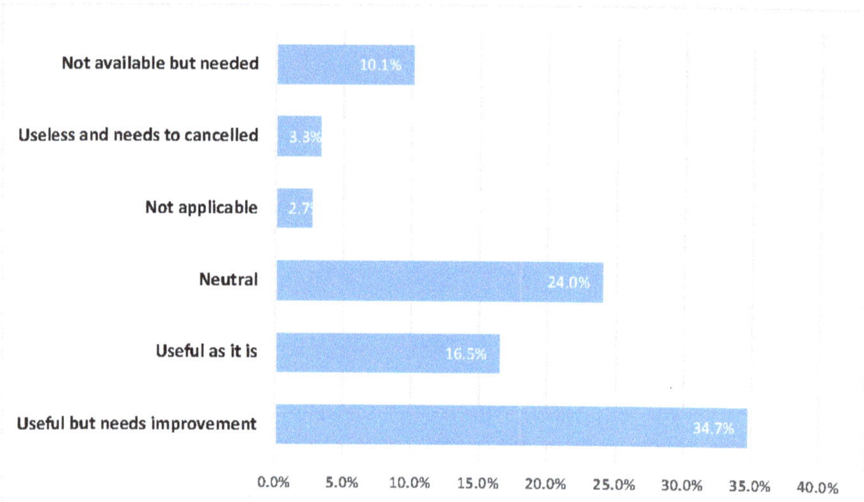

Figure 14. Trainees' satisfaction with the training programs.

In the same direction, we also tried to understand some key factors with a significant impact on trainees' perception towards the quality of the Training Programs.

With regard to the satisfaction rate of the trainees, 34.7%, 31%, 22%, 25.3%, 25.5%, and 25.3% considered the program, academic half-day, bedside teaching, grand round, bedside round with a senior, and bedside round with a consultant to be good, respectively (see Table 9, below).

Table 9. Training Program aspects that are valued by residents.

Key Aspect	Rate %
Program	34.70%
Academic Half-day	31%
Bedside teaching	22%
Grand round	25.30%
Bedside round with a senior	25.50%
Bedside round with a consultant	25.20%

It is important to emphasize that residents commented that academic half-day is a key aspect of satisfaction, which implies that research and training beyond professional service are recognized as top priorities by trainees. This is a key finding that needs to be exploited further through strategic consultation by enhancing the research and academic life and practice of residents. Trainees selected attitude, interest in learning, and interest in helping others as top priorities while rating features of other residents working in the program.

More than half (58%) were very likely to recommend the program in which they work to others, while 12.2% were unlikely to do so (see Figures 15 and 16 below). This seems to be a strong core component of trainees' satisfaction. In future research, we plan to study the value components for his likelihood behavior for recommendation using a Structural Equation Modeling research tool. In the Section 5, we also provide our constructive interpretation of the key findings of our research towards a progressive model for Trainees' Satisfaction in Medical Education.

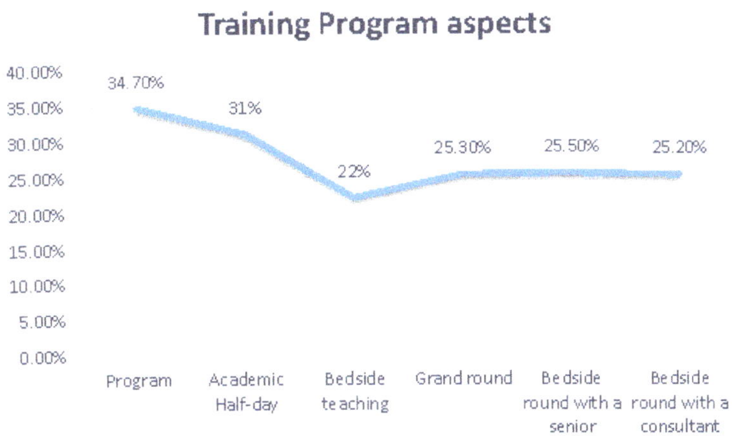

Figure 15. Training program aspects that are valued by residents.

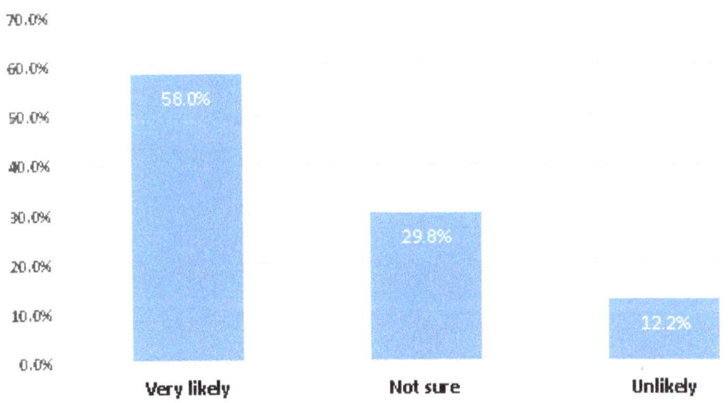

Figure 16. Trainees' responses to whether they would recommend their training program to others.

These absolute numbers can also be used for future benchmarking research. They can serve as managerial tools for enhancement and also for monitoring and control of the value proposition of the medical Training Programs supervised by the SCFHS.

In response to the question on whether they would choose another center for residency training after joining the residency program, nearly one third of the trainees (30.5%) agreed they would (see

Figure 17 below). This is potentially another factor that links the satisfaction rate of trainees to their perceptions towards their Training Center. It seems that one third of respondents do not build a strong bond with their selected Training Center. This finding requires further investigation. We must reveal the factors behind this attitude held by residents. The absolute rate of almost one third is rather high and means that several residents are not satisfied with their current Training Center.

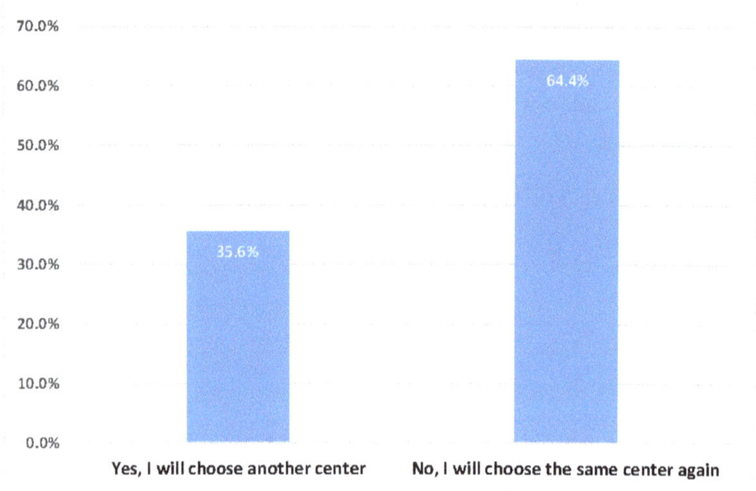

Figure 17. Trainees' responses to whether they would choose another center for their residency training after having joined their program.

More males than females consider it "very likely" that they will recommend their program to other trainees, while more females consider it "unlikely" that they will do so (see Figures 18–20 below).

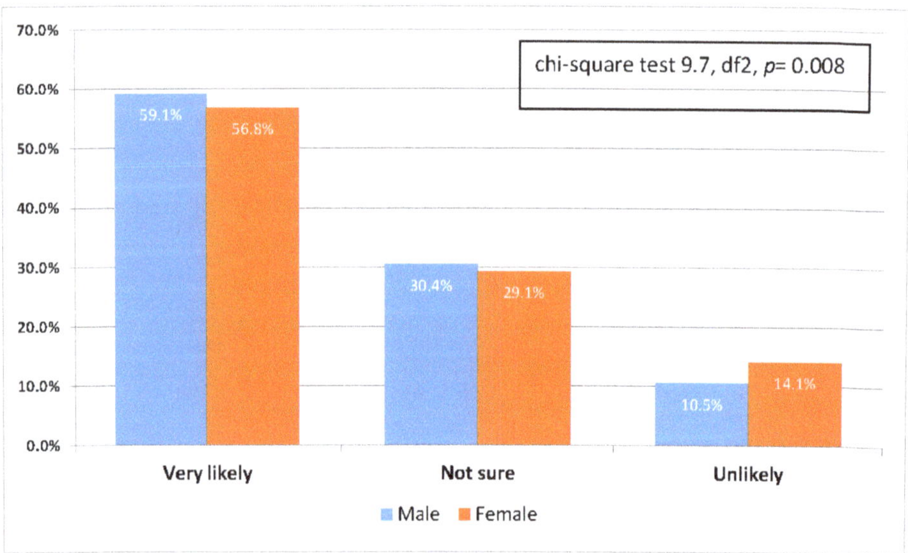

Figure 18. Gender differences regarding the likelihood of respondents recommending their Training Program to other trainees.

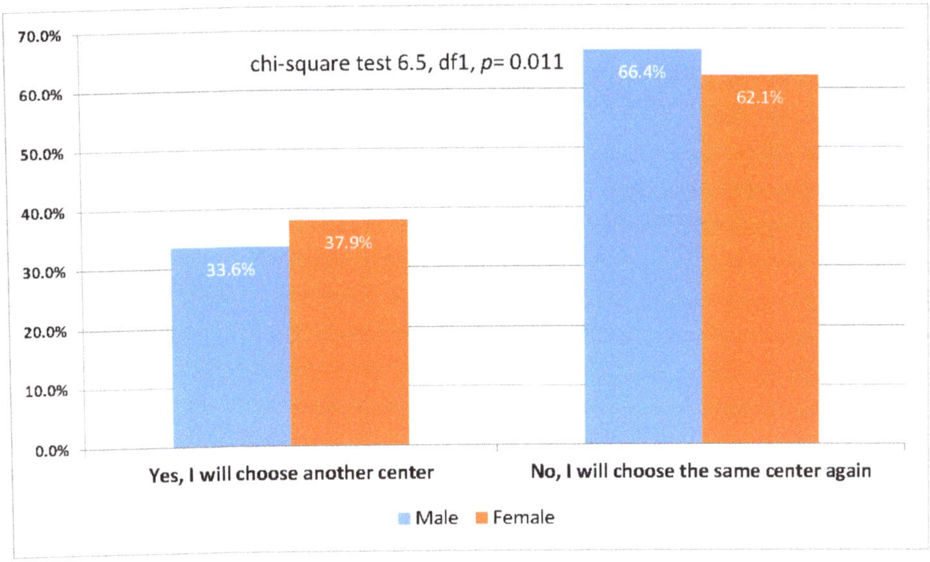

Figure 19. Gender differences in responses to the question: In retrospect (after you have joined the program), would you choose another center for your residency training?

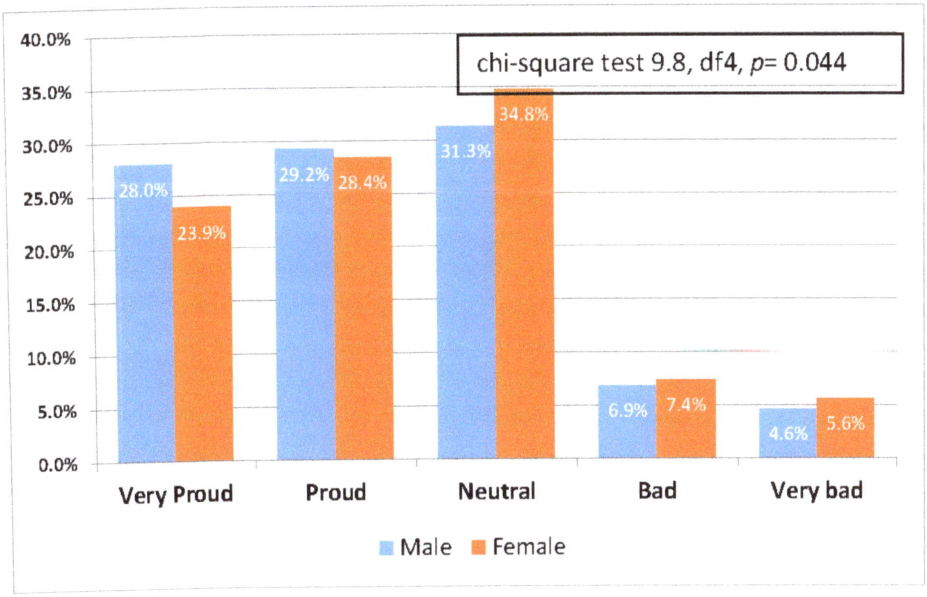

Figure 20. Gender differences in responses to the question: How "proud" or "bad" do you feel about being a trainee in your (current Training Center)?

Female residents seem to be more likely to switch Training Centers than males. In Figure 19 below, there is a graphical representation of this finding.

Overall, residents feel proud to have been members of their current Training Center. There were no huge gender-based differences found. It is also important in our future research to understand the reasons why about 10% of residents are unhappy in their current Training Centers. It is a key priority of our future research to identify unhappy residents and to provide flexible mechanisms for their support and psychological enhancement (see Figure 20 below).

In this section, we provided some initial assessments of the attitudes of residents relating to their Training Program. It seemed that there is a rather fair satisfaction level of residents towards their Training Programs, and a lot of space for improvements. We also suggest that we should use the various rates of the relevant questions as benchmarks for future analysis, monitoring, and control. They can also serve as strategic objectives from improvements. We do believe it is feasible to set the increase of relevant satisfaction components for the Training Programs by 5–10% for the next two years as an objective through reflective strategic actions carried out by the SCFHS.

In the next section, we focus on one more key objective of our research study, which is to investigate the degree of burnout of residents as expressed by themselves in our survey. We intend to interpret the key facts of this situation and to make significant recommendations in the Discussion section.

4.6. Special Analysis of Trainees' Satisfaction Related to Burnout

Burnout is always referred to the literature as a key component of residents' dissatisfaction in their medical Training Programs. It is a factor with diverse direct and indirect psychological effects, and also linked with depression and anxiety. In this section, we provide our key findings related to various aspects of burnout (see Table 10, below).

Table 10. Overview of burnout rates.

Burnout	Rate
Always	28.2%
Sometimes	38.5%
Rarely	15.6%
It depends	13.7%
Never	4.0%

In response to the question "How frequently do you feel "burned out" and unable to cope anymore?", 28.2% of trainees believe they "always" feel this way, 38.5% "sometimes" have this feeling, 15.6% feel it rarely, 13.7% said it depends, while approximately 4% "never" feel burned out (see Figure 21 below). This is a major finding of our research that poses critical questions for managerial decisions and key responsive actions. The too high rate of burnout seems to be one of the negative catalysts for the snapshot of the trainees' satisfaction in residents' Training Programs. Almost two third of residents feel that they experience a significant rate of burnout in their lives that has a key impact on their professional conduct and their lives. We intend in the near future to run a survey based on structural equation modeling in order to understand the cause and effect relations of burnout and also to measure the impact of the high level of burnout to different aspects of residents' personal and professional lives. We also present some key recommendations in the Discussion section.

Concerning the burnout rate of residents, there is not a gender-based uniform pattern. In Figure 22, there is a detailed overview of gender differences related to burnout. Once again, the high rate of burnout for males and females alike is a major finding of our survey and needs further study. It seems to be one of the core components of trainees' dissatisfaction. The SCFHS must focus on this finding and initiate actions and plans to provide residents with psychological relief and increase their courage and physical capabilities.

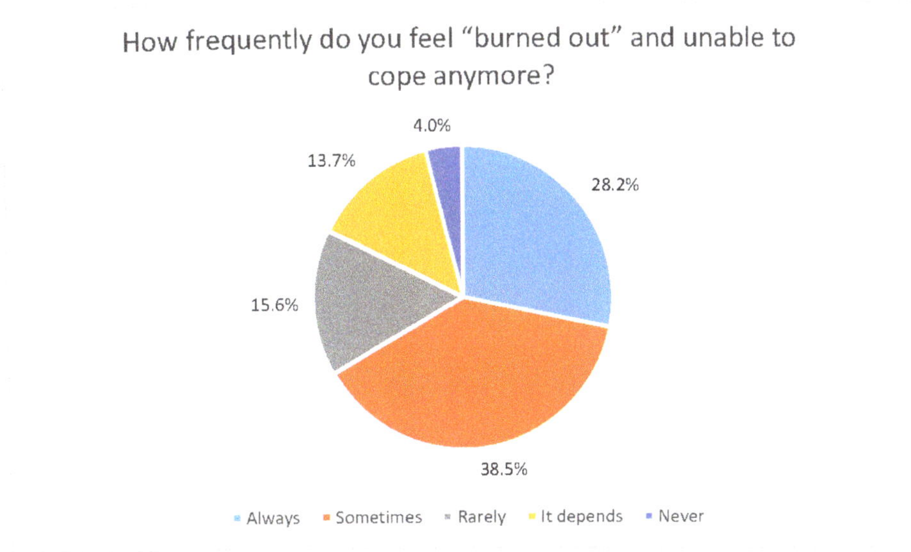

Figure 21. Responses to the question: How frequently do you feel "burned out" and unable to cope anymore?

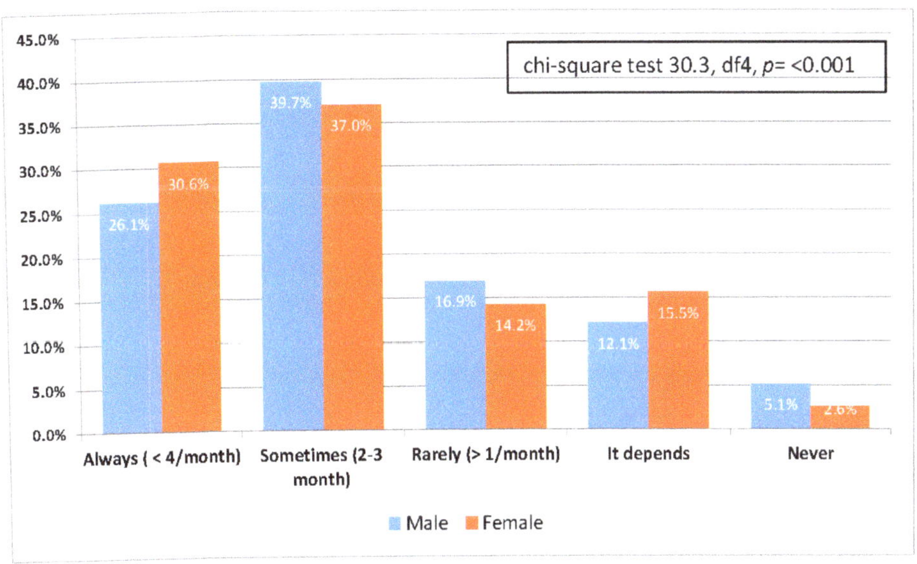

Figure 22. Gender differences in responses to the question: How frequently do you feel that you are "burned out" and "cannot cope anymore" due to work stress?

In Figure 23, we also summarize a very delicate factor of trainees' satisfaction related to sexual harassment. The percentage is low related to international benchmarks but it seems that females experience verbal sexual harassment twice as frequently as males. This is also another significant key direction for future research.

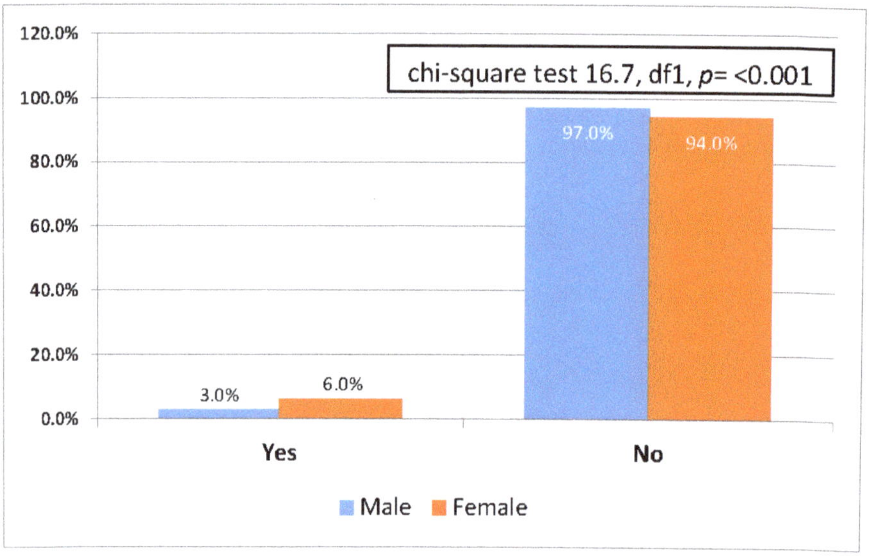

Figure 23. Gender differences in experience of verbal sexual harassment.

In the next section, we provide our key interpretations of the findings and we also contribute theoretically with a progressive model for Trainees' Satisfaction. We must provide the following significant statements for the end of this Analysis of findings section:

- Some of the collected data and their relevant analysis are only for internal use in the SCFHS. In this research paper, we focus on data and findings that can be shared publicly to promote knowledge in the domain of Sustainable Health
- Our intention in this research is to build a systematic, trusted set of benchmarks for the study of residents' satisfaction in their medical Training Programs. In future research, we intend to further study the cause and effect relationships for the key findings of this research. This seems to be a limitation factor for the current study but our current contribution remains unique, bold, and significant. Our study is one of the first to analyze resident population issues related to satisfaction at such a large scale.

5. Discussion, and Interpretation of the Key Findings of Our Research

In this section, we try to synthesize the key findings of our research. We also try to provide some key implications of our research.

5.1. Key Findings and Interpretations

In this section, we will provide an overview of the key findings of our research together with some key interpretations and propositions. This section is organized as follows. First, we present the main findings of our research. Then we discuss them, and we interpret them towards reflective actions and propositions for the SCFHS.

Analysis of Overall Job satisfaction: This is a major finding of our research. We propose to set the average benchmark of Trainees Satisfaction to 58% as a working managerial benchmark for enhancement initiatives. We also recommend that this benchmark must be monitored with a targeted increase by 5% to 10% per year for the next three years. Actions and guidelines towards this quality enhancement are proposed in the next section. The key aspects of overall job satisfaction are presented below:

- The absolute overall job satisfaction for Residents related to Medicine is 58.6%, a number which definitely allows for significant improvement and sets an initial benchmark
- The overall satisfaction for Pharmacy is similar to the one related to Medicine and equals 58.8%
- The highest rate is for the Applied Health Science trainees at almost 70%, while
- The lowest rate is related to Dentistry at 56.6%.

Analysis of Core Components (Q1–Q7) of Job satisfaction: This is a major finding of our research. We propose to set any of the core components Q1 to Q7 as working managerial benchmarks. In the table below, we emphasize that there is a strategic need for an integrated initiative to increase the rate of job satisfaction of trainees related to their academic program and the relationship with the SCFHS with a target of 15% increase per year in the next three years. A set of actions are provided in the next section. We will also proceed to conduct further research on this major finding of our current research. Currently we are designing and implementing a new survey based on structural equation modeling aiming to investigate the key hermeneutic factors for all these core parameters of Job Satisfaction. We provide some of the key aspects for the core components of job satisfaction below:

- Fellows in residency programs have an almost 8% higher satisfaction compared to junior and senior residents.
- Female participants highly rate their satisfaction gained from their interactions and co-existence with other residents and colleagues in their center (76.6%).
- They also appreciate the administrative components in their center at a fair rate (64.5%) and they are satisfied with their specialty at a good rate (62.1%).
- Their perceptions towards their Training Center and their training program at the center are also rather fair (53.3% and 56.8%, respectively).
- It seems that there are many margins for improvement of their perceptions towards their academic activities in the center, as well as their interaction and integration with the SCFHS (41% and 47.8%, respectively).
- Male participants in our survey have similar attitudes to the seven components with some minor differences. Compared to women, they rate their overall satisfaction (+1.5%), their satisfaction related to administration in their Training Center (+1.7%), and their satisfaction towards academic activities in the center (+1.2%) slightly more highly.
- Males are happier with their specialty (+4.3%) and also they value their relationship with the SCFHS more highly (+2.4%).
- Similarly to females, males seem to need further developments and enhancement in their academic activities in the center.
- The rather low degree of satisfaction of residents related to the academic activities in the training institution must initiate a debate on a strategic plan for quality enhancement.

Some additional comments that also highlight key aspects of the interaction and integration of the trainees with the training center are summarized below:

- The trainees' satisfaction level was highest towards their program director (69.7%), followed by their satisfaction with the chief resident (69%);
- They were least satisfied with their department head (56.8%).
- More than half (51.4%) of the trainees were very likely to recommend the program in which they worked to others.
- Throughout the training, 96.1% of the trainees experienced burnout and only 3.9% reported that they did not feel it.
- Over half of the respondents were proud of their current Training Center, while one-tenth (11.7%) stated that they would choose another specialty if given a chance.
- The most common abusive behavior reported in our study sample was verbal abuse (35.8%).

- In total, 33.1% of the respondents participated in research, with 40.9% were mainly involved in collecting data and 36.3% were involved in proposal preparation.

These are another major findings of our research. We propose that the burnout rate must also be a benchmark for the SCFHS quality assurance initiative and the residency programs. The current high rate is unacceptable, and we propose systematic actions for the decrease of the absolute value by 15% per year for the next three years. We introduce also introduce one more benchmark related to research involvement, which is currently at 35%, with a recommendation for it to be enhanced by 15% per year in the next five years (see Table 11, below).

Table 11. Key Benchmarks for the Job Satisfaction of residents.

Key Benchmarks Introduced in Our Study	Current Rate	Goal Set for the Next 3 Years
Overall Job satisfaction of trainees	58%	+5–10% per year
Q1: Academic Activities in the Center	41.5%	+15% per year
Q2: Residents and Colleagues in the Center	76%	+5–10% per year
Q3: Administrative Components in the Center	65%	+5–10% per year
Q4: Training Program in the Center	58%	+5–10% per year
Q5: Specialty	64%	+5–10% per year
Q6: Training Center	52%	+15% per year
Q7: SCFHS	49%	+15% per year
Burnout Rate	90%	−15% per year
Verbal Abuse	35%	−10% per year
Research Involvement	33%	+15% per year

Note: Colored numbers refer to significant findings (for improvement or interpretation).

Our key findings can be compared with similar studies. We need though to communicate in a bold way that our research is likely the first covering such a (see Table 12, below) large population for the KSA and one of the very few with such a representative sample in residency programs. Our key findings can be associated with similar studies, even in different populations. For example, the job satisfaction rate observed in our sample can be compared with the research work of Alsubaiea and Isouard [7], in which they concluded that the job satisfaction rate that they found was fair. Also, our key findings can be associated with works like Chen et al. [8] showing that job satisfaction is associated with numerous factors including working conditions, financial rewards, and the doctor's relationships with patients. The key contribution of our work is that the job satisfaction revealed two problematic contexts: the academic program and the Training Center, as well as the need for collaboration with the supervising organization for the residency programs. This is a finding that needs further investigation, and we intend to proceed to a new research study soon.

Our research study also dealt with variables found in the research studies of Bawakid et al. [9] indirectly, since our key findings also support the need to focus on the continuous support and improvement of the working environment. We also noted that burnout is a key obstacle preventing high job satisfaction and efficiency. We also believe that participation in research and research involvement can also be compared with the key findings in the work of Soto–Rubio et al. [12], which emphasized that high emotional intelligence has a positive favorable effect on job satisfaction.

Our finding also that residents are really interested in the Academic program and the Training Center can be also compared with key findings in the research of Aoyagi et al. [13]. Our work also highlighted significant components for dissatisfaction of residents that overlaps somewhat with a study conducted by Liu et al. [14]. Our research also confirmed the key findings of [15], in which a list of factors for job satisfaction includes satisfaction towards a supervisor and satisfaction towards the administration.

Some additional elaboration on our key findings can be provided: The prevalence of burnout in medical residencies, according to the international literature, ranges from 27% to 75%, depending on the specialty [23]. However, our results revealed that 96.1% of the trainees had experienced burnout and only 3.9% reported that they did not feel it throughout the course of training. Contrary to our results, a study among 59 Dutch psychiatry residents four teaching hospitals found that only 7% of their respondents met the criteria for burnout, which is very low [24]. Additionally, in Brazil, the prevalence of burnout was 27.9% among medical residents, which is also low [25].

In 2012, a survey was conducted to determine the extent of burnout in the United States, and 68.2% of medical students were found to have high and intermediate levels of burnout [26]. In a study in Canada, 79.2% of dermatologists were found to have a moderate-to-high burnout rate [27]. Another study conducted at the Massachusetts General Hospital, United States, revealed that burnout was prevalent in 28% of medicine and psychiatry residents [28]. Additionally, a survey conducted in the United States found that 69% of residents working in 20 different programs in 2014 experienced burnout [29].

According to a narrative review conducted by Dyrbye et al., which included articles published between 1990 and 2015, 45–56% of residents experienced burnout [30]. Moreover, Jagsi et al. performed a meta-analysis of 35 studies in 2014, which revealed that there is a mean prevalence of 33.1% of students and residents who reported experiencing sexual harassment [31]. Eight out of ten other studies reported that between 45% and 93% of residents had experienced some form of inappropriate behavior during their residency training on at least one occasion [32]. The most common abusive behavior reported in our study sample was verbal abuse, followed by physical, sexual (verbal), and sexual (physical) abuse (4.5%, 4.4%, and 1.8%, respectively). Fnais et al. performed a meta-analysis of 51 studies across multiple disciplines and countries in 2014 and revealed that 59.4% of trainees have experienced bullying, with verbal harassment being the most common form of abuse (63%) [32].

In Figure 24 below, we synthesize the various findings of our research and we present an initial framework for the Trainees satisfaction as a basis for advanced decision making.

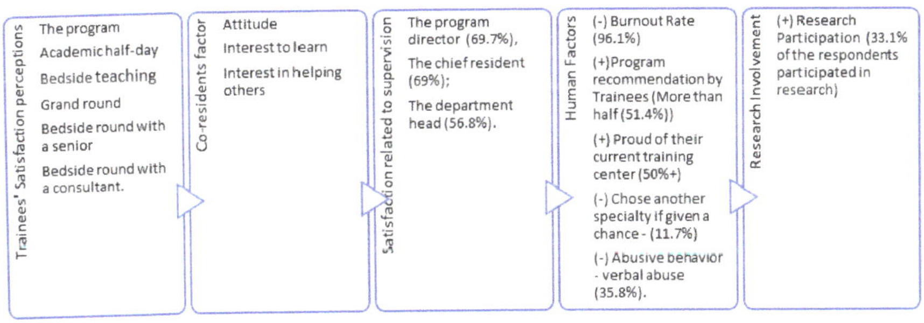

Figure 24. An integrative model for understanding Trainee satisfaction with Residency Training Programs.

We provide a 5-pillar model abstraction for the interconnected variables of Trainee satisfaction in residence programs. This can be used as the basis for more sophisticated enhancement of the quality of training and decision making. It can be also used as the basis for specifying key performance indicators focused on trainee satisfaction.

Table 12. Respondents per Survey.

Reference	Authorship	Details of the Survey
[7]	Alsubaie et al.	Involved 2362 nurses in Saudi Arabia Includes 4 studies.
[8]	Chen et al.	Study was conducted of 39 doctors from five township hospitals in Guangxi Zhuang, China
[9]	Bawakid et al.	Study of 237 FPs working in primary healthcare centers under the Saudi Ministry of Health that was conducted in two regions (Jeddah and Eastern region).
[10]	Zhang SX et al.	A sample of 304 healthcare staff (doctors, nurses, radiologists, technicians, etc.) in China.
[11]	Leskovic et al.	Conducted in spring 2013 (n = 556) and spring 2020 at the peak of the pandemic in Eastern Europe (n = 781) in Slovenia.
[12]	Soto-Rubio et al.	Involved 125 nurses in Spain.
[13]	Aoyagi et al.	Narrative synthesis which showed study estimates that ranged from 23.1% to 95.8% rates of high willingness to work, depending on context. Meta-analyses of specific factors showed that for male HCWs, physicians, and nurses, full-time employment, perceived personal safety, awareness of pandemic risk and clinical knowledge of influenza pandemics, role-specific knowledge, pandemic response training, and confidence in personal skills were statistically significantly associated with increased willingness to work
[14]	Liu et al.	Questionnaires were administered to 897 doctors from Eleven tertiary public hospitals in Shanghai, China.
[15]	Platis et al.	A questionnaire was distributed to 246 nurses in Greece.
[16]	Chao et al.	Involved 344 licensed professionals in 1 rural regional hospital in Taiwan.
[17]	Maissiat et al.	This study was conducted with 242 employees of a municipality of Rio Grande do Sul, Brazil.
[18]	Pandey et al.	This study involved 177 accredited social health activists in India.
[19]	Labrague et al.	This study involved 261 frontline nurses in the Philippines.

Each of these pillars is related positively or negatively with significant components of satisfaction.

- • • *Trainees' Satisfaction perceptions*: In the first pillar, the Training Program itself together with modes of academic development and supervision are valued as positive factors for residents' perceptions. The following areas have been highlighted as critical for the trainees' perceptions towards the program and supervision.
    - → *The program*
    - → *Academic half-day*
    - → *Bedside teaching*
    - → *Grand round*
    - → *Bedside round with a senior*
    - → *Bedside round with a consultant.*
- • • *Co-residents factor*: another critical component for trainees' satisfaction is related to co-resident features, with very positive contribution to satisfaction from the following dimensions:
    - → *Attitude*
    - → *Interest in learning*
    - → *Interest in helping others*

- • • *Satisfaction related to supervision*: the third significant component of trainees' satisfaction is related to the professional relationship and appreciation of supervision.
  → *The program director (69.7%),*
  → *The chief resident (69%);*
  → *The department head (56.8%).*
- • • *Human Factors*: The human factors integrate different aspects of human personality and behavior. The key component of dissatisfaction is the burnout rate and also verbal abuse and willingness to choose another specialty if given a chance. Positive enablers are the feelings of being proud to be part of a Training Center and the likelihood to recommend their center based on their experience and satisfaction.
  → *(-) Burnout Rate (96.1%)*
  → *(+) Program recommendation by Trainees (More than half (51.4%))*
  → *(+) Proud of their current Training Center (50%+)*
  → *(-) Would choose another specialty if given a chance—(11.7%)*
  → *(-) Abusive behavior—verbal abuse (35.8%).*
- • • *Research Involvement*: The last but not least important component of satisfaction is related to the quality and the significance of undertaken research during resident training.
  → *(+) Research Participation (33.1% of the respondents participated in research)*

5.2. Strategic Propositions and Managerial Implications

As depicted in the previous section, continuous improvement on residency programs and related job satisfaction can be orchestrated through reflective actions and initiation of a strategic plan. The benchmarks that presented can be used as a managerial tool for enhanced job satisfaction. Furthermore, here are some additional strategic propositions:

- The multidimensional enhancement of the academic activities in Training Centers is required. For this purpose, we intend to initiate a new survey for the required actions and the suggestions of administrators and trainees. A task force can deal in a professional way with the enhancement of the academic and research activities of residents in the academic programs.
- The cultivation of a trusted relationship between the residents and the SCFHS also needs to be promoted. It is necessary to update communication channels and to increase the awareness of residents for the added value and its contribution to their careers. The SCFHS must provide continuous communication channels and reinforcement services to residents.
- The enhancement of the Training Centers and a continuous improvement process and strategy over time are needed. It is a recommendation to use the benchmarks of this study as managerial tool for enhancing satisfaction rates over time for the next 5 years by 5–10% per year. This is a bold requirement that needs further investigation, planning, and implementation.
- It seems from all the different disciplines that there is a need for focused strategic actions targeted towards:
  - The multidimensional enhancement of the academic activities in a Training Center. For this purpose, we intend to initiate a new survey focusing on the required actions and the suggestions of administrators and trainees.
  - The cultivation of a trusted relationship between the residents and the SCFHS. It is necessary to update communication channels and to increase the awareness of residents to increase the added value and its contribution to their careers.
  - The enhancement of the Training Centers and a continuous improvement process and strategy over time. It is a recommendation to use the benchmarks of this study as managerial

tool for enhancing satisfaction rates over time for the next 5 years by 5–10% per year. This is a bold requirement that needs further investigation, planning, and implementation.
  - The development of an institution wide initiative for the enhancement of the Training Centers in all of their aspects. In the next section of our survey, we provide numerous additional qualitative features of the perceptions of the residents towards their Training Center. We do believe that one of the main findings of our research is related to the average perceptions of respondents towards their Training Center and their academic activities. These two aspects jointly summarize most of the core components for the residents' experiences and behavior.

- From this short overview of satisfaction with the Training Program, it is highly recommended to use the current benchmarks of the Useful as it is, and the Useful but needs improvement rate responses as an initial point for improving the satisfaction rate over the next two years by 20%.
- It is necessary to enhance the research and academic life and practices of residents.
- The too high rate of burnout seems to be one of the negative catalysts for this snapshot of the trainees' satisfaction in residents' Training Programs. Almost two third of residents feel a significant rate of burnout in their lives, with a key impact on their professional conduct and their lives. We intend in the near future to run a survey based on structural equation modeling in order to understand the cause and effect relations of burnout and also to measure the impact of the high level of burnout on different aspects of residents' personal and professional lives. We also present some key recommendations in the Discussion section.
- It seems that one third of respondents do not build a strong bond with their selected Training Center. This finding requires further investigation. We must reveal the interpretive factors for this attitude amongst residents. The absolute rate at almost one third is rather high and means that several residents are not satisfied with their current Training Center.
- It is also important in our future research to understand the reasons for making about 10% of residents unhappy in their current Training Centers. It is a key priority of our future research to identify unhappy residents and to provide flexible mechanisms for their support and psychological enhancement (see Figure 19 above).

In the next section, we conclude our research with key statements and also provide the key directions for future research.

6. Conclusions, Limitations, and Future Research

Our research study is a systematic effort to understand residents' attitudes related to job satisfaction. In such a complicated research environment, we tried at a first stage to obtain a comprehensive picture of the phenomenon.

Our research study is one of the first in KSA as well as worldwide, with a special emphasis on a thorough understanding of issues affecting the job satisfaction of residents at a major scale. In comparison to other studies, we have obtained a significant achievement by obtaining almost 4000 questionnaire results from a major portion of the whole population of residents in the KSA. A key limitation of our research is that all findings and key interpretations refer to the current snapshot of the professional and academic conduct of trainees. Thus, the generalization of findings must be made while taking into consideration that we only studied residents from the Kingdom of the Saudi Arabia.

One more limitation of this survey is that we intended to develop a thorough trusted initial benchmark for various components of trainees' satisfaction. We do believe that the outcome of our intellectual effort and research methodology enabled summarizing a clear overview of trainees' satisfaction benchmarks at present. We have to admit though that especially in this study, we did not provide the same effort towards analyzing the cause and effect relations of the trainee satisfaction phenomenon. We plan to shortly run a new survey based on structural equation modeling to further inform our key findings. The ultimate objective is also to update our recently published set of Key

Performance Indicators (KPIs) for the Quality of Residents Training with an additional set of KPIs related to residents' satisfaction.

One additional limitation of our study is the disclosure of significant information. Some of the aspects and findings of our survey must be used only for internal use of the Saudi Commission of the Health Specialties, since they refer to indirect evaluation and assessment of various training centers supervised by the SCFHS. Finally, we have also to communicate that this research study refers to a certain period, and thus all the findings and conclusions must be interpreted within the given time. Any effort to extend the validity of the findings for the future must be made with special care and conditions. We do believe though that all the key findings have direct implications on the launching of revised policies for the enhancement of the Training Programs and trainee satisfaction.

We also contributed significantly by summarizing key hermeneutic factors for the job satisfaction of trainees and we also introduced some key benchmarks for the measurement of its rate. Our work contributes to the literature of job satisfaction in medical training by introducing new measurable benchmarks.

It also serves as a key methodological approach for the introduction of reflective actions in different areas of residency including the academic activities, the Training Program, the Training Center, the collegiality, the psychological load of residents, and their interactions with the administration and the residency supervising bodies.

It is also a bold contribution to Sustainable Health literature, since it promotes a vision for enhanced skills and competencies of health specialists with a balance of life and work. It is also linked to the discussion and the debate on Sustainable Development Goals as introduced by the United Nations. The Sustainable Development Goal #3—Establish Good Health and Well-Being—must also extend to health specialists. The promotion of good mental health and wellbeing of health experts must be a priority of modern societies. Within this context, our work is also directly related to the Sustainable Development Goal #4—Provide Quality Education. Our main interest in this study is to understand the conditions and requirements that make residents happy and efficient in their training and professional conduct. Last but not least, our work also contributed to the Sustainable Health Vision through the SDG #8 Create Decent Work and Economic Growth. It is a key responsibility of the SCFHS to promote the professional capacity of residents so that the Future Saudi Society will enjoy having high-skilled health experts with a strong sense of social responsibility and high efficiency.

In the near future, we intend to further extend our theoretical model in the following directions: (i) Delivery of a research study based on Structural Equation Modeling to understand the contribution of each of the factors to the Quality of Residents' Training and the effectiveness of knowledge creation [50]; (ii) the design and implementation of an innovative Artificial Intelligence enabled ecosystem for personalized training based on key assumptions of our model [51]; (iii) focused research on the key determinants and hermeneutic variables of Human Factors affecting residents' education; [iv] the integration of smart cities research and smart healthcare strategies [52,53] under the vision of the Kingdom 2030 Digital Transformation of Healthcare.

Author Contributions: The authors contributed equally to this research work and they were all involved in an integrated way in the various stages of this research including conceptualization, methodology, software, validation, formal analysis, investigation, resources, data curation, writing—original draft preparation, writing—review and editing, visualization, supervision, project administration, and funding acquisition. All authors have read and agreed to the published version of the manuscript.

Funding: This research received no external funding.

Acknowledgments: The authors would love to thank all of their colleagues in the Saudi Commission for Health Specialties and all of the participants in the survey.

Conflicts of Interest: The authors declare no conflict of interest.

References

1. Khoja, T.; Rawaf, S.; Qidwai, W.; Rawaf, D.; Nanji, K.; Hamad, A. Health care in Gulf Cooperation Council Countries: A review of challenges and opportunities. *Cureus* **2017**, *9*, e1586. [CrossRef] [PubMed]

2. Howe, A.; Smajdor, A.; Stöckl, A. Towards an understanding of resilience and its relevance to medical training. *Med. Educ.* **2012**, *46*, 349–356. [CrossRef] [PubMed]
3. Gregory, S.; Demartini, C. Satisfaction of doctors with their training: Evidence from UK. *BMC Health Serv. Res.* **2017**, *17*, 851. [CrossRef] [PubMed]
4. Dyrbye, L.; Shanafelt, T. A narrative review on burnout experienced by medical students and residents. *Med. Educ.* **2016**, *50*, 132–149. [CrossRef] [PubMed]
5. De Cavalcande Almeida, G.; de Souza, H.R.; de Almeida, P.C.; de Calvancante Almeida, B.; Almeida, G.H. The prevalence of burnout syndrome in medical students. *Arch. Clin. Psychiatry* **2016**, *43*, 6–10.
6. Rodrigues, H.; Cobucci, R.; Oliveira, A.; Cabral, J.V.; Medeiros, L.; Gurgel, K.; Gonçalves, A.K. Burnout syndrome among medical residents: A systematic review and meta-analysis. *PLoS ONE* **2018**, *13*, e0206840. [CrossRef]
7. Alsubaie, A.; Isouard, G. Job Satisfaction and Retention of Nursing Staff in Saudi Hospitals. *Asia Pac. J. Health Manag.* **2019**, *14*, 68–73. [CrossRef]
8. Chen, Q.; Yang, L.; Feng, Q.; Tighe, S.S. Job Satisfaction Analysis in Rural China: A Qualitative Study of Doctors in a Township Hospital. *Scientifica* **2017**, 1964087. [CrossRef]
9. Bawakid, K.; Rashid, O.A.; Mandoura, N.; Shah, H.B.U.; Mugharbel, K. Professional Satisfaction of Family Physicians Working in Primary Healthcare Centers: A Comparison of Two Saudi Regions. *J. Fam. Med. Prim. Care* **2018**, *7*, 1019–1025.
10. Zhang, S.X.; Liu, J.; Afshar Jahanshahi, A.; Nawaser, K.; Yousefi, A.; Li, J.; Sun, S. At the Height of the Storm: Healthcare Staff's Health Conditions and Job Satisfaction and Their Associated Predictors during the Epidemic Peak of Covid-19. *Brain Behav. Immun.* **2020**, *87*, 144–146. [CrossRef]
11. Leskovic, L.; Erjavec, K.; Leskovar, R.; Vukovic, G. Burnout and Job Satisfaction of Healthcare Workers in Slovenian Nursing Homes in Rural Areas during the Covid-19 Pandemic. *Ann. Agric. Environ. Med.* **2020**. [CrossRef]
12. Soto-Rubio, A.; Giménez-Espert, M.D.C.; Prado-Gascó, V. Effect of Emotional Intelligence and Psychosocial Risks on Burnout, Job Satisfaction, and Nurses' Health during the COVID-19 Pandemic. *Int. J. Environ. Res. Public Health* **2020**, *17*, 7998. [CrossRef] [PubMed]
13. Aoyagi, Y.; Beck, C.R.; Dingwall, R.; Nguyen-Van-Tam, J.S. Healthcare workers' willingness to work during an influenza pandemic: A systematic review and meta-analysis. *Influenza Respir. Viruses* **2015**, *9*, 120–130. [CrossRef] [PubMed]
14. Liu, J.; Yu, W.; Ding, T.; Li, M.; Zhang, L. Cross-sectional survey on job satisfaction and its associated factors among doctors in tertiary public hospitals in Shanghai, China. *BMJ Open* **2019**, *9*, e023823. [CrossRef] [PubMed]
15. Platis, C.; Reklitis, P.; Zimeras, S. Relation between job satisfaction and job performance in healthcare services. *Procedia Soc. Behav. Sci.* **2015**, *175*, 480–487. [CrossRef]
16. Chao, M.-C.; Jou, R.-C.; Liao, C.-C.; Kuo, C.-W. Workplace stress, job satisfaction, job performance, and turnover intention of health care workers in rural Taiwan. *Asia Pacific J. Public Health* **2015**, *27*, NP1827–NP1836. [CrossRef]
17. da Silvera Maissiat, G.; Lautert, L.; Dal Pai, D.; Patri tavares, J. Work context, job satisfaction and suffering in primary health care. *Rev. Gaúcha Enferm.* **2015**, *36*, 42–49. [CrossRef]
18. Pandey, J.; Singh, M. Donning the mask: Effects of emotional labour strategies on burnout and job satisfaction in community healthcare. *Health Policy Plan.* **2016**, *31*, 551–562. [CrossRef]
19. Labrague, L.J.; De los Santos, J.A.A. Fear of Covid-19, psychological distress, work satisfaction and turnover intention among frontline nurses. *J. Nurs. Manag.* **2020**. [CrossRef]
20. Low, Z.X.; Yeo, K.A.; Sharma, V.K.; Leung, G.K.; McIntyre, R.S.; Guerrero, A.; Lu, B.; Lam, C.C.S.F.; Tran, B.X.; Nguyen, L.H.; et al. Prevalence of burnout in medical and surgical residents: A meta-analysis. *Int. J. Environ. Res. Public Health* **2019**, *16*, 1479. [CrossRef]
21. Levin, K.H.; Shanafelt, T.D.; Keran, C.M.; Busis, N.A.; Foster, L.A.; Molano, J.R.V.; O'Donovan, C.A.; Ratliff, J.B.; Schwarz, H.B.; Sloan, J.A.; et al. Burnout, career satisfaction, and well-being among US neurology residents and fellows in 2016. *Neurology* **2017**, *89*, 492–501. [CrossRef] [PubMed]
22. Baer, T.E.; Feraco, A.M.; Sagalowsky, S.T.; Williams, D.; Litman, H.J.; Vinci, R.J. Pediatric resident burnout and attitudes toward patients. *Pediatrics* **2017**, *139*, e20162163. [CrossRef] [PubMed]
23. Moir, F.; Yielder, J.; Sanson, J.; Chen, Y. Depression in medical students: Current insights. *Adv. Med. Educ. Pract.* **2018**, *9*, 323. [CrossRef] [PubMed]

24. Raj, K.S. Well-being in residency: A systematic review. *J. Grad. Med. Educ.* **2016**, *8*, 674–684. [CrossRef]
25. Dyrbye, L.N.; Burke, S.E.; Hardeman, R.R.; Herrin, J.; Wittlin, N.M.; Yeazel, M.; Satele, D.V. Association of clinical specialty with symptoms of burnout and career choice regret among US resident physicians. *JAMA* **2018**, *320*, 1114–1130. [CrossRef]
26. Ripp, J.A.; Privitera, M.R.; West, C.P.; Leiter, R.; Logio, L.; Shapiro, J.; Bazari, H. Well-being in graduate medical education: A call for action. *Acad. Med.* **2017**, *92*, 914–917. [CrossRef]
27. Tawfik, D.S.; Profit, J.; Morgenthaler, T.I.; Satele, D.V.; Sinsky, C.A.; Dyrbye, L.N.; Shanafelt, T.D. Physician burnout, well-being, and work unit safety grades in relationship to reported medical errors. *Mayo Clin. Proc.* **2018**, *93*, 1571–1580. [CrossRef]
28. Spataro, B.M.; Tilstra, S.A.; Rubio, D.M.; McNeil, M.A. The toxicity of self-blame: Sex differences in burnout and coping in internal medicine trainees. *J. Women Health* **2016**, *25*, 1147–1152. [CrossRef]
29. Busireddy, K.R.; Miller, J.A.; Ellison, K.; Ren, V.; Qayyum, R.; Panda, M. Efficacy of interventions to reduce resident physician burnout: A systematic review. *J. Grad. Med. Educ.* **2017**, *9*, 294–301. [CrossRef]
30. Rotenstein, L.S.; Ramos, M.A.; Torre, M.; Segal, J.B.; Peluso, M.J.; Guille, C.; Mata, D.A. Prevalence of depression, depressive symptoms, and suicidal ideation among medical students: A systematic review and meta-analysis. *JAMA* **2016**, *316*, 2214–2236. [CrossRef]
31. Baker, K.; Sen, S. Healing medicine's future: Prioritizing physician trainee mental health. *AMA J. Ethics* **2016**, *18*, 604. [PubMed]
32. Jalal, S.R.; Osman, A.; Azizi, S. Additional factors influencing resident satisfaction and dissatisfaction. *Adv. Med. Educ. Pract.* **2017**, *8*, 769. [CrossRef] [PubMed]
33. Vargas-Vera, M.; Lytras, M.D. Exploiting semantic web and ontologies for personalised learning services: Towards semantic web-enabled learning portals for real learning experiences. *Int. J. Knowl. Learn.* **2008**, *4*, 1–17. [CrossRef]
34. Lytras, M.; Tsilira, A.; Themistocleous, M. Towards the semantic e-learning: An ontological oriented discussion of the new research agenda in e-learning. In Proceedings of the American Conference on Information Systems, Tampa, FL, USA, 4–6 August 2003; p. 388.
35. Krebs, R.; Ewalds, A.L.; van der Heijden, P.T.; Penterman, E.J.M.; Grootens, K.P. Burn-out, commitment, personality and experiences during work and training; Survey among psychiatry residents. *Tijdschr. Psychiatr.* **2017**, *59*, 87–93. [PubMed]
36. Gouveia, P.A.D.C.; Ribeiro, M.H.C.; Aschoff, C.A.M.; Gomes, D.P.; Silva, N.A.F.D.; Cavalcanti, H.A.F. Factors associated with burnout syndrome in medical residents of a university hospital. *Rev. Assoc. Med. Bras.* **2017**, *63*, 504–511. [CrossRef] [PubMed]
37. Dyrbye, L.N.; West, C.P.; Satele, D.; Boone, S.; Tan, L.; Sloan, J.; Shanafelt, T.D. Burnout among U.S. medical students, residents, and early career physicians relative to the general U.S. population. *Acad. Med.* **2014**, *89*, 443–451. [CrossRef]
38. Shoimer, I.; Patten, S.; Mydlarski, P.R. Burnout in dermatology residents: A Canadian perspective. *Br. J. Dermatol.* **2018**, *178*, 270–271. [CrossRef]
39. Porter, M.; Hagan, H.; Klassen, R.; Yang, Y.; Seehusen, D.A.; Carek, P.J. Burnout and resiliency among family medicine program directors. *Fam. Med.* **2018**, *50*, 106–112. [CrossRef]
40. Chaukos, D.; Chad-Friedman, E.; Mehta, D.H.; Byerly, L.; Celik, A.; McCoy, T.H., Jr.; Denninger, J.W. Risk and resilience factors associated with resident burnout. *Acad. Psychiatry* **2017**, *41*, 189–194. [CrossRef]
41. Holmes, E.G.; Connolly, A.; Putnam, K.T.; Penaskovic, K.M.; Denniston, C.R.; Clark, L.H.; Rubinov, D.D.; Meltzer-Brody, S. Taking care of our own: A multispecialty study of resident and program director perspectives on contributors to burnout and potential interventions. *Acad. Psychiatry.* **2017**, *41*, 159–166. [CrossRef]
42. Dyrbye, L.N.; Thomas, M.R.; Huschka, M.M.; Lawson, K.L.; Novotny, P.J.; Sloan, J.A.; Shanafelt, T.D. A multicenter study of burnout, depression, and quality of life in minority and nonminority US medical students. *Mayo Clin. Proc.* **2006**, *81*, 1435–1442. [CrossRef] [PubMed]
43. Jagsi, R.; Griffith, K.A.; Jones, R.; Perumalswami, C.R.; Ubel, P.; Stewart, A. Sexual harassment and discrimination experiences of academic medical faculty. *JAMA* **2016**, *315*, 2120–2121. [CrossRef] [PubMed]
44. Karim, S.; Duchcherer, M. Intimidation and harassment in residency: A review of the literature and results of the 2012 Canadian Association of Internsand Residents National Survey. *Can. Med. Educ. J.* **2014**, *5*, e50–e57. [CrossRef] [PubMed]

45. Fnais, N.; Soobiah, C.; Chen, M.H.; Lillie, E.; Perrier, L.; Tashkhandi, M. Harassment and discrimination in medical training: A systematic review and meta-analysis. *Acad. Med.* **2014**, *89*, 817–827. [CrossRef] [PubMed]
46. Ten Cate, O.; Scheele, F. Competency-based postgraduate training: Can we bridge the gap between theory and clinical practice? *Acad. Med.* **2007**, *82*, 542–547. [CrossRef]
47. Kjaer, N.K.; Kodal, T.; Shaughnessy, A.F.; Qvesel, D. Introducing competency-based postgraduate medical training: Gains and losses. *Int. J. Med. Educ.* **2011**, *2*, 110–115. [CrossRef]
48. Scheele, F.; Teunissen, P.; Van Luijk, S.; Heineman, E.; Fluit, L.; Mulder, H.; Meininger, A.; Wijnen-Meijer, M.; Glas, G.; Sluiter, H.; et al. Introducing competency-based postgraduate medical education in the Netherlands. *Med. Teach.* **2008**, *30*, 248–253. [CrossRef]
49. Toussaint, N.D.; McMahon, L.P.; Dowling, G.; Soding, J.; Safe, M.; Knight, R.; Fair, K.; Linehan, L.; Walker, R.G.; Power, D.A. Implementation of renal key performance indicators: Promoting improved clinical practice. *Nephrology* **2015**, *20*, 184–193. [CrossRef]
50. Naeve, A.; Yli-Luoma, P.; Kravcik, M.; Lytras, M.D. A modelling approach to study learning processes with a focus on knowledge creation. *Int. J. Technol. Enhanc. Learn.* **2018**, *1*, 1–34. [CrossRef]
51. Spruit, M.; Lytras, M. Applied Data Science in Patient-centric Healthcare. *Telemat. Inform.* **2018**, *35*, 2018. [CrossRef]
52. Visvizi, A.; Lytras, M.D. Editorial: Policy Making for Smart Cities: Innovation and Social Inclusive Economic Growth for Sustainability. *J. Sci. Technol. Policy Mak.* **2018**, *9*, 1–10.
53. Visvizi, A.; Lytras, M.D. *Transitioning to Smart Cities: Mapping Political, Economic, and Social Risks and Threats*; Elsevier: New York, NY, USA, 2019.

Publisher's Note: MDPI stays neutral with regard to jurisdictional claims in published maps and institutional affiliations.

sustainability

MDPI

Article

University Students' Perception, Evaluation, and Spaces of Distance Learning during the COVID-19 Pandemic in Austria: What Can We Learn for Post-Pandemic Educational Futures?

Tabea Bork-Hüffer [1], Vanessa Kulcar [2], Ferdinand Brielmair [1], Andrea Markl [1], Daniel Marian Immer [2], Barbara Juen [2], Maria Hildegard Walter [2] and Katja Kaufmann [1,3,*]

1 Institute of Geography, University of Innsbruck, 6020 Innsbruck, Austria; tabea.bork-hueffer@uibk.ac.at (T.B.-H.); Ferdinand.Brielmair@student.uibk.ac.at (F.B.); Andrea.Markl@uibk.ac.at (A.M.)
2 Institute of Psychology, University of Innsbruck, 6020 Innsbruck, Austria; Vanessa.Kulcar@uibk.ac.at (V.K.); Daniel.Immer@uibk.ac.at (D.M.I.); Barbara.Juen@uibk.ac.at (B.J.); Maria.Walter@uibk.ac.at (M.H.W.)
3 Institute for Comparative Media and Communication Studies, Austrian Academy of Sciences and University of Klagenfurt, 1010 Vienna, Austria
* Correspondence: Katja.Kaufmann@uibk.ac.at

Abstract: The COVID-19 pandemic caught societies worldwide unprepared in 2020. In Austria, after a lockdown was decreed on 16 March 2020, educational institutions had to switch to a patched-up distance learning approach, which has been largely maintained to date. This article delivers empirical insights from an interdisciplinary mixed-methods research study that investigated university students' perceptions of and experiences with distance learning as well as their educational (home) spaces during the pandemic in Innsbruck, Austria. It combines results from a quantitative survey conducted with 2742 students in early 2021 with a qualitative multi-method and longitudinal research study that accompanied 98 students throughout four data-collection phases in 2020. Results show a significant improvement since spring 2020 with both teachers and learners adjusting to the distance learning formats and the use of digital tools, yet students urgently desired a return to face-to-face teaching and university life, particularly for its social benefits. Strikingly, more than half of the participants wanted to maintain the option of overall distance education after the pandemic. Based on the perspectives of students, it is appropriate to demand significant changes in post-pandemic education adapted to the era of the post-digital, for which this article gives short-term as well as medium-term recommendations.

Keywords: COVID-19; young adults; hybrid learning; remote teaching; educational spaces; tertiary education; higher education; Austria; mixed methods; post-digital

Citation: Bork-Hüffer, T.; Kulcar, V.; Brielmair, F.; Markl, A.; Immer, D.M.; Juen, B.; Walter, M.H.; Kaufmann, K. University Students' Perception, Evaluation, and Spaces of Distance Learning during the COVID-19 Pandemic in Austria: What Can We Learn for Post-Pandemic Educational Futures? *Sustainability* **2021**, *13*, 7595. https://doi.org/10.3390/su13147595

Academic Editors: Miltiadis D. Lytras, Abdulrahman Housawi and Basim Alsaywid

Received: 6 June 2021
Accepted: 2 July 2021
Published: 7 July 2021

1. Introduction

Worldwide, 90% of all pupils and students (1.5 billion) were allegedly affected by the closure of educational institutions effected as a response to stall the spread of the COVID-19 pandemic in 2020 [1]. Austria, too, was affected by wide-reaching restrictions imposed early-on on educational institutions, including the province of Tyrol, home of the ski resort of Ischgl, which later rose to fame as a pan-European COVID-19 hotspot. Between mid-March 2020 and March 2021, overall three nationwide so-called "strict lockdowns" (1st lockdown from 16 March till April 2020, 2nd lockdown from 17 November till 6 December 2020, 3rd lockdown from 26 December till 7 February) were enacted, which encompassed stay-at-home orders, social distancing measures, and the extensive closure of non-essential businesses, in order to contain the spread of SARS-CoV-2. Higher education in Austria switched to distance learning starting with the first lockdown in March 2020. Since then, the tertiary educational institutions have not returned to in-person forms of teaching. Besides for slight relaxations during summer 2020 when infection numbers had dropped significantly, distance learning was maintained until the time of writing this article in June 2021.

In this article, we present results from a comprehensive interdisciplinary mixed-methods research study, which investigates the complexity of perceptions, evaluations, practices, experiences, and spatial contexts of distance learning for students in the Tyrolean capital of Innsbruck in Austria throughout the first year of the COVID-19 pandemic. The article focuses on empirical answers to the following research questions: how did students perceive and evaluate distance learning to date? Which spatial—i.e., social, material and digital—contexts have influenced distance learning? What are the lessons learned for the short-term design of distance education during the pandemic and a mid-term provision of post-pandemic education adapted to the era of the post-digital [2]? The data collected with students on distance learning during the COVID-19 pandemic that we use to answer these questions were collected as part of a mixed-methods research project collaboratively conducted by the Institutes of Geography and Psychology at the University of Innsbruck in Tyrol. The study applied a sequential design in which the qualitative results from a longitudinal qualitative multi-method study that included 98 university students were used to design a follow-up quantitative survey with 2742 students from the University of Innsbruck.

After introducing existing research and our methodology, we present the background to the understanding of the empirical results regarding the types of courses visited and the digital media used for distance learning during the pandemic. Next, we show the results of the study regarding students' perceptions, evaluation of distance learning, the socio-material home spaces in which they pursued distance learning, their social media use, and the effects of these variables on distance learning. Eventually, we discuss recommendations for pandemic education and for distance education beyond the pandemic that is more socially sustainable and responsive to the realities of the post-digital.

2. Effects of the COVID-19 Pandemic on the Spatiality of University Students' Education and Learning

The worldwide closure of educational institutions due to the COVID-19 pandemic sparked interest among many researchers in how the forced conversion to distance education and learning affected teachers and learners. Indeed, numerous national and international studies were conducted in 2020 that examined various aspects of the transition to and implementation of distance education and learning. Many of the studies explore either academic work and life [3–5] or students' mental health [6–11], or a combination of the two topics [12,13].

Comparative studies on the psychological effects of the pandemic suggest that young people, and particularly young adults, are especially vulnerable to the mental health effects of the pandemic [14,15]. This vulnerability partly stems from their developmental phase. Young adults in their qualification periods are in a formative transition phase of their lives, which is marked by insecurities and changes [16,17], even more so with a pandemic around [18]. The additional challenges during the COVID-19 crisis can therefore easily overwhelm them. As their studies, and therefore also the universities, play a central role in students' lives, changes in how the education system works affect all other aspects of their everyday experiences. Social contacts among students decrease [19], and some move back in with their parents [20] or find themselves separated from families and friends while studying abroad due to closed borders [21]. In addition, they are under pressure to keep up their performance while having difficulties finding motivation and concentrating during the time of crisis [22]. As the young adults' lives profoundly change and their mental health is severely affected, various studies suggest that the mid-term consequences will likely be felt for some years after a withdrawal of pandemic-related measures [23–26].

Furthermore, current research sheds light on the dimension of sustainability in the context of the pandemic. Studies on sustainability education [27,28] as well as on sustainable online learning and teaching [29–31] have highlighted challenges and possibilities for sustainable development in the context of the pandemic. Especially social distancing and temporary lockdowns are reported to have a significant impact on the everyday life and education of millions of students. Sustainable online learning, which includes access to and the possibility of actively participating in online education during the pandemic, is considered to be a decisive factor for successful teaching and learning and, furthermore, a key factor for maintaining and increasing social sustainability. Social sustainability is therefore at risk if students have limited access, or even no access at all, to online education.

Lastly, economic impacts play a significant role, and they directly affect tertiary education institutions. In February 2021, 436,982 people were registered as unemployed with the Austrian public employment agency AMS [32] (p. 1), and a further 71,941 participated in training courses [32] (p. 3), out of a total Austrian population of almost 9 million [33]. The cohort of those aged 15–24 years was already more affected by unemployment before the crisis and remains one of the most affected groups with a lack of prospects on the job market. As a result, student enrolment, including in the tertiary education institutions of Tyrol, is increasing. Meanwhile, students already enrolled feel that their performance has been dropping due to the mental-health impacts of the crisis [22]. Universities are challenged to accommodate these different groups and interests and at the same time need to care for students' mental health to avoid negative long-term effects—all while building up and maintaining a teaching system almost completely based on distance learning.

In studies on students' experiences of the COVID-19 pandemic, the spatiality of education and learning, however, is hardly considered. This should come as a surprise as the COVID-19 crisis has a very palpable spatial dimension. Distance learning per se implies spatiality, both through the relocation of aspects of learning and teaching to the students' home spaces and their entanglements with digital learning spaces, and through the fact that students no longer experience circulation between their home spaces and the university as a space of education due to this outsourcing. Gobbi and Rovea [34] argue, in this context, that spatial changes due to distance learning go hand in hand with a reduction of the university as a joint (inter)actional and socio-material space to scattered digital spaces on one or more devices. These devices are embedded in the everyday living environments, the home spaces, of the students [35], which is why students experience a new kind of homemaking. Furthermore, while the home space, with various functionalities already intersecting in it, becomes the (new) individual educational and learning space, this does not mean that students find themselves alone in this new assemblage. Rather, they are embedded in altered relational networks [36], in which people, objects, and routines remain closely connected [35] and sociality, including the collective of the university, is maintained by digital means [37]. Following Bork-Hüffer et al. [38], one can speak of cON/FFlating educational and learning spaces that are simultaneously shaped by entangled socio-material and techno-social relations and spaces. They remain socially shaped by multiple perceptions, meanings, values, and ideologies of students, teachers [36], and peer-groups. In line with this, the entangled material, social, and technological dimensions of educational and learning spaces must be taken account of in the analysis of distance learning and its impact on students.

The importance of space to education and learning was already emphasized as part of the spatial turn in the geographies of education [39]. Without a doubt, during the pandemic, new and altered formal and informal educational and learning spaces are emerging in which learners in turn have learning experiences that are particularly central to the social and cultural geographies of education. Given the importance of the affective-emotional experience of space, i.e., of (mediated) sense of place [40–42], it can be argued that parallel to the (new) educational and learning spaces, new manifestations of a sense of educational and learning place also emerge. Subsequently, these new spatially shaped learning experiences condition a constant reproduction and interpretation of students' identities through socio-spatial-technological practices [36], which is also inscribed in the formation of a sense of educational and learning place [43,44]. Student identity, or a student identity specific to each individual, should also be interpreted in the context of the living situation, the home space, because it can influence students' experiences as such both positively and negatively and has an impact on identity formation within social activities, learning environments, friendship networks, and other socio-cultural factors [45].

3. Methods

The data that are the basis of the analysis presented in this article were collected as part of a sequential mixed-methods research design, in which the insights from a qualitative multi-method project (study part 1) were used to design a quantitative survey instrument (study part 2). While the data collection was sequential, an integrated data analysis followed in which both forms of data were merged.

Part 1, the qualitative part of the study, built upon an analysis of data collected as part of the ongoing research project COV-IDENTITIES at the Institute of Geography of the University of Innsbruck, Austria. The project applies a qualitative multi-method and longitudinal approach, accompanying young adults, among them university students, since early April 2020 throughout the COVID-19 pandemic. The aim of the COV-IDENTITIES project is to explore changes in young adults' socio-material and techno-social everyday spaces and practices (including study, work, contacts, leisure) over the course and different phases of the COVID-19 pandemic. The COV-IDENTITIES dataset includes comprehensive data related to distance learning. It combines written narratives with smartphone-based methods (mobile instant messaging interviews) [46] and in-depth interviews.

Written narratives offer a qualitative approach to elicit individual experiences with and reflections on complex change processes as well as potentially traumatic experiences [47]. The approach gives young people room for subjective descriptions and interpretations of their feelings and experiences [48,49]. Mobile instant messaging interviews (MIMIs) offer comprehensive in situ insights into the very concrete daily practices and spaces of selected participants [50]. Using established messenger apps (here WhatsApp), research participants are contacted by researchers at regular (in our case, two-hourly) intervals and are asked about their current practices and spaces, and also about socio-material and techno-social elements with which they engage during those practices, against the backdrop of their daily routines. In contrast to standardized mobile experience-sampling approaches, when using MIMIs, researchers engage with participants in the digital instant messaging spaces where they explore together the participants' momentary experiences [46]. MIMIs thus allow researchers to gain deep insights into the complex everyday spaces and practices of research subjects in situ, even under conditions of curfews and social distancing. Research subjects can use a wide range of multimedia elements (text messages, photos, videos, voice messages, GIFs, emojis, screenshots) for their responses. To maximize the variety of perspectives in our qualitative study, additional participants from the pool of narrative data were invited to participate in in-depth interviews, to facilitate even more comprehensive insights into different perceptions, evaluations, experiences, and strategies with distance education.

Longitudinal data collection using narratives and MIMIs took place in four phases (DCs) in 2020:

- DC-1: 1–7 April: 1st lockdown
- DC-2: 20–27 April: during stepwise relaxation of the 1st lockdown
- DC-3: 2–14 June: full relaxation of the 1st lockdown
- DC-4: 16–23 November: 2nd lockdown.

The in-depth interviews were conducted in parallel in July, August, and November 2020. Data collection is planned to continue throughout 2021.

The sample encompassed students who were enrolled at universities in Innsbruck at the time of DC-1. The participants were aged between 18 and 36 years, and 65% were female (average at DC-1). As part of the COV-IDENTITIES project, 340 written narratives and 44 full-day MIMIs with university students were collected throughout the four DCs. The exact number of participants includes, for narratives in DC-1: $n = 98$, DC-2: $n = 93$, DC-3: $n = 82$, DC-4: $n = 67$; and for MIMIs, DC-1: $n = 13$; DC-2: $n = 12$, DC-3: $n = 8$, DC-4: $n = 11$. An additional ten follow-up in-depth interviews with students were conducted that focused on distance learning only. The participants in the interviews were selected with the aim of collecting a maximum variety of perspectives, ranging from students who had previously reported strongly negative experiences with distance learning to those who had reported strongly positive ones.

The qualitative data were analyzed with the help of MaxQDA using qualitative content analysis, which is an appropriate strategy for analyzing data from all three methods (narratives, MIMIs, interviews), e.g., [47]. The WordCloud used in Section 4.5 was produced in MaxQDA (Berbi Software, Berlin, Germany) and shows all software (including distance learning software) that was mentioned in the written narratives in different sizes according to the number of times they were named in narratives.

Based on the insights gained from the qualitative part 1, in part 2, a quantitative survey was designed to expand the findings of the qualitative part, to be conducted with a larger population of students at the University of Innsbruck. This survey was based on the results of the qualitative data set, as well as on outcomes of previous surveys conducted by this workgroup, which were designed to elicit information about the mental health and practices of students during the COVID-19 crisis. The resulting questionnaire included questions about different criteria relevant to students' perceptions and experiences of distance learning. These included contextual factors, such as living conditions, spaces where students take part in distance learning, and technical equipment; personality factors, including motivation and the ability to work independently; and subjective workload and information on the concrete implementation of students' courses. Additionally, data about participants' satisfaction with distance learning and performance, measured as the number of exams taken and scores achieved, were retrieved. The survey was sent to students at the University of Innsbruck on behalf of the vice-rector of teaching and studies. For the students, this recruitment by university officials indicated the importance of the survey for the future shaping of their own studies, and thereby likely increased their motivation to participate and answer all questions truthfully. The survey was open for replies from 19 February to 8 March 2021. The survey was presented in a multiple-choice and Likert response format, and students could also choose not to answer individual questions, to avoid forced answers.

In total, 2742 students completed the survey, which equals 10% of the total number of students at the University of Innsbruck in 2020 [51]. Most participants were between 20 and 30 years old (79%). Females were slightly overrepresented, making up 66% of the survey's participants (whereas females constitute 53% of all students at the University of Innsbruck). Males made up 32.5%, while 1.5% did not assign themselves to either of these genders, or did not want to answer the question.

To explore students' satisfaction with distance learning, we used six items asking for their satisfaction with distance learning in general, the workload, the content, their learning success, and the university's communication. The items were rated on a 5-point Likert scale (1 = 'I do not agree at all' to 5 = 'I agree completely'). A scale was created by calculating the mean. The internal consistency was satisfactory (Cronbach's $\alpha = 0.82$). To explore students' subjective performance during the distance learning semester, we used three items asking about changes in the number of exams and ECTS compared to before the distance learning started. The items were rated on a three-point Likert scale (-1 = fewer exams/ECTS; 0 = no significant changes; 1 = more exams/ECTS). Cronbach's $\alpha = 0.88$ was satisfactory. We calculated the mean of both items, with values <0 indicating decreased performance and values >0 increased performance.

For this article, the quantitative data set was analyzed using IBM SPSS Statistics, Version 26 (IBM, Armonk, NY, USA). We conducted descriptive analyses, independent t-tests, analyses of variance, and correlation analyses. Participants had the opportunity to select "no answer", so sample sizes varied for different tests. Effects were interpreted as significant with $p < 0.05$. We used Bonferroni correction when we ran multiple tests to avoid error inflation.

Eventually, an integrated analysis stage took place in which both data sets were analyzed with regard to findings for the main research questions presented above, and findings in the quantitative survey were cross-checked with the qualitative data and vice versa. Fully building upon the advantage of the mixed-methods study, the results are closely interwoven in the following results section. All data are used there to give substantiated and broad insight into students' perceptions, evaluations, and spaces of distance learning during the COVID-19 pandemic.

4. Results

4.1. Types of Courses Attended and Types of Distance-Learning Practices

Figure 1 summarizes the types of courses that the students reported having attended during the winter term, which lasted from 1 October 2020 till 1 February 2021. Most of the courses were teacher- and classroom-focused formats (lectures (VO, Vorlesungen), lectures with exercises (VU, Vorlesungen mit Übungen), proseminars (PS, Proseminare), seminars (SE, Seminare), or orientation courses (SL, Studienorientierungslehrveranstaltungen)), but a considerable percentage also attended classes that include practical, field, or laboratory experiences in normal, non-pandemic study terms (exercise courses (UE, Übungen), field trips (EX, Exkursionen), field courses (EU, Exkursionen mit Übungen), practical training sessions (PR, Praktika), work groups (AG, Arbeitsgemeinschaften)). Overall, 35% of the students reported in the quantitative survey that they had taken part in at least one course that took place in a physical, face-to-face format during the winter term 2020/2021.

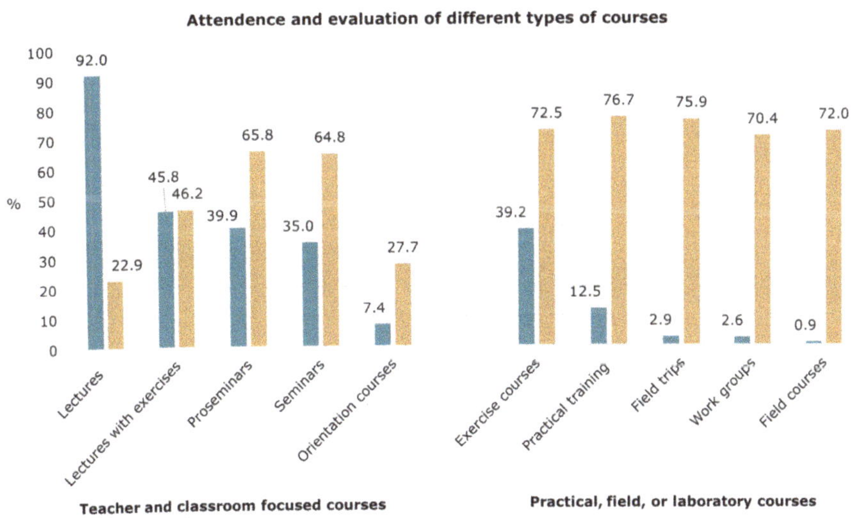

Figure 1. Types of courses that students attended during the winter term 2020/21 (green) and types of courses that suffered most from the conversion to distance learning (brown). Types of courses that suffered most are filtered by students taking the courses during the winter term 2020/21. Lectures $n = 2524$, lectures with exercises $n = 1256$, proseminars $n = 1094$, seminars $n = 959$, orientation courses $n = 202$, exercise courses $n = 1075$, practical training sessions $n = 343$, field trips $n = 79$, work groups $n = 71$, field courses $n = 25$ (multiple choice, source: quantitative survey, February 2021).

During the pandemic, the tasks pursued most often during distance learning, i.e., following lectures online or working on tasks given independently at home, did not include direct teacher–student interaction, as Figure 2 shows. This is generally not surprising, as a large part of learning tasks for students involves listening to lectures or self-study even during regular (non-pandemic) study terms. However, the use of online forums for teacher–student exchange and live question–answer sessions were new or at least largely increased types of interaction that sprung up during the pandemic. Furthermore, 34.9% reported that during the winter semester 2020/2021 they had worked in parts of their courses in small groups, and 21.8% that they had done so in large groups, indicating that interactive formats were maintained in some course types.

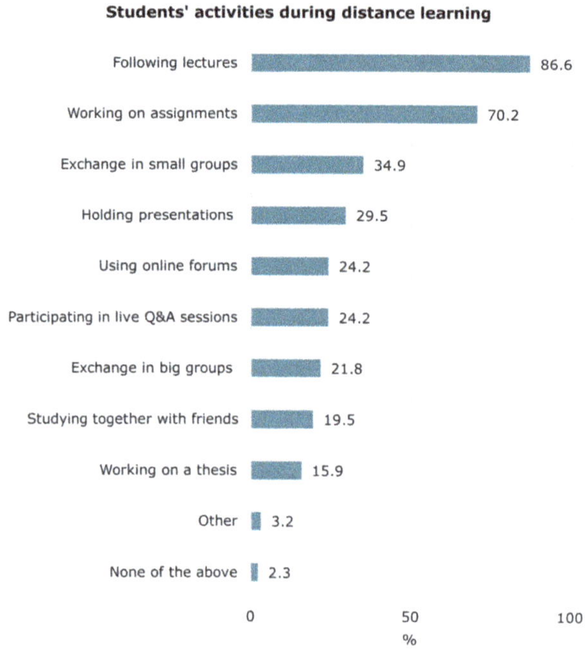

Figure 2. Types of activities that students conducted during the winter term 2020/2021 (multiple choice, source: quantitative survey, February 2021).

4.2. Hardware and Software Used for Distance Learning

To access the Internet, as the most important link between teachers and students, as well as in their learning and working process during distance learning, the students used various types of hardware. They primarily used their laptops/computers, smartphones, e-book readers, game consoles, and televisions for this purpose. Laptops and stand-alone PCs were used, as they were before the pandemic began, for word processing; running, applying, and learning course-relevant and helpful software; research; and communication. New additions to distance education included attending classes via online (learning) platforms and videoconferencing software, as well as accessing and watching instructional videos created by instructors for learners and posted online. Students used their smartphones in addition to their computers either alternately or simultaneously. These were often used for personal purposes, but they were also used in distance learning. Students used their smartphones to communicate with instructors and other students via various messenger services and email apps. However, smartphones were also used to retrieve and watch instructional videos and to participate in video conferences in virtual classrooms. In some cases, apps were used to help students organize and complete work assignments and to gather information.

Some students also indicated that they used their TVs to attend courses (primarily lectures) and to watch instructional videos on a larger screen and in a more comfortable setting, such as on the couch at home. Game consoles were also used, providing a range of software that students used both to communicate and to access instructional videos.

Due to the closure of libraries and the resulting limited access to literature, some students also used their e-book readers for research purposes and to read course-relevant literature, provided that the relevant literature was also available in digital form.

In addition to the hardware, students mentioned a variety of software they used during the COVID-19 pandemic. Figure 3 provides an overview of the software, categorized by their primary function in distance education. It should be noted here that in addition to the software used across the university, students also used specialized software specific to their discipline (e.g., ArcGis, MATLAB, SPSS). Challenges particularly related to the use and availability of specialized software are explored in more detail in Section 4.3.

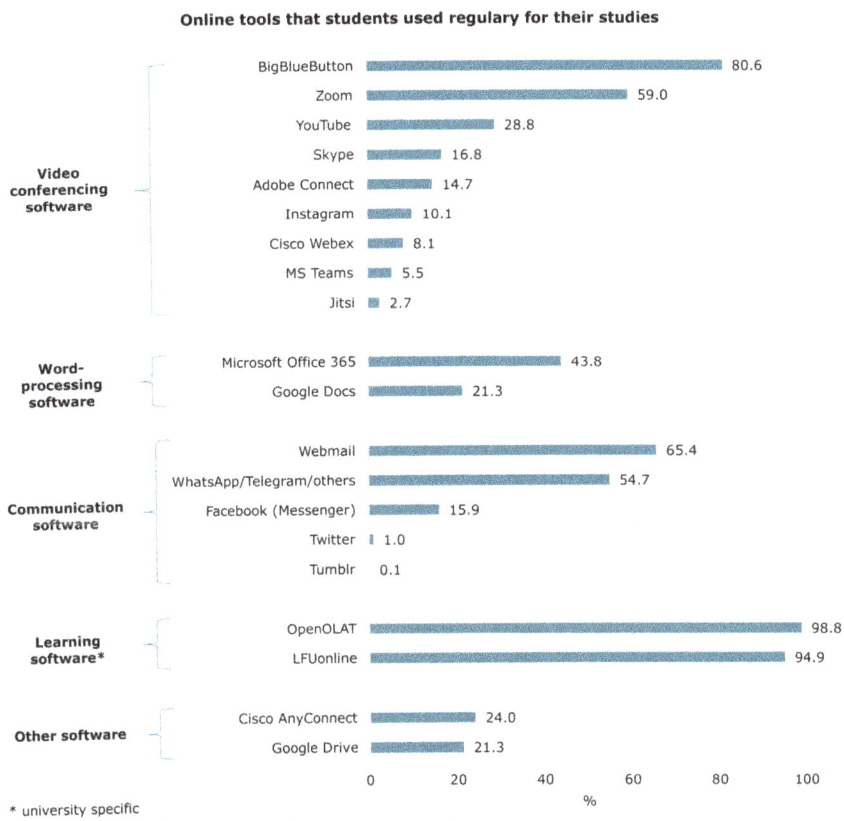

Figure 3. Software used in distance education across institutions (multiple choice; source: quantitative survey, February 2021).

As can be seen from Figure 3 showing the variety of software used across the university, students used a number of different kinds of video (conferencing) software, most of which was chosen by instructors for distance learning. It should be mentioned here that based on the insights from the qualitative longitudinal part, we can conclude that in the summer term 2020, at the beginning of the COVID-19 pandemic, many students reported a higher number of different kinds of video (conferencing) software used in distance learning than in the following winter semester 2020/21. With regard to word-processing software, some students stated that especially the possibility of simultaneous online editing of text files proved to be helpful while working on group tasks. For communication within the group, students used social-media software such as the Facebook and WhatsApp platforms and messenger services, as well as e-mail software such as Webmail. In addition to Webmail, students also used the (learning) software of Innsbruck's universities to communicate with their instructors, conduct research, and access information about course content and organization.

Overall, students generally used digital media more frequently and over a longer period of time since the switch to distance learning (see also Section 4.3). A more detailed description of the challenges with digital media follows in the next subsection.

4.3. Experiences with and Evaluation of Distance Learning during the Pandemic

Students had very different experiences in connection with the conversion to distance learning. Overall, the type of course chosen clearly had an impact on the perception of distance learning, as Figure 1 shows. The courses that contained practical elements were considered to be the most negatively affected by the conversion to a distance learning format. Still, students taking practical, field, or laboratory courses during the winter term 2020/21 did not differ from students not taking such courses in their satisfaction with distance learning ($t(2688) = 1.83$, $p = 0.067$). Additionally, there were no differences in performance ($t(2193) = -0.54$, $p = 0.589$).

The qualitative study highlights major individual differences, for example in the perceived challenges, individual structural conditions of distance learning (see also Section 4.4), attitudes toward distance learning, and in the personal appropriation of digital media (see also Section 4.5). For the purpose of comparison, the students' experiences are reduced and categorized in the following to challenges (negative experiences) and potentials (positive experiences) of distance learning. It should be noted that the qualitative data made clear that some elements of a course could have been perceived positively and other elements negatively. Figure 4 quantifies the perception of elements of distance learning based on the results of the quantitative survey.

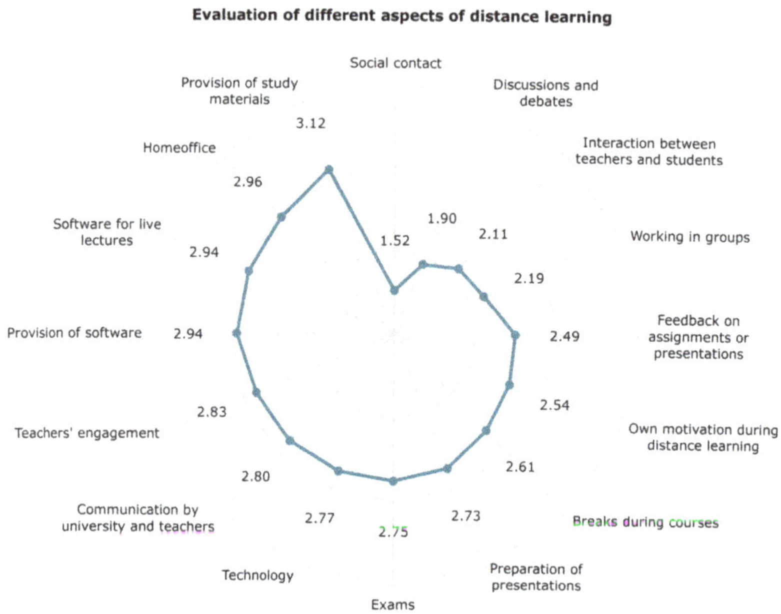

Figure 4. Evaluation of social, communicational, organizational, technical, and personal dimensions of distance learning ("How well do the following aspects work in distance learning in general?", single choice, 1 = 'very poorly', 2 = 'poorly', 3 = 'well', 4 = 'very well', source: quantitative survey, February 2021).

4.3.1. Negative Experiences: Challenges of Distance Learning

In the quantitative and qualitative data, students reported negative experiences with distance education, citing a variety of reasons why they were dissatisfied with distance education in different situations. One aspect that resonated through most responses was the lack of social exchange and contact in the context of distance education. Students missed the interaction with instructors (74.4%) and fellow students (91.9%) during class, as well as casual interactions (85.5%) (see also Figure 4). Those who missed these interactions reported decreased satisfaction with distance learning and decreased performance (see Table 1).

Table 1. Group differences between students missing interactions during distance learning and students not missing them.

	n	M	SD	*n*	M	SD	T	d	*p*
	Missing			Not Missing					
	Satisfaction with Distance Learning								
Interactions with instructors	2017	3.32	0.80	695	3.84	0.76	14.91	0.66	<0.001
Interactions with other students	2491	3.40	0.80	221	4.05	0.78	11.54	0.81	<0.001
Casual interactions	2318	3.41	0.81	394	3.74	0.81	7.55	0.41	<0.001
	Performance during Distance Learning								
Interactions with instructors	1623	−0.11	0.58	588	0.11	0.60	7.82	0.38	<0.001
Interactions with other students	2011	−0.07	0.59	200	0.18	0.61	5.52	0.42	<0.001
Casual interactions	1883	−0.07	0.59	328	0.07	0.61	3.92	0.24	<0.001

Note. *n* = sample number for each response option, M = mean, SD = standard deviation, T = *t*-test, d = Cohen's d, *p* = significance level.

Many found that social exchange, which is required and encouraged in face-to-face teaching, is more difficult in online formats. Debates and discussions during class were missed or perceived as poor by a considerable number of the students (78.0%). Group and team assignments and projects were perceived as more difficult in distance education than in face-to-face education. Participant Emma (24, narrative, DC-2), for example, explains that *"an assignment [is] significantly more time-consuming online than it [would be] in the university, there is often a lack of exchange during group work."*

Furthermore, the anonymity, particularly during lectures and on online learning platforms, was criticized:

> *"The most negative thing about distance learning, in my opinion, is the lack of contact with fellow students. Sure, you can [use] Skype and write emails, but I only do that with those I already know. In some courses I don't know anyone and since the webcams are never turned on, I don't have any faces to go with the names and this anonymity is quite strange and I feel quite uncomfortable because you can't make any contacts that way."*
>
> (Annika, 24, narrative, DC-4)

With 37.1% evaluating feedback on their own work and presentations by teachers as poor or very poor, giving feedback during distance learning did not work well for a considerable share of the participants. The students also stated that the exchange with lecturers and fellow students before, after, and between the courses is important to them and has a significant influence on their well-being and potential learning effects. In the quantitative survey, 85.5% noted that they missed the informal opportunities for discussion and exchange outside the classroom sessions.

The results of the quantitative survey show that students were relatively satisfied with the engagement of teachers, their communication, and their provision of learning material as well as the overall communication by the University (see Figure 4). The qualitative longitudinal data indicate that this evaluation has strongly improved since the beginning of the pandemic. Especially in the weeks of the first lockdown in March and April of the summer semester 2020, during the acute changeover phase to distance learning (survey phases DC-1 and DC-2), students experienced difficulties due to lack of information about courses, lack of (clear) communication from lecturers, lack of information about examination regulations and formats, especially from some lecturers, but also from institutes, examination offices, and the University:

> *23.04.20, 17:10—Interviewer: How would you describe your current emotional state?*
>
> *23.04.20, 17:13—Rebecca: Upset. I got an email from the internship office a few minutes ago that kills me. Everything is so uncertain in my studies and all we keep getting is empty promises and actually no real information. I'm extremely worried whether I'll be able to graduate this semester as planned and that just upsets me immensely [...] Actually, I'm slowly running out of energy to continue and still study because I just don't know what for and I'm not learning anything because we don't get any fixed exam dates.*
>
> (Rebecca, 26, MIMI chat log, DC-2)

However, even after the first lockdown and a hasty conversion to distance learning, in the qualitative data, students continued to note the lack of availability of some teachers, the perceived lack of technical and didactic skills of some teachers, and the feeling that some teachers exerted too much pressure on students. This included, for example, the use of software that was unsuitable for the implementation of the respective teaching mode, an insufficient use of functional possibilities of software, and incomprehensible sound and video recordings. Students criticized an insufficient adaptation of the teaching concepts to digital circumstances in some courses, or the replacement of teaching by the sole provision of materials (PowerPoint slides or literature).

In addition, in the qualitative data collection, students provided data about the psychological impacts of distance learning, particularly regarding motivation and concentration problems, but also about a loss of joy in studying, increased stress, feelings of being overwhelmed, and depression. They noted that during courses without active involvement, their concentration decreased, and more so than in face-to-face teaching formats:

> *"I find long seminars too exhausting, because even with breaks my concentration in front of the laptop quickly drops, especially when I don't have anything to do actively"*
>
> (Lotta, 23, narrative, DC-4)

In addition, some students reported feeling stressed by a perceived pressure to be constantly available online. However, the opposite effect—a lack of engagement and resulting decreased stress due to their physical distance from the educational institution of the university—was also mentioned by some students. Some of these students eventually raised concerns as to what the point of their study would be, when no intensive personal involvement with course content was necessary in order to successfully complete courses. Some students found the large amount of instructional and work material burdensome and, in some cases, overwhelming. While many students described familiarizing themselves with digital software as incidental and self-evident, previously unknown digital software, or the lack of opportunity to try it out prior to using it caused some study participants to feel stressed and anxious before presentations and exams. Feeling overwhelmed with materials in the distance learning situation and worrying about presentations or exams led some students to decide not to take exams, which prevented them from successfully completing some of their courses. A few students reported being severely overwhelmed and described having experienced depressive episodes during the pandemic.

In addition to the negative effects of distance learning on students' mental health, distance learning also had a negative impact on their physical health. For example, in the qualitative study, students communicated that they had suffered from headaches, back pain, sore eyes, and discomfort associated with working with digital devices. They also raised the issue of increased and quicker-onset of fatigue associated with screen-based work.

Moreover, participants in the qualitative data collection described a variety of technical difficulties, especially associated with the (internet) connection, software, and hardware. They reported connection problems and slow internet transmission rates, which interfered with smooth distance learning. Students with poor internet connection quality were both less satisfied with distance learning ($r = 0.27$, $p < 0.001$, $n = 2712$) and reported reduced performance ($r = 0.13$, $p < 0.001$, $n = 2211$). Other students criticized the software used for distance learning and found it not to be functional or user-friendly. Some reported that they were distracted by the large amount of software, which, however, also included media used for purposes other than distance learning (see also Section 4.5). Due to the high dependency on software, technical failures were all the more detrimental to productive distance learning. Problems with microphones were found to be particularly disruptive, as microphones provided the only opportunity for direct verbal exchange. Since some (special) software only works on suitable hardware, additional problems arose for some students.

In the quantitative survey, 38.1% reported that they were forced to purchase new hardware for distance learning purposes. The reason given for this was that most laptops are not designed for computing energy-intensive (special) programs and the university did not grant access to the devices located within the university buildings for reasons of COVID-19 protection measures. Funding and obtaining suitable hardware presented a challenge for some students, and they described relying on their parents for financial and material support because of this. Participants who had to make additional purchases for distance learning were less satisfied (M = 3.25, SD = 0.81, $n = 1030$, $t(2710) = -10.62$, $p < 0.001$, d = 0.41) and reported reduced performance (M = −0.12, SD = 0.63, $n = 867$, $t(1722.32) = -4.23$, $p < 0.001$, $d = 0.20$) compared with students not having to make such purchases (satisfaction M = 3.58, SD = 0.81, $n = 1682$; performance M = −0.00, SD = 0.57, $n = 1344$). Others reported that they felt compelled to disregard COVID-19 conditional spacing rules and collaborate (sometimes in groups) on an available powerful computer because a new hardware purchase was not financially economical and/or possible.

4.3.2. Positive Experiences: Potentials of Distance Learning

Both in the qualitative and quantitative study, students described various positive elements and circumstances in the context of distance learning (see Figure 5).

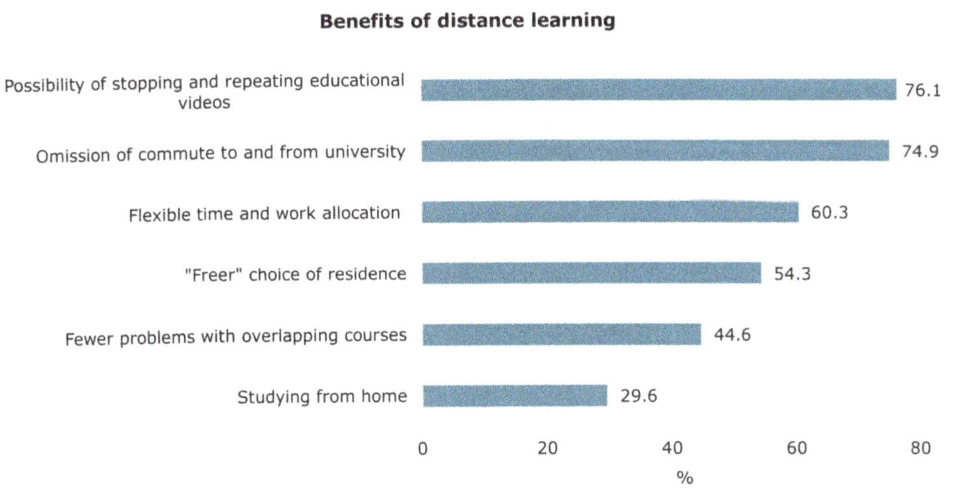

Figure 5. Advantages of distance learning that students reported during the winter term 2020/2021 (multiple choice, source: quantitative survey, February 2021).

The students who experienced advantages of distance learning were more satisfied with the situation and reported a better performance during the winter term 2020/21 (see Table 2).

Table 2. Group differences between students experiencing advantages of distance learning and students not experiencing them.

	n	M	SD	*n*	M	SD	T	d	*p*
	Advantage Experienced			**Not Experienced**					
	Satisfaction with Distance Learning								
Possibility of studying at one's own pace	2067	3.54	0.79	645	3.18	0.86	9.58	0.45	<0.001
Omission of commute	2028	3.56	0.81	684	3.14	0.77	11.80	0.53	<0.001
Flexible time and work allocation	1637	3.67	0.76	1075	3.13	0.81	17.63	0.69	<0.001
"Freer" choice of residence	1474	3.61	0.80	1238	3.27	0.81	10.76	0.42	<0.001
Fewer problems with overlapping courses	1206	3.70	0.75	1506	3.26	0.82	14.73	0.56	<0.001
Studying solely from home	802	3.78	0.79	1910	3.32	0.80	13.94	0.58	<0.001
	Performance during Distance Learning								
Possibility of studying at one's own pace	1656	−0.02	0.59	555	−0.14	0.60	4.18	0.20	<0.001
Omission of commute	1659	−0.00	0.59	552	−0.19	0.60	6.28	0.32	<0.001
Flexible time and work allocation	1311	0.06	0.58	900	−0.21	0.58	10.85	0.47	<0.001
"Freer" choice of residence	1185	0.01	0.58	1026	−0.12	0.60	5.01	0.22	<0.001
Fewer problems with overlapping courses	971	0.08	0.58	1240	−0.15	0.59	9.26	0.39	<0.001
Studying solely from home	647	0.11	0.61	1564	−0.11	0.57	-8.10	0.38	<0.001

Note. *n* = sample number for each response option, M = mean, SD = standard deviation, T = *t*-test, d = Cohen's d, *p* = significance level.

The digital accessibility of recordings and videos of lectures was found to be very useful by many students. Many also liked the option of being able to stop and repeat videos for a better understanding of the content. The qualitative data revealed that this accessibility of recordings also enabled the students to organize their learning phases independently of the lecture times, which was appreciated, among others, by working students and students with children:

> *"I see opportunities (...) in the fact that it can be a relief for students who still have to work on the side, because a freer time management is possible. Also, you are very flexible where you are and so you can study from anywhere."* (Elena, 21, Narrative, DC-4)

In the quantitative data there was no significant effect due to children being present in the same household on satisfaction ($t(2678) = 1.18$, $p = 0.237$, $n = 2680$) or performance ($t(229.39) = 0.76$, $p = 0.450$, $n = 2185$). In the interpretation of this result, closures of childcare services should be factored in.

Other students appreciated recorded courses because they facilitated distributing daily screen time throughout the week and helped them avoid packed distance learning days. In addition, others said that distance learning allowed them to take multiple courses with the same or overlapping course times, making it possible for them to study somewhat quicker than they would have in the regular face-to-face teaching system. Especially for students with multiple study subjects this can be beneficial. Students studying more than one subject reported increased performance during the distance learning semester (M = 0.09, SD = 0.60, $n = 495$) compared to students with only one subject, who reported decreased performance (M = −0.09, SD = 0.58, $n = 1602$, $t(2095) = 5.97$, $p < 0.001$, d = 0.31). There was no difference in satisfaction ($t(2567) = 0.33$, $p = 0.742$, $n = 2569$). The degree aspired to (Bachelor, Master, Diploma, teacher, PhD/Doctorate) did not have significant effects on satisfaction with distance learning ($F(4, 2691) = 2.14$, $p = 0.074$, $n = 2695$). There was a significant effect on performance ($F(4, 228.43) = 9.39$, $p < 0.001$, Welch-test, $n = 2201$). Students training to be teachers ($n = 194$) reported a stronger increase in their performance during distance learning than Bachelor, Master, or Diploma students (all $p < 0.001$).

Since physical attendance at the university campus was no longer required or possible, some students chose to leave Innsbruck and return to their home countries/towns and/or parental homes (cf. Section 4.3). Many participants reported that the elimination of necessary travel and commuting resulted in noticeable time savings, a reduction in travel costs, and their attendance at more classes:

> *04/23/20, 11:24—Rebecca: What I like is the online lectures. This allows me to attend significantly more lectures because I have less time issues with driving back and forth. I have lectures on Engineering, Botany, and at [the] Geiwi [campus]—it's really hard to be on time a lot of the time.* (Rebecca, 26, MIMI chat log, DC-2)

While some students, especially at the beginning of the pandemic, struggled with not having work or internships, income from work, or having to make dual conversions from study and internships or work to digital formats, others also appreciated a better work or internship/study balance through distance learning:

> *05.04.20, 11:21—Lucy: Honestly, there are only advantages for me with the current situation in terms of university. Since I work full-time as a teacher and do my studies virtually on the side, I have more leeway to arrange my tasks. The professors are also more relaxed with the submissions and I find that very pleasant. I also like the fact that you don't have to be present at the seminars and that more emphasis is placed on personal responsibility.* (Lucy, 27, MIMI chat log, DC-1)

In this context, some students spoke of increased "convenience" in their studies. Students also rated the experience gained with the home-office work model, and their familiarization with useful learning, working, and communication software, as potentially valuable experiences for the future. Some students found these new opportunities a relief in principle.

Creative and interactive teaching formats and experiences with new digital tools also received particularly positive feedback in the narratives and MIMIs. After all, 49.1% of the students enjoyed using more digital media for distance learning, in comparison to 32.9% who disliked the increased use of digital media. Those who appraised the increased use of digital media as positive reported a higher satisfaction with distance learning (M = 3.70, SD = 0.76, n = 1335) than those who appraised it as negative (M = 3.09, SD = 0.79, n = 891; $t(2224) = 18.32$, $p < 0.001$, d = 0.79). Additionally, approval of the increased media use was associated with increased performance (M = 0.04, SD = 0.60, n = 1086) but disapproval was associated with decreased performance (M = −0.17, SD = 0.57, n = 749, $t(1833) = 7.49$, $p < 0.001$, d = 0.36).

The participants in the qualitative longitudinal part also noted that they could see a positive development in distance learning over the course of 2020 and that both they and the teachers had slowly adapted or become accustomed to distance learning. Even though many of the challenges and problems in distance education (mentioned above) could not be completely eliminated in the winter term 2020/21, in the quantitative survey, 37.2% found that distance learning had improved slightly and 19.1% found it had improved greatly in comparison to the first semester afflicted by COVID-19, i.e., the summer term 2020.

4.4. Places of Stay and Socio-Material Educational Spaces during the Pandemic

At the core of the COV-IDENTITIES project is the study of students' everyday spaces during the pandemic. A variety of factors influenced the home and educational spaces beyond the technical challenges already described in Section 4.3, and thus also students' learning conditions. These include living arrangements, household structures, mobilities and changes of residence during the pandemic, and the related social and material learning environments and the changes thereof.

Looking at the socio-material home spaces, when asked with whom they lived together at the time of the quantitative survey in early 2021, 40.2% reported that they lived with their parents, 23.1% in shared apartments with flatmates, 19.6% with their partners, 8.8% alone, and 4.9% in student accommodation. Combinations of answers were noted in the "other" section (e.g., living together with partners and flatmates in shared apartments). Students living alone in particular reported suffering due to isolation and stay-at-home-orders—especially during the first lockdown (DC-1 and DC-2). At the beginning of 2021, after nearly one year of pandemic and distance learning, the living situation still had significant effects on satisfaction with distance learning ($F(4, 625.16) = 10.02$, $p < 0.001$,

Welch corrected, $n = 2620$). In particular, students living in shared apartments reported significantly lower satisfaction than all other students ($p < 0.001$ for living with partner or parents, $p = 0.031$ for living alone, Games–Howell corrected). There was no significant effect of the living situation on performance ($F(4, 2133) = 2.00$, $p = 0.093$, $n = 2138$).

At the same time, the qualitative and quantitative data indicate that there has been some flux in places of stay. In this context, the qualitative study points to unusually frequent, mostly temporary, changes of residence as well as, in some cases, to relocations, which are related to the pandemic. Overall, the quantitative survey revealed that 24.4% changed their places of residence once and 13.5% several times during the first year of the pandemic, while 62% did not relocate. There were no effects of relocation on satisfaction with ($t(2334.98) = 1.51$, $p = 0.132$, $n = 2712$) or performance during distance learning ($t(2209) = -0.52$, $p = 0.606$, $n = 2211$). Some of those who relocated decided to leave Innsbruck to be with their families. The qualitative longitudinal study shows that the study participants decided to return to their families especially during the lockdown in spring 2020 and the subsequent summer semester 2020 (in the survey DC-1, DC-2, already significant decrease in DC-3). A smaller proportion of the qualitative study participants also spent the second lockdown in autumn 2020 with families. Almost all students who made this decision describe positive psycho-social effects from returning to their families, especially in DC-1 and DC-2. They also emphasize the availability of recreational and green spaces, especially gardens, but also positive effects of an often-rural location that allows easier access to nature during lockdowns compared to the city:

> *"After 6 weeks of staying in Innsbruck, I decided to start my journey home and go back to the Allgäu [region in Southern Germany] to my family. [...] I really enjoy the exchange of ideas, cooking and eating together with my family and playing with my nephew and niece in the big garden. Even though I have to accept a two-week quarantine again when crossing the border, I find it much easier under these circumstances."* (Theo, 28, narrative, DC-2)

Rarely, however, negative aspects of moving back in with the family are also described, such as the resurgence of family conflicts, sometimes with accompanying (re)negotiation of parent–child or sibling relationships, and a lack of peace and quiet for learning. We additionally asked if people felt they had a stable circle of friends in Innsbruck, the city of their studies. There were no effects of the stable circle of friends on satisfaction with distance learning ($t(2639) = 0.53$, $p = 0.594$, $n = 2641$) or on performance ($t(2152) = -1.25$, $p = 0.212$, $n = 2154$).

Other study participants reported psychologically stressful socially conflictive phases during the pandemic, which ultimately also had an impact on their learning (see also Section 4.3). These include conflicts with flatmates in shared apartments and with partners—in the case of the latter—both due to spatial separation and to excessive spatial proximity, especially during the lockdowns. These conflicting social situations were for some students also the reasons they moved back to their families, left or changed flat-sharing communities, moved in with their partners, or moved away from them. However, there are also examples in the narratives that emphasize the mutual support between students that some experienced in shared living arrangements:

> *"I get along quite well with distance learning, but also only because my two flatmates study the same as I do and so we can do a lot together and support and motivate each other. I think without the two of them it would be much harder for me to find the motivation to stick with it. However, it's already working out much better this term than last summer term."* (Alicia, 20, narrative, DC-4)

Spatial confinement, lack of physical space for learning, and lack of sufficiently quiet learning environments are other factors that negatively influenced distance learning (see also Figure 6):

Figure 6. Snapshot taken by participant during a MIMI: In the pandemic, the master's thesis is written in the shared kitchen instead of the university library (Stella, 25, MIMI material, DC-2).

> *"However, I do notice how it stresses me out when my flatmates are also here sometimes and both do their distance learning in the kitchen-living room of our shared apartment. Our rooms are very small (about 10 square meters) so there's not really enough room for a desk."* (Aaron, 26, Narrative, DC-4)

The lack of work and study space in the university premises, especially the library, is highlighted frequently in the narratives and MIMIs:

> *23.04.20, 13:00—Interviewer: Does the location (now at home in the home office) change your activity?*
>
> *23.04.20, 13:01—Alma: Yes, I am unfortunately not as focused at home as I would be otherwise when I am working at the library. And I don't have new books available. Many online libraries are overloaded or not working properly.*
>
> *23.04.20, 13:03—Interviewer: What distracts you most from your tasks at home?*
>
> *23.04.20, 13:10—Alma: My consoles (XBOX and Nintendo Switch) and my TV. Short breaks become longer breaks as a result.* (Alma, 25, MIMI chat log, DC-2)

However, the quantitative survey, which asked about a change of the location of study before and after the pandemic, showed that the increase in students who worked at home, at 21.1%, was not as large as might have been expected (see Figure 7). This reflects that many students were also learning at home during normal times. Still, those who had previously used alternative places of study (informal physical spaces of learning), such as a library, friends' places, or when travelling prior to the pandemic, were no longer able to do so during the pandemic and the related restrictions on the use of public learning spaces. Students who reported changes in their location of study (n = 1285, M = 3.30, SD = 0.81) were less satisfied with the distance learning situation than those who did not (n = 1427, M = 3.60, SD = 0.80, $t(2710) = 9.64$, $p < 0.001$, $d = 0.37$). Additionally, their performance decreased during the distance learning (n = 1133, M = −0.12, SD = 0.60), while the performance of students whose learning location did not change stayed constant (n = 1078, M = 0.03, SD = 0.58, $t(2208.89) = 5.80$, $p < 0.001$, d = 0.25).

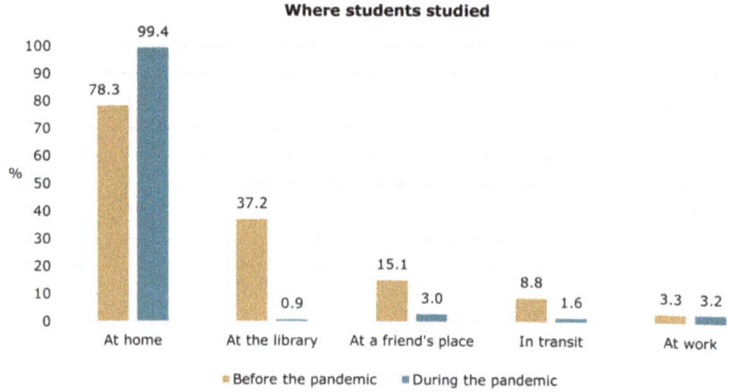

Figure 7. Locations of study before and during the pandemic as percentages (multiple choice, source: quantitative survey, February 2021).

4.5. Techno-Social Spaces and Polymedia Use during the Pandemic

Everyday spaces of students are characterized by an entanglement of techno-social and socio-material contexts (see Section 2). The qualitative part of the project sheds light not only on the use of digital media for distance learning but also on their further use for private purposes in the broadest sense. This further use of digital media is still highly relevant to distance learning because, on the one hand, it has made a central positive contribution—especially psycho-social, but also in terms of physical health, income, leisure time, civic engagement—as reported by participants in the qualitative study part. On the other hand, the MIMIs, in which students' everyday practices were surveyed for a full day at each survey phase, illustrate that distance learning was very often accompanied by private, parallel media use. Some students used several programs simultaneously, often on multiple screens and devices. With their polymedia concept, Madianou and Miller [52] point out how such converged media use has various social and emotional consequences for the users.

The following WordCloud illustrates the diversified use of digital media (see Figure 8). The WordCloud immediately highlights the central role that social media played for the participants—social media such as WhatsApp, Instagram, Skype, and YouTube were *also* used in distance learning, but primarily for private purposes.

Figure 8. WordCloud of software used by students for private purposes and distance learning during the pandemic (visualized are all instances of software named in the narratives; COV-IDENTITIES project).

The resulting blurring of private and learning contexts, of university and home, caused more trouble for some students in the context of distance learning, promoting stress and the feeling of being overwhelmed, see [38]:

> *"I do everything university-related in the living room, so now I will always associate the living room with university."* (Lukas, 20, Narrative, DC-3)

4.6. Students' Aspirations and Recommendations for Distance Learning

In the qualitative narratives and interviews, students were asked about their aspirations and recommendations for making distance learning more enjoyable and productive. These were then quantified by letting the participants of the quantitative survey choose of which recommendations and aspirations they approved. Many of the answers pointed to elements of teaching, which were not only relevant during a pandemic but also beyond (see also Figure 9).

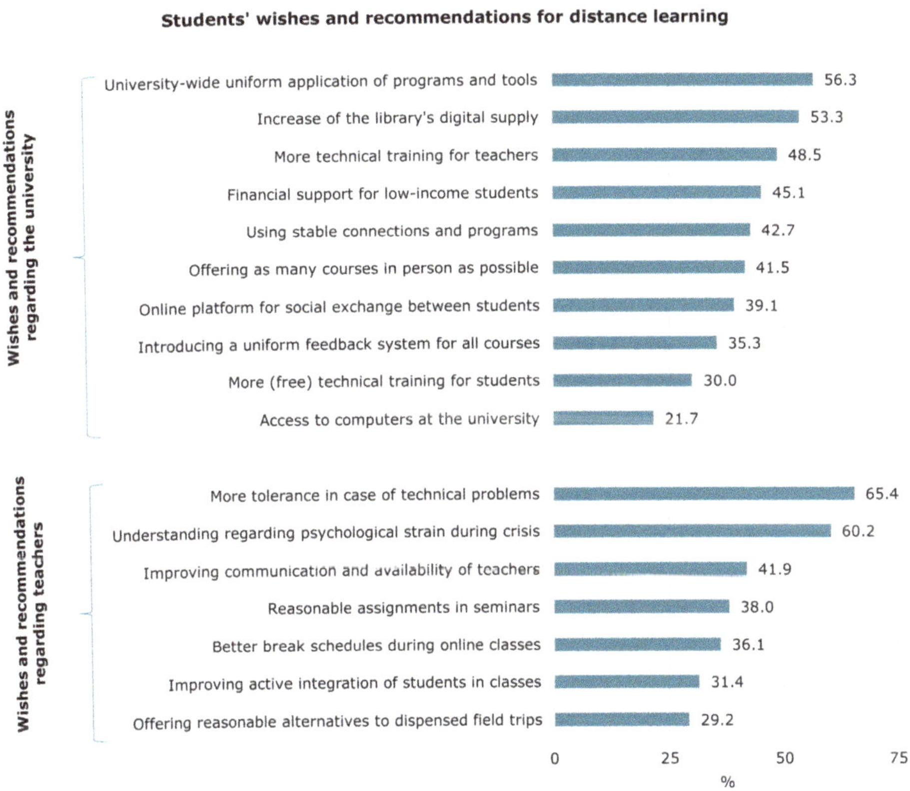

Figure 9. Students' wishes and recommendations regarding digital media for distance learning (multiple choice, source: quantitative survey, February 2021).

Despite the advantages that distance learning could have for them, students wanted to experience as much face-to-face teaching as possible during the pandemic situation. In addition, many students expressed the desire for adequate alternatives for field and laboratory experiences, such as field trips and practical training sessions, which would be cancelled due to the pandemic. Pure assignment processing instead of practical exercises and interactive teaching formats was not seen as a suitable and acceptable substitute. Furthermore, a majority of students wanted to see a continuation and expansion of the practice of recording courses and making instructional videos available to students. Also relevant for regular teaching were the expressed wishes for more online courses, more blended learning formats, and the introduction of an evaluation or feedback system that accompanies the teaching period in order to allow teachers to implement possible improvements in course quality while courses are still running. With regard to the design of (distance) learning courses, students also wished for more interactivity, more room for discussion, a reduction of computer-based assignment processing, and the aforementioned avoidance of assignment processing as an alternative to face-to-face courses. The desire for more breaks was also frequently expressed. In the quantitative survey, 56.2% reported that not taking (enough) breaks was an issue for them. Ronja emphasized their importance in a qualitative interview:

> *"Breaks during seminars are very important. [laughs] Oh my God, that's a huge thing that a lot of people forget when they're sitting in front of the computer, that you need breaks because you just don't have anyone to give you direct feedback. Breaks! Incredibly important! I think it's even worse to listen in front of the computer than anywhere else. That means taking a short break every half hour, preferably every 45 min, so that you can just take a breath, get fresh air, etc."* (Ronja, 27, interview, November 2020)

In the qualitative data collection, lecturers were asked for more reachability, more commitment and creativity, consideration, and tolerance for students with technical problems, as well as clearly formulated and comprehensible assignments. In addition, the desire for obligatory (technical) training for teachers was mentioned in order to compensate for what students perceived as a lack of technical competence on the part of some teachers. All these wishes are also directed at regular teaching and are also formulated in the following narrative excerpt by Paul (24, DC-4):

> *"Lecturers [should] be encouraged to be more creative in online teaching. The frontal teaching without interactive elements by technically limited lecturers, which is often propagated as being without alternative, could be loosened up by training, technical support (possibly students with an affinity for technology) and best-practice examples. Intermedia formats are already being used successfully by a few lecturers."*

The universities and institutes were asked to continue with a continuous flow of information about important innovations, changes, and developments in connection with teaching. In addition, students would like to see uniform pandemic-related regulations across institutes with regard to classroom teaching, distance learning, alternatives for internships and excursions, and the use of software.

For example, some students wanted more access to (technical) training for themselves and their instructors in order to better master the challenges that come with (distance) learning. In addition, there was a desire for further development of software so that it both offers more features and is less error-prone due to so-called bugs (errors in the software code/operation). Additionally, free access to course-relevant software would address some of the difficulties students face. A continuation of digital formats, as well as the use of uniform software across institute borders, is just as urgently recommended as is the rental of hardware or financial subsidies from the university to students for necessary hardware purchases. Especially during a pandemic, students who do not have the necessary hardware would appreciate having access to powerful computers either physically in the university building or digitally via remote access from home.

In the quantitative survey, students were asked to choose elements of distance learning which they would like to keep even after the pandemic (see Figure 10), with a large majority stating their preference for maintaining online access to all materials—since this question did not inquire as to how many perceived they already had good access prior to the pandemic, it is not certain how much change students would like here. Significantly, a high percentage of students wanted to keep the recording of classes, and slightly more than half were interested in maintaining opportunities for distance learning overall.

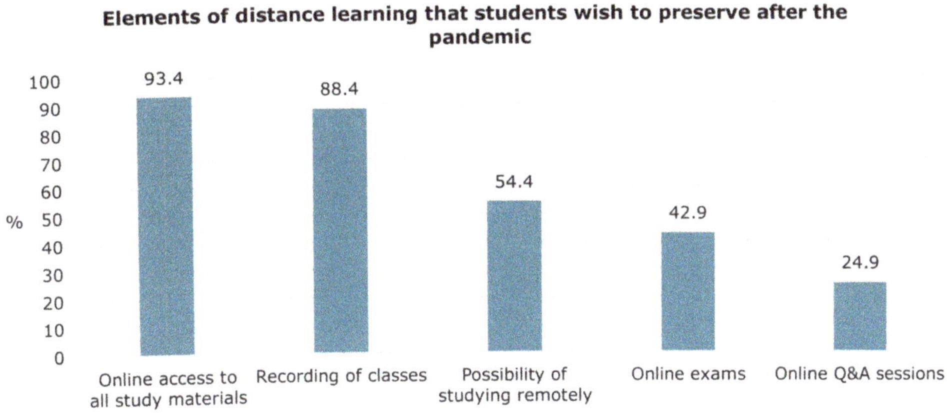

Figure 10. Elements of distance learning that students wish to maintain even after a return to presence teaching (multiple choice, source: quantitative survey, February 2021).

5. Discussion

Despite the challenges perceived by students during the past year of pandemic education, the current generation of students is able to continue with many elements of their studies through the opportunities that distance learning offers. A large proportion of the students in our study have managed to continue and even successfully complete their university education in this past year characterized by the COVID-19 pandemic. Nevertheless, it is probably more appropriate to speak of them as persevering during this time. For them, education during the crisis period meant being much more self-reliant and having to deal very flexibly with a potpourri of different virtual-teaching formats.

In terms of pandemic education, our research participants reported a significant improvement since the beginning of the pandemic in spring 2020. Especially at the beginning of the pandemic, the change often meant self-study—lists of teaching materials and literature were sent out, teaching staff first had to get used to virtual-teaching formats, and the software used caused difficulties. The satisfaction with distance learning improved with both teachers and learners adjusting to the distance learning formats.

Through the evaluation of the qualitative and quantitative data, moreover, it became clearly visible that the students of the Innsbruck universities were satisfied with the information provided by the universities via websites and social media, and that the information and communication skills of the universities were rated positively. Aristovnik et al. [13] state in their study that students worldwide rate their respective universities with similar degrees of positivity in this regard. This points to the importance of efficient communication on the part of university management and argues for maintaining a communication strategy that keeps students informed about developments at universities.

Nonetheless, as in other countries e.g., [53–57], the students in our Austrian study perceived a variety of challenges. Many participants urgently desired a return to face-to-face teaching and university life as well as direct face-to-face contact with teachers, their classmates, and other fellow students. The limitations of adequately replacing practical, field and laboratory courses during conditions of distance learning were strongly criticized. For some, technical difficulties posed a barrier, especially poor or overloaded internet connections and insufficiently powerful computers. Restless learning environments and the feeling of not being able to escape the constant virtual presence of university in the private room were also among the everyday realities of the pandemic. Young people were confronted with many uncertainties during this time, while contact persons were often less tangible than in normal times. This widening of inequalities because of the pandemic promises to be particularly problematic and poses an impediment to reaching the Sustainable Development Goal of quality education for all [29,58]; those students who do not have sufficient psycho-social support from families, who have no (adequate) learning space at home, who lack the financial means (e.g., for access to hardware and software), and those who have already suffered from psychological stress and pre-existing conditions are left further behind. The psychological effects of the pandemic and distance learning on students described in this study, especially the increased risk of depression, are in line with the study results of Huckins et al. [10] (see also [59]). Accordingly, students wish for more understanding of mental health problems during this crisis.

Another problematic aspect of the current distance learning relates to difficulties in time management as the shift to distance learning not only leads to spatial changes, but also to a lack of institutionally framed time management [34]. Students reported problems in taking adequate breaks and in achieving a time structure while learning from their homes. This suggests they need more support through both having temporal frameworks set by their teachers or universities, and education on how to improve at self-management and setting up their own time structures.

However, most students clearly would like to see the continuation and, in most cases, significant further development of those gains from digitization from the first pandemic year: blended learning formats, the widespread use of online learning-management systems, and the use of creative digital tools [29]. Strikingly, in the quantitative survey, even more than half of the participants expressed their wish to maintain the option of overall distance education even after the pandemic. The primary reasons are, especially, greater flexibility, time savings, and more options to adjust learning to one's own speed and abilities; new forms of educational and learning spaces seem to empower students when organizing their own learning. Especially students who work during their studies or who have children to care for seem to profit from the increased flexibility, which thus can help lessen social inequalities and make university education more accessible. Based on the perspectives of students, it is appropriate to demand a significant change in post-pandemic education. After all, since the switch to online formats affected not only the education but also the work sectors, it seems that enhanced digitization of university education is urgently needed to prepare graduates for the realities of a working life in the 21st century.

6. Conclusions and Recommendations

Although the advent of the "post-digital" [2]—an era in which the digital has become a natural, banal, and ubiquitous part of everyday life—has long since been declared, many universities in Austria and beyond have (knowingly) been lagging behind in adapting education to these times. This condition has played out painfully for students, teachers, and eventually society during the COVID-19 pandemic. Based on the results of our mixed-methods study we conclude by pointing out recommendations for action in dealing with this situation during continued higher education during the pandemic. We also sketch pathways for post-pandemic educational futures that are socially more sustainable in preparing young people for their lives in the post-digital.

In the face of an ongoing and uncertain pandemic situation, the following short-term recommendations for lecturers are aimed at helping to improve remote-teaching processes in order to support students in dealing with managing their university education during these challenging times [29,60]. Lecturers are recommended, where possible, to reduce computer-based assignment processing, insert more frequent breaks in classes (at least every 30–45 min), and to make teaching more interactive as well as involve students more closely. Particularly, lectures can be recorded and provided online to provide more time and spatial flexibility to students. Students' parallel media use for private purposes poses an increasing challenge during online classes—just as during presence teaching, when students bring their mobile devices to class. Our results indicate, however, that this challenge has increased all the more during distance learning. As measures aimed at reducing private media usage during distance learning, we recommend using interactive and changing teaching formats, establishing and reminding students of netiquette, raising students' awareness of negative effects on mental health and learning outcomes, and encouraging the use of the camera function (if necessary, temporarily). The loss of opportunities for in-class contact could be partially mitigated through offering students to meet teachers in the virtual classroom before or after each class and the extension of online consultation hours. To support contact between students, teachers are further encouraged to implement teaching methods that promote interaction. Even though group work is perceived as being more difficult to conduct digitally, group assignments can still provide a first step toward creating closer contacts between students.

While lecturers are the first point of contact for most students during distance learning, universities as institutions can and should help to support all parties involved based on the following short-term recommendations for action for general teaching. University management and administration are recommended to educate students and lecturers on health-promoting measures in the home office to reduce physical discomfort and prevent fatigue and stress, which is detrimental to quality learning [28,61]. Both students and lecturers need to be made more aware of the physical and mental health strains of distance teaching and learning, so as to increase their understanding and tolerance of the shortcomings of others. The use of standardized software in teaching across administrative-unit boundaries needs to be fostered. Universities need to find solutions to provide adequate alternatives to field, laboratory, and practical exercises. If this is not possible, they are recommended to provide intensive catch-up formats for such classes once attendance teaching resumes. To support social exchange between (particularly first-year) students, (a) student online platform(s) for exchange and/or joint learning should be established. Admission to these platforms should be made subject to prior agreement to the adherence to netiquette in these platforms. Universities should make teachers aware of restrictions that individual students potentially face during the pandemic resulting from socio-material learning environments, mental stress, and technical challenges. Exceptions for individual students should be permitted and developed to ensure accessibility to education for disadvantaged students. Resumption of face-to-face teaching as soon as legally—and from a public-health perspective—possible again, should be announced with sufficient advance notice to allow students who have returned to live with their families to come back to the universities' locations and rent accommodation in time.

With advancing vaccination rates, post-pandemic educational futures become more tangible, and the need for empirically well-grounded discussions on such futures increases. Based on our extensive study, we see the implementation of the following medium-term measures as essential to make teaching future-proof and distance learning more learner centered and sustainable beyond the phase of the pandemic [62]: the continuation of the digitization of university processes, literature, and library services must be strongly fostered. Based on this study, the retention of online course offerings, the establishment of hybrid courses, and the strong encouragement of creative and innovative online and blended teaching methods are recommended [27]. Current distance learning formats should not be adopted unchanged, but prevailing problems should be removed, such as the lack of social contacts and time structure. The technical possibilities of (new) learning software are continuously developing and expanding. They offer new opportunities but also present challenges for lecturers. Regular and compulsory training for teachers in these new opportunities is, in our opinion, the only option to keep university teaching up-to-date with the young generations' expectations of education in the post-pandemic, post-digital age. For teachers to be able to reserve sufficient time to acquire and update their digital skills, it will be necessary for universities to offer corresponding teaching reductions. After all, the acquisition of digital skills is an important pillar of tertiary education, preparing students for an increasingly digitized labor market and society. As has been long noted by (critical) education scholars, pupils and students need skills that prepare them for a responsible and active use of digital media, allowing them to become mature digital participants in society (e.g., [63,64]). This is an essential component of socially sustainable education and society in the post-digital.

Author Contributions: Conceptualization: T.B.-H., B.J., M.H.W., F.B. and K.K.; methodology: T.B.-H., K.K., F.B., D.M.I., B.J. and M.H.W.; validation: T.B.-H. and V.K.; formal analysis: F.B., T.B.-H., V.K. and D.M.I.; investigation: T.B.-H., F.B., V.K., D.M.I. and A.M.; resources: T.B.-H., B.J. and M.H.W.; data curation: T.B.-H. and V.K.; writing—original draft preparation: T.B.-H., F.B., V.K., K.K. and A.M.; writing—review and editing: K.K., T.B.-H., V.K., A.M. and F.B.; visualization: D.M.I., V.K., F.B. and T.B.-H.; Supervision: T.B.-H., B.J. and M.H.W.; project administration: T.B.-H., B.J. and M.H.W.; funding acquisition: T.B.-H., B.J. and M.H.W. All authors have read and agreed to the published version of the manuscript.

Funding: This research was funded by the University of Innsbruck, project number 338221, and by Förderkreis 1669—WissenschafftGesellschaft, grant number 329327.

Institutional Review Board Statement: Ethical clearance was provided for the COV-IDENTITIES project by the Board for Ethical Questions in Science of the University of Innsbruck, Innsbruck, Austria, on 1 April 2020.

Informed Consent Statement: Informed consent was obtained from all subjects involved in the COV-IDENTITIES study.

Data Availability Statement: The quantitative data are not yet publicly available, as the data analysis and processing are still ongoing. The qualitative data are not publicly available due to privacy and ethical reasons.

Acknowledgments: The authors thank the student participants and the student researchers involved in this project. Further, the authors thank the anonymous reviewers as well as the Open Access Fonds of the University of Innsbruck for their generous support of the publication of this article.

Conflicts of Interest: The authors declare no conflict of interest.

References

1. UNESCO. Global Monitoring of School Closures Caused by COVID-19. 2020. Available online: https://en.unesco.org/covid19/educationresponse (accessed on 20 May 2021).
2. Cascone, K. The aesthetics of failure: 'Post-digital' tendencies in contemporary computer music. *Comput. Music J.* **2000**, *24*, 12–18. [CrossRef]
3. Händel, M.; Bedenlier, S.; Gläser-Zikuda, M.; Kammerl, R.; Kopp, B.; Ziegler, A. Do Students have the Means to Learn during the Coronavirus Pandemic? Student Demands for Distance Learning in a Suddenly Digital Landscape. *PsyArXiv* **2020**. [CrossRef]
4. Edelhauser, E.; Lupu-Dima, L. Is Romania Prepared for eLearning during the COVID-19 Pandemic? *Sustainability* **2020**, *12*, 5438. [CrossRef]
5. University of Potsdam. Ergebnisbericht zu PotsBlitz "Online-Lehre 2020" im SoSe. 2020. Available online: https://pep.uni-potsdam.de/media/PotsBlitz/Berichte/PotsBlitz_Gesamtbericht.pdf (accessed on 23 June 2021).
6. Händel, M.; Stephan, M.; Gläser-Zikuda, M.; Kopp, B.; Bedenlier, S.; Ziegler, A. Digital readiness and its effects on higher education students' socio-emotional perceptions in the context of the COVID-19 pandemic. *J. Res. Technol. Educ.* **2020**. [CrossRef]
7. Zimmermann, M.; Bledsoe, C.; Papa, A. The Impact of the COVID-19 Pandemic on College Student Mental Health. A Longitudinal Examination of Risk and Protective Factors. *PsyArXiv* **2020**. [CrossRef]
8. Wang, C.; Zhao, H. The Impact of COVID-19 on Anxiety in Chinese University Students. *Front. Psychol.* **2020**, *11*, 1168. [CrossRef] [PubMed]
9. Tang, W.; Hu, T.; Yang, L.; Xu, J. The role of alexithymia in the mental health problems of home-quarantined university students during the COVID-19 pandemic in China. *Personal. Individ. Differ.* **2020**, *165*, 110131. [CrossRef]
10. Huckins, J.F.; daSilva, A.W.; Wang, W.; Hedlund, E.; Rogers, C.; Nepal, S.K.; Wu, J.; Obuchi, M.; Murphy, E.I.; Meyer, M.L.; et al. Mental Health and Behavior of College Students During the Early Phases of the COVID-19 Pandemic. Longitudinal Smartphone and Ecological Momentary Assessment Study. *J. Med. Internet Res.* **2020**, *22*, e20185. [CrossRef]
11. Kaparounaki, C.K.; Patsali, M.E.; Mousa, D.-P.V.; Papadopoulou, E.V.K.; Papadopoulou, K.K.K.; Fountoulakis, K.N. University students' mental health amidst the COVID-19 quarantine in Greece. *Psychiatry Res.* **2020**, *290*, 113111. [CrossRef]
12. Sahu, P. Closure of Universities Due to Coronavirus Disease 2019 (COVID-19). Impact on Education and Mental Health of Students and Academic Staff. *Cureus* **2020**, *12*, e7541. [CrossRef]
13. Aristovnik, A.; Keržič, D.; Ravšelj, D.; Tomaževič, N.; Umek, L. Impacts of the COVID-19 Pandemic on Life of Higher Education Students. A Global Perspective. *Sustainability* **2020**, *12*, 8438. [CrossRef]

14. Salari, N.; Hosseinian-Far, A.; Jalali, R.; Vaisi-Raygani, A.; Rasoulpoor, S.; Mohammadi, M.; Rasoulpoor, S.; Khaledi-Paveh, B. Prevalence of stress, anxiety, depression among the general population during the COVID-19 pandemic: A systematic review and meta-analysis. *Glob. Health* **2020**, *16*, 57. [CrossRef]
15. Xiong, J.; Lipsitz, O.; Nasri, F.; Lui, L.M.W.; Gill, H.; Phan, L.; Chen-Li, D.; Iacobucci, M.; Ho, R.; Majeed, A.; et al. Impact of COVID-19 pandemic on mental health in the general population: A systematic review. *J. Affect. Disord.* **2020**, *277*, 55–64. [CrossRef]
16. Arnett, J.J. *Emerging Adulthood: The Winding Road from the Late Teens through the Twenties*, 2nd ed.; Oxford University Press: Oxford, UK, 2014.
17. Barlett, C.; Barlett, N. The young and the restless: Examining the relationships between age, emerging adulthood variables, and the Dark Triad. *Personal. Individ. Differ.* **2015**, *86*, 20–24. [CrossRef]
18. Kaufmann, K.; Straganz, C.; Bork-Hüffer, T. City-life No More? Young Adults' Disrupted Urban Experiences and their Digital Mediation under COVID-19. *Urban Plan.* **2020**, *5*, 324–334. [CrossRef]
19. Elmer, T.; Mepham, K.; Stadtfeld, C. Students under lockdown: Comparisons of students' social networks and mental health before and during the COVID-19 crisis in Switzerland. *PLoS ONE* **2020**, *15*, e0236337. [CrossRef] [PubMed]
20. Son, C.; Hegde, S.; Smith, A.; Wang, X.; Sasangohar, F. Effects of COVID-19 on college students' mental health in the United States: Interview survey study. *J. Med. Internet Res.* **2020**, *22*, e21279. [CrossRef]
21. Conrad, R.; Rayala, H.; Menon, M.; Vora, K. Universities' Response to Supporting Mental Health of College Students during the COVID-19 Pandemic. 2020. Available online: https://www.psychiatrictimes.com/view/universities-response-supporting-mental-health-college-students-during-covid-19-pandemic (accessed on 23 June 2021).
22. Besser, A.; Flett, G.L.; Zeigler-Hill, V. Adaptability to a sudden transition to online learning during the COVID-19 pandemic: Understanding the challenges for students. *Scholarsh. Teach. Learn. Psychol.* **2020**. [CrossRef]
23. Dale, R.; Budimir, S.; Probst, T.; Stippl, P.; Pieh, C. Mental health during a COVID-19 lockdown over the Christmas period in Austria. *SSRN Electron. J.* **2021**. [CrossRef]
24. Schlack, R.; Neuper, L.; Hölling, H.; De Bock, F.; Ravens-Sieberer, U.; Mauz, E.; Wachtler, B.; Beyer, A.-K. Auswirkungen der COVID-19-Pandemie und der Eindämmungsmaßnahmen auf die psychische Gesundheit von Kindern und Jugendlichen. *J. Health Monit.* **2020**, *5*, 23–34.
25. Zhou, S.; Zhang, L.; Wang, L.; Guo, Z.; Wang, J.; Chen, J.; Liu, M.; Chen, X.; Chen, J. Prevalence and socio-demographic correlates of psychological health problems in Chinese adolescents during the outbreak of COVID-19. *Eur. Child Adolesc. Psychiatry* **2020**, *29*, 749–758. [CrossRef] [PubMed]
26. Ravens-Sieberer, U.; Kaman, A.; Erhart, M.; Devine, J.; Schlack, R.; Otto, C. Impact of the COVID-19 pandemic on quality of life and mental health in children and adolescents in Germany. *Eur. Child Adolesc. Psychiatry* **2021**. [CrossRef]
27. Navarro-Espinosa, J.A.; Vaquero-Abellán, M.; Perea-Moreno, A.-J.; Pedrós-Pérez, G.; Aparicio-Martínez, P.; Martínez-Jiménez, M.P. The Higher Education Sustainability before and during the COVID-19 Pandemic: A Spanish and Ecuadorian Case. *Sustainability* **2021**, *13*, 6363. [CrossRef]
28. Wolff, L.-A. Sustainability Education in Risks and Crises: Lessons from COVID-19. *Sustainability* **2020**, *12*, 5205. [CrossRef]
29. Chu, A.M.Y.; Liu, C.K.W.; So, M.K.P.; Lam, B.S.Y. Factors for Sustainable Online Learning in Higher Education during the COVID-19 Pandemic. *Sustainability* **2021**, *13*, 5038. [CrossRef]
30. Leon, I.; Sagarna, M.; Mora, F.; Otaduy, J.P. BIM Application for Sustainable Teaching Environment and Solutions in the Context of COVID-19. *Sustainability* **2021**, *13*, 4746. [CrossRef]
31. Collado-Valero, J.; Rodríguez-Infante, G.; Romero-González, M.; Gamboa-Ternero, S.; Navarro-Soria, I.; Lavigne Cerván, R. Flipped Classroom: Active Methodology for Sustainable Learning in Higher Education during Social Distancing Due to COVID-19. *Sustainability* **2021**, *13*, 5336. [CrossRef]
32. Arbeitsmarktservice Österreich. Übersicht Über den Arbeitsmarkt Februar 2021 [Employment Market Overview February 2021]. Available online: https://www.ams.at/arbeitsmarktdaten-und-medien/arbeitsmarkt-daten-und-arbeitsmarkt-forschung/berichte-und-auswertungen (accessed on 23 June 2021).
33. Statistik Austria. Bevölkerung [Population]. Available online: https://www.statistik.at/web_de/statistiken/menschen_und_gesellschaft/bevoelkerung/index.html (accessed on 23 June 2021).
34. Gobbi, A.; Rovea, F. Distance teaching and teaching 'as' distance. A critical reading of online teaching instruments during and after the pandemic. *Teoría Educ. Rev. Interuniv.* **2021**, *33*, 71–87. [CrossRef]
35. Card, P.; Thomas, H. Student housing as a learning space. *J. Geogr. High. Educ.* **2018**, *42*, 573–587. [CrossRef]
36. House-Peters, L.A.; Del Casino, V.J.; Brooks, C.F. Dialogue, inquiry, and encounter: Critical geographies of online higher education. *Prog. Hum. Geogr.* **2019**, *43*, 81–103. [CrossRef]
37. Bork-Hüffer, T.; Mahlknecht, B.; Kaufmann, K. (Cyber)Bullying in Schools—When Bullying Stretches Across cON/FFlating Spaces. *Child. Geogr.* **2021**, *19*, 241–253. [CrossRef]
38. Bork-Hüffer, T.; Mahlknecht, B.; Markl, A. Kollektivität in und durch cON/FFlating spaces: 8 Thesen zu Verschränkungen, multiplen Historizitäten und Intra-Aktionen in sozio-materiell-technologischen (Alltags-)Räumen. *Z. Kult. Kollekt.* **2020**, *6*, 131–170.
39. Kraftl, P.; Andrews, W.; Beech, S.; Ceresa, G.; Holloway, S.L.; Johnson, V.; White, C. Geographies of education: A journey. *Area* **2020**, 1–9. [CrossRef]
40. Tuan, Y.-F. Place: An Experiential Perspective. *Geogr. Rev.* **1975**, *65*, 151–165. [CrossRef]

41. Massey, D.B. For space. *Geogr. Ann.* **2005**, *89*, 83–85.
42. Bork-Hüffer, T. Mediated sense of place: Effects of mediation and mobility on the place perception of German professionals in Singapore. *New Media Soc.* **2016**, *18*, 2155–2170. [CrossRef]
43. Qian, J.; Zhu, H. Chinese urban migrants' sense of place: Emotional attachment, identity formation, and place dependence in the city and community of Guangzhou: Chinese urban migrants' sense of place. *Asia Pac. Viewp.* **2014**, *55*, 81–101. [CrossRef]
44. Marcu, S. Emotions on the move: Belonging, sense of place and feelings identities among young Romanian immigrants in Spain. *J. Youth Stud.* **2012**, *15*, 207–223. [CrossRef]
45. Holton, M.; Riley, M. Student geographies and homemaking: Personal belonging(s) and identities. *Soc. Cult. Geogr.* **2016**, *17*, 623–645. [CrossRef]
46. Kaufmann, K.; Peil, C. The mobile instant messaging interview (MIMI): Using WhatsApp to enhance self-reporting and explore media usage in situ. *Mob. Media Commun.* **2020**, *8*, 229–246. [CrossRef]
47. Vandebosch, H.; Green, L. *Narratives in Research and Interventions on Cyberbullying among Young People*; Springer International Publishing: Cham, Switzerland, 2019.
48. Atkinson, R. *The Life Story Interview*; SAGE: Thousand Oaks, CA, USA, 1998.
49. Pabian, S.; Erreygers, S. Generating personal stories on negative online peer interactions through a photo elicitation method. In *Narratives in Research and Interventions on Cyberbullying among Young People*; Vandebosch, H., Green, L., Eds.; Springer: Cham, Switzerland, 2019; pp. 77–95.
50. Kaufmann, K.; Peil, C.; Bork-Hüffer, T. Producing in situ Data in a Pandemic with Mobile Instant Messaging Interviews (MIMIs). *Int. J. Qual. Methods* **2021**, in press.
51. Universität Innsbruck. Auf Einen Blick [At a Glance]. Available online: https://www.uibk.ac.at/universitaet/profil/dokumente/uni-in-zahlen-2020.pdf (accessed on 23 June 2021).
52. Madianou, M.; Miller, D. Polymedia: Towards a new theory of digital media in interpersonal communication. *Int. J. Cult. Stud.* **2013**, *16*, 169–187. [CrossRef]
53. Allam, S.N.S.; Hassan, M.S.; Mohideen, R.S.; Ramlan, A.F.; Kamal, R.M. Online Distance Learning Readiness During COVID-19 Outbreak Among Undergraduate Students. *Int. J. Acad. Res. Bus. Soc. Sci.* **2020**, *10*, 642–657. [CrossRef]
54. Al-Kumaim, N.H.; Alhazmi, A.K.; Mohammed, F.; Gazem, N.A.; Shabbir, M.S.; Fazea, Y. Exploring the Impact of the COVID-19 Pandemic on University Students' Learning Life: An Integrated Conceptual Motivational Model for Sustainable and Healthy Online Learning. *Sustainability* **2021**, *13*, 2546. [CrossRef]
55. Cicha, K.; Rizun, M.; Rutecka, P.; Strzelecki, A. COVID-19 and Higher Education: First-Year Students' Expectations toward Distance Learning. *Sustainability* **2021**, *13*, 1889. [CrossRef]
56. Khan, M.A.; Vivek; Nabi, M.K.; Khojah, M.; Tahir, M. Students' Perception towards E-Learning during COVID-19 Pandemic in India: An Empirical Study. *Sustainability* **2021**, *13*, 57. [CrossRef]
57. Coman, C.; Țîru, L.G.; Meseșan-Schmitz, L.; Stanciu, C.; Bularca, M.C. Online Teaching and Learning in Higher Education during the Coronavirus Pandemic: Students' Perspective. *Sustainability* **2020**, *12*, 10367. [CrossRef]
58. Leal Filho, W.; Brandli, L.L.; Lange Salvia, A.; Rayman-Bacchus, L.; Platje, J. COVID-19 and the UN Sustainable Development Goals: Threat to Solidarity or an Opportunity? *Sustainability* **2020**, *12*, 5343. [CrossRef]
59. Gómez-Galán, J.; Martínez-López, J.Á.; Lázaro-Pérez, C.; Sarasola Sánchez-Serrano, J.L. Social Networks Consumption and Addiction in College Students during the COVID-19 Pandemic: Educational Approach to Responsible Use. *Sustainability* **2020**, *12*, 7737. [CrossRef]
60. Joia, L.A.; Lorenzo, M. Zoom In, Zoom out: The Impact of the COVID-19 Pandemic in the Classroom. *Sustainability* **2021**, *13*, 2531. [CrossRef]
61. Li, H.; Hafeez, H.; Zaheer, M.A. COVID-19 and Pretentious Psychological Well-Being of Students: A Threat to Educational Sustainability. *Front. Psychol.* **2021**, *11*, 628003. [CrossRef] [PubMed]
62. Sá, M.J.; Serpa, S. The COVID-19 Pandemic as an Opportunity to Foster the Sustainable Development of Teaching in Higher Education. *Sustainability* **2020**, *12*, 8525. [CrossRef]
63. Dorsch, C.; Kanwischer, D. Mündigkeit in einer Kultur der Digitalität—Geographische Bildung und "Spatial Citizenship" [Maturity in a culture of digitality—Geographical education and 'spatial citizenship']. *Z. Didakt. Ges.* **2020**, *11*, 23–40.
64. Gryl, I.; Jekel, T. Spatially informed citizenship education as an approach for global understanding. In *Geography Education for Global Understanding*; Demirci, A., Miguel González, R.D., Bednarz, S.W., Eds.; Springer: Cham, Switzerland, 2018; pp. 43–56.

sustainability

Article

Multilevel Antecedents of Organizational Speed: The Exemplary Case of a Small Italian R&D Organization

Lara Agostini * and Anna Nosella

Department of Management and Engineering, University of Padova, 35131 Padova, Italy; anna.nosella@unipd.it
* Correspondence: lara.agostini@unipd.it

Abstract: The COVID-19 outbreak has compelled many organizations to adapt to a rapidly changing environment. In this context, the aim of this article is to unveil how a small R&D organization has been able to rapidly take advantage of the opportunities offered by the COVID-19 outbreak and to understand the factors that have enabled organizational speed. Results of the qualitative analysis of this exemplary single case show that a combination of factors at the individual level (i.e., characteristics of the CEO and collaborators), organizational level (i.e., structure, resources, operative systems and processes, culture), and partner level (i.e., characteristics of the partner portfolio) is required to go through a very fast recognition–decision–execution process.

Keywords: organizational speed; dynamic capability; ambidexterity; COVID-19; R&D organization

Citation: Agostini, L.; Nosella, A. Multilevel Antecedents of Organizational Speed: The Exemplary Case of a Small Italian R&D Organization. *Sustainability* **2021**, *13*, 7502. https://doi.org/10.3390/su13137502

Academic Editors: Miltiadis D. Lytras, Abdulrahman Housawi and Basim Alsaywid

Received: 25 May 2021
Accepted: 2 July 2021
Published: 5 July 2021

1. Introduction

The COVID-19 outbreak, which the World Health Organization declared a pandemic on 11 March 2020, has disrupted all business environments. Compared with other contagious diseases, COVID-19 appears to be both more deadly and more contagious, thus leading scientists to consider it "the worst health crisis of our times" [1]. Consequently, all COVID-19 health policies have been directed towards decreasing the replication of the disease and saving lives. This situation is posing serious problems for all individuals, organizations, and countries; looking at organizations, the economic shutdown has put pressure on manufacturing, decreased demand, and contracted industrial profits [2], shocking organizations and putting economic sustainability at risk, which has called for attention on this issue [3,4]. However, at the same time, it might also offer a potential opportunity for some organizations, especially for R&D labs operating in the healthcare industry domain, which has a direct impact on human health and wellbeing.

The topic of how organizations adapt to changes in their external environment is an issue around which there is recent and still open debate. Although the literature has dedicated increasing attention to this matter in recent decades, organizations continue to fail in doing it effectively, especially in the case of discontinuous change [5]. The changes that COVID-19 has brought about are not only discontinuous, but also unexpected, which further contributes to making this challenge tougher. Indeed, an organization needs to be flexible enough to move and reconfigure its organizational resources and processes to cope with environmental changes [6], which is far from being an easy undertaking, especially in certain situations where the key is responding *quickly* to environmental challenges [7]. In this regard, very recently, Dykes et al. [8] conceptualized *organizational speed* as "the dynamic Gestalt-like capacity of an organization to quickly identify, assemble, reconfigure, modify, and deploy its organizational processes and activities" [8] (p. 270), which seems to be exactly what firms need in such unexpected situations as the COVID-19 outbreak. They explain that this dynamic capability encompasses three dimensions, namely recognition speed, decision speed, and execution speed, emphasizing the holistic perspective they use to study organizational change, and explicitly calling for further investigation on this underexplored concept, with a particular focus, among others, on antecedents [8].

In addition, to the best of our knowledge, strategic issues, related to the broader theme of dynamic capabilities that organizational speed is part of, have mainly been investigated in manufacturing contexts, with some exceptions focusing on service industries (e.g., [9]), thus leaving the domain of R&D organizations unexplored.

On such bases, this article aims to answer the following research questions: How can an R&D organization rapidly take advantage of the opportunities offered by the COVID-19 outbreak? What are the factors that enable the organizational speed of an R&D organization in the face of the COVID-19 outbreak?

For this purpose, we examine a single yet exemplary case of a small Italian R&D organization that has faced these challenges rapidly by launching several new explorative and exploitative projects to respond to the open questions posed by the COVID-19 outbreak.

This research is particularly relevant because it allows lessons to be learned after a crisis-like situation, so as to take advantage of the positive experience of an organization that was able to turn this difficult situation into an opportunity [4]. This study contributes, from a theoretical and practical standpoint, to the debate surrounding how small R&D organizations can promptly react to the disruption this new pandemic has brought about and turn challenges into opportunities. From a theoretical perspective, we elaborate on the concept of organizational speed, offering an overarching view on the recognition–decision–execution process in times of rapid change and associated antecedents at the individual, organizational, and partner levels. In addition, we contribute to the recent stream exploring the connection between the triple bottom-line dimensions of environmental, social, and economic sustainability and the COVID-19 pandemic, with a particular focus on economic sustainability. From a practical perspective, we make managers of small R&D organizations aware of how they could be prepared to face such disruptive situations, where the only chance to survive is by transforming challenges into opportunities.

The remainder of this article is structured as follows: in the next section, we briefly lay out the theoretical bases for our study, and then we illustrate the methodology that we followed and present the results of our analysis, before discussing them in light of existing studies and highlighting implications for theory and practice.

2. The COVID-19 Outbreak

After the first cases of pneumonia in China (Wuhan City, Hubei Province) at the end of 2019, the rapid escalation of coronavirus disease 2019 (COVID-19) led the World Health Organization declare it a global pandemic on 11 March 2020.

It is widely recognized that the COVID-19 pandemic crisis embraces the health, economic, and social domains. In this context, economists, scientists in the pharmaceutical, epidemiology and biology fields, and policy-makers have led the scene of ongoing debates, but the deep impact on work and organizations calls for a managerial and organizational perspective to complete the picture [10].

While being acquainted with all of the serious difficulties that people and organizations are facing, we must also acknowledge the opportunities that such a changing context offers to organizations [11]. Unquestionably, the COVID-19 pandemic has left no alternatives to organizations but to reconsider how to adapt to the new situation in the most suitable way [12]. Therefore, the capacity to rapidly respond to changes has been considered the key to survival, which is assured not only by financial assets in this scenario [13], further confirming the relevance of successfully adjusting to a mutable environment in all of its facets [14].

However, the ability to change is not enough, but rather speed is central; indeed, in a pandemic, "time is so valuable and essential, that the question of costs is far less important than the ability to get to a solution sooner" [15] (p. 410). Based on the rationale that organizations that have transformed to survive through previous crises have been able to change what they do quickly [16], authors of recent publications in the COVID-19 domain stress the prominent role of speed, both in engaging with new ideas and taking actions (e.g., [11,15,17]). Organizational capabilities related to such abilities are not simply associated with the redeployment of resources, but rather they require other mechanisms, which contributes to increasing the complexity and demands further research [18].

Despite this evidence, "we know little about the novel organizational and change aspects that the disease represents beyond our established and comfortable researching boundaries" [10] (p. 263).

On such premises, we aim to take advantage of the unique chance to carry out phenomenon-driven research, i.e., shaping research as the phenomenon occurs [10,19], which provides the perfect conditions for conducting research on organizational speed in the COVID-19 pandemic scenario.

3. Theoretical Background

One of the key issues in the strategic management field relates to modifications in the environment and the subsequent organizational adaptation on the part of firms through their strategic changes [20]. The literature has widely recognized the importance of strategic change, especially in dynamic environments, and has associated it with the reconfiguration of the resources and capabilities an organization possesses (e.g., [21]). In this context, agility and flexibility are the most used terms by the literature to refer to the ability organizations have to adapt to the changing environment and satisfy market demand [22]. However, the two terms (i.e., flexibility and agility) are often used interchangeably, blurring the boundaries between them [22]. Recently, some articles have tried to shed light on this issue, highlighting that, since "agility emphasizes speed and flexibility as the primary attributes of an agile organization" [23], flexibility seems to be just a component of agility, which is a more encompassing concept. In other words, agility has been recognized by some authors as a natural evolution of flexibility since fast market changes and global competition, which characterize the current competitive arena, have also emphasized the need for speed: flexibility is thus "an agility capability, among other capabilities such as responsiveness or speed" [22] (p. 1138). Along the same lines, Singh et al. [24] and, more recently, Baškarada and Koronios [25] further argue that agility enables a company to respond to the external *stimuli* producing changes along two independent dimensions: magnitude of change and rate of variety change, which respectively refer to flexibility and speed relative to the competitors. Besides flexibility and speed, Walter [26], in her up-to-date literature review, also identifies responsiveness and competency as the main agility capabilities, all falling under the broader umbrella of dynamic capabilities [26]. Our article is theoretically grounded in this literature and focuses on one of the dimensions of agility, namely *organizational speed*, which, in the face of the COVID-19 scenario, seems to be a winner and a needed capability for companies not deeply investigated by the literature so far [25]. Indeed, organizational speed can be considered one of the components of agility, even though it emphasizes rapidity, based on the idea that "for action to occur—and occur quickly when conditions warrant—the firm must be alert to the need for action, quickly decide what to do, and have the capability to execute the action" [8] (p. 272). Organizational speed is thus composed of three main dimensions: *recognition speed*, which is the speed with which an organization recognizes opportunities and challenges; *decision speed*, which is the speed with it reaches decisions to act; and *execution speed*, which is the speed with which it mobilizes resources, processes, and activities for the implementation of an initiative [8]. While so far studies on organizational speed are scant, studies on agility have mainly embraced the manufacturing industry, leaving R&D organizations and laboratories under-investigated; only recently some articles, focused on the broader concept of dynamic capabilities, have started investigating R&D and technology-based organizations. To provide an overview of the studies on this issue in the R&D context, we performed a search in ISI Web of Science (WoS) combining the keywords "R&D lab*" OR "innov* lab*" OR "research lab*" OR "research cent*" OR "R&D organi*" OR "technology-based" with the keyword "dynamic capabilit*", both in the Topic (title, abstract, keywords). Results yielded 90 articles, which, after filtering by WoS categories (Management, Business), document type (Article, Review), and language (English), decreased to 58. After reading the full text, we identified 17 articles that fit well with the domain of interest. Table 1 exhibits some details of these articles, showing some first evidence concerning the paucity of literature in this area.

Articles can be divided into two main groups where the former contains the articles that assess the impact of dynamic capabilities on firm performance, whereas the latter contains those that investigate the antecedents. Overall, articles in the first group prove that different dynamic capabilities, such as R&D and marketing capabilities [27], absorptive, innovative, and adaptive capabilities (e.g., [28]), and organizational capability [29,30], play a positive role in enhancing different dimensions of firm performance.

Table 1. Overview of the literature on dynamic capabilities in R&D and technology-based organizations.

Reference	Purpose	DC Investigated	Main Evidence	Setting	Methodology
do Nascimento Welter et al. [31]	To identify the instruments and organizational mechanisms that provide the development of the innovative capacities of companies	Innovative capacity	The process of developing the innovative capacity of companies was studied, and this development occurred through behaviors and skills, routines, and processes and mechanisms of learning and knowledge governance that underpin the development of the product, process, and behavioral dimensions.	Small technology-based institutions in Brazil	Qualitative—multiple case study
Deligianni et al. [32]	To examine the nature of the relationship between technological competence and product innovation, and the moderating effects of the entrepreneur's political competence and prior start-up experience	Entrepreneurial political capability	Technological competence and entrepreneurial competences are key contingencies that influence resource orchestration efficiency in the context of new venture innovation.	New technology-based ventures in Greece	Quantitative
Salehi et al. [33]	To understand how the roles played by network actors evolve during the development and commercialization process of an emerging technology and what operational and dynamic capabilities are developed by actors through collaboration	DC in general	Actors developed sensing capabilities in the pre-collaboration stage, which drove joint new product development. During the collaboration, seizing capabilities were developed where resource commitment and alignment of resources among actors were essential. Capabilities gained through commercialization and large-scale production were predominantly transforming capabilities where actors realigned their structure and had a positive impact on capability development in the wider network.	Technology-based start-ups in a network context	Qualitative—single case study

Table 1. *Cont.*

Reference	Purpose	DC Investigated	Main Evidence	Setting	Methodology
Deakins and Bensemann [34]	To investigate the nexus between entrepreneurial learning and the innovation process	Entrepreneurial learning	Entrepreneurial learning has a critical role in the innovation process, enabling TBSFs to overcome resource constraints and challenges in a lean contextual environment such as New Zealand.	Technology-based new firms (TBNFs) in the agri-business sector in New Zealand	Qualitative interviews
Jensen and Clausen [35]	To examine the origins of exploration and exploitation capabilities in NTBFs	Exploration and exploitation capabilities	There is an effect of exploitation and exploration behavior on capability emergence, and this effect is mediated through routines for deliberate learning.	New technology-based firms (NTBFs) in Norway	Quantitative
Cho et al. [36]	To propose an assessment framework for research and development (R&D) innovation capabilities	R&D-based innovation capabilities	Conceptualization of R&D-based innovation capabilities as composed of management, innovation, planning, and implementation capabilities; a framework with indices for each component is presented and tested.	SMEs in technology-based industries (chemistry, electronics, telecommunications, and semiconductors)	Conceptual + Qualitative
Löfsten [30]	To analyze the organizational capabilities among NTBFs and examine how these are linked to the firms' long-term survival	Organizational capabilities	Business experience (work experience, education, multidisciplinary) of the CEO positively influences firm survival, whereas financing and having international customers do not.	NTBFs in Sweden	Quantitative
Hutterer et al. [37]	To investigate how a research center is able to fulfil contradictory demands by scientific and industrial stakeholders	Ambidexterity	The fundamental abilities for managing contradictory demands are located on an individual level, and it is argued that especially autonomous, well-educated people and their competences of self-organization enable the research center to be ambidextrous.	Large research center in the mechatronic sector	Qualitative—single case study

Table 1. *Cont.*

Reference	Purpose	DC Investigated	Main Evidence	Setting	Methodology
Ortín-Ángel and Vendrell-Herrero [38]	To compare the evolution of firms' total factor productivity between university spin-offs and NTBFs from a capability perspective	DC in general	University spin-offs have lower initial substantive capabilities but greater dynamic capabilities than independent NTBFs.	University spin-off and NTBFs in Spain	Quantitative
Biedenbach and Müller [28]	To explore how absorptive, innovative, and adaptive capabilities within early project phases affect project and portfolio performances in pharmaceutical and biotechnology R&D organizations	Absorptive, innovative, and adaptive capabilities	The results show effects of absorptive, innovative, and adaptive capabilities on short- and long-term project performance and portfolio performance. Absorptive and adaptive capabilities are the primary contributors to the performance outcome, whereas innovative capabilities are a minor contributor.	Pharmaceutical and biotechnology R&D organizations	Qualitative-quantitative interviews
Brinckmann and Hoegl [39]	To examine how teamwork capability and relational capability of the entrepreneurial team affects the development of new firms	Relational capability and teamworking capability	The study finds that the founding team's initial relational capability is important for the development of NTBFs, whereas the founding team's initial teamwork capabilities is not.	NTBFs in Germany	Quantitative
Strehle et al. [40]	To investigate the impact of organizational learning as an endogenous growth driver for technology-based new ventures	Strategic planning, financial planning, and evaluation, human resource planning, and evaluation, product development, marketing and sales	Eight bundles of management control systems, which are used as proxies for the emergence of the dynamic capabilities strategic planning, financial planning, and evaluation, human resource planning, and evaluation, product development, marketing and sales, and partnering, are positively associated with the growth of the venture.	Technology-based new ventures	Qualitative

Table 1. *Cont.*

Reference	Purpose	DC Investigated	Main Evidence	Setting	Methodology
Thal and Shahady [29]	To examine perceptions regarding the practice of innovation in the US Air Force's science and technology (S&T) community	Innovation management as a form of organizational capability	The primary reasons the S&T community pursues innovation are a desire to be state-of-the-art, to use technology better, and to respond to the customer. However, innovation was not well integrated into their business and technology strategies, which may result in ad hoc innovation efforts that are incongruent. Therefore, the S&T community may be better served by providing its workforce with the organizational processes to better facilitate game-changing innovation.	Air Force Research Laboratory	Qualitative interviews
Wu and Wang [41]	To analyze how firms transform resources into performance	Resource integration capability, resource reconfiguration capability, learning capability, ability to respond to the rapidly changing environment	DC serve as a link for transforming internal and external resources (specialized know-how, capital, operational management capability, reputation, cooperative alliance experience) first into firm competitiveness and then into financial performance.(Dynamic capabilities increase with firm resources and with the willingness of support firms to cooperate, and, in turn, they serve to increase firm competitiveness and then financial performance.)	Technology-based firms in Taiwan	Quantitative

Table 1. *Cont.*

Reference	Purpose	DC Investigated	Main Evidence	Setting	Methodology
Andries and Debackere [42]	To look at how the existing literature at the company level can inform us about adaptation in new technology-based companies	Adaptation	Model that proposes initial as well as later-acquired human, technological, financial, and networking resources as possible enablers for business model adaptation, which consists of different episodes, characterized by uncertainty or ambiguity.	New technology-based firms	Conceptual
Atuahene-Gima et al. [43]	To investigate the relationship between the marketing strategy innovativeness and new product performance	Marketing strategy innovativeness (MSI)	The team's extra industry relationships and market dynamism enhanced the impact of MSI on new product performance. In contrast, top management team's intra-industry relationships, financial relationships, and technology dynamism hindered the impact of MSI on new product performance.	Technology-based firms in China	Quantitative
Kor and Mahoney [27]	To examine the effects of the dynamics, management, and governance of R&D and marketing resource deployments on firm-level economic performance	R&D and marketing capabilities	A history of increased efforts in developing and maintaining marketing capabilities is an enduring source of competitive advantage. Moreover, if existing R&D capabilities are not renewed, in a high-velocity business world, economic rents from prior firm-level capabilities dissipate rapidly. Management experience moderates this effect.	Technology-based entrepreneurial firms in the medical, surgical, and dental instruments industry	Quantitative

Articles in the second group, encompassing contributions that are more recent, start to look at how dynamic capabilities are formed, thus examining antecedents of specific dynamic capabilities, such as ambidexterity [37], exploration and exploitation capabilities [35], or innovative capacity [31]. Overall, antecedents are found at three main levels, namely the individual, organizational, and partner levels. At the individual level, skills [31] are mentioned as important antecedents, with particular reference to the attitude towards adaptability and a strong previous knowledge base [35]. At the organizational level, routines and processes and mechanisms of learning and knowledge governance seem to serve to innovative capacity [31]; in particular, routines in systematizing existing knowledge and experience are given particular emphasis for their potential to transform that knowledge into future capabilities [35]. Finally, partnerships seem to contribute to develop sensing and seizing capabilities [33].

It is clear that the literature on how dynamic capabilities are formed is scant and that there is a lack of an overarching view on antecedents at different levels. In addition, it mainly focuses on new technology-based firms, leaving the other R&D organizations unexplored. Moreover, there is only one article [41] specifically dealing with the ability to respond to a rapidly changing environment, but only as one of the components through which the impact of dynamic capabilities on firm competitiveness and performance is tested. Therefore, the dimension of *speed* has not received the attention it deserves.

To bridge this gap, this article aims to understand how an R&D organization can rapidly take advantage of the opportunities offered by the COVID-19 outbreak and to unveil the factors that enable the organizational speed of an R&D organization in the face of the COVID-19 outbreak.

4. Methodology

We employed a qualitative methodological approach for its ability to capture evolutionary, relational, temporal, and cultural complexities and offer detailed and contextualized descriptions of actions and interactions associated with the innovation domain [44]. We relied on the recent article by Elsahn et al. [45] that aims to provide directions for more rigorous and transparent case studies in the field of technology and innovation management. Our study follows an approach that adopts a variance ontology with a process epistemology [46] in trying to describe innovation processes in terms of stages, phases, or cycles identifying causal relationships. Following the suggestion to select multiple cases or an exemplary single case theoretically proposing a strong justification, we selected the exemplary case of a small R&D organization, Polo GGB (Pole of Innovation in Genomics, Genetic and Biology), because it represents a "revelatory case" [47] of rapid reaction to the challenges posed by the COVID-19 outbreak. Moreover, it is set in Italy, which was the first country (after China) where the virus spread dramatically.

The rapidity Polo GGB shows in quickly adapting to the challenges posed allowed us to carry out an in-depth analysis of a complex phenomenon, analyzing it from different perspectives and precisely while things were happening. This is particularly important to increase internal validity of results and to avoid the risk that informants do not remember relevant events precisely [48]. Actually, this approach aims to understand the intricacies and complexity of the matter object of study by allowing researchers to unveil meaningful features of real-life events [49]. In addition, this responds to the recent push for research designs that are more focused on specific empirical phenomena based on a rich and contextualized qualitative approach [50]. Providing such a detailed analysis is time-consuming and implies a significant involvement of respondents, which was not simple at all considered the contingent situation related to the COVID-19 pandemic and pressure on organizations. This contributes to our opting for a single exemplary case where the object of analysis is likely to be detectable, as recommended by Pettigrew [51]. Indeed, when the COVID-19 pandemic exploded, the chosen R&D organization was strongly motivated and committed to launch new products and services in the market as soon as possible, which let us observe what it did and what allowed it to do so. Therefore, the purpose is to develop new understanding and knowledge in a context of novel arrangements (i.e., the COVID-19 pandemic), exploring an under-investigated topic (i.e., organizational speed) from a pragmatic point of view and in line with what the literature suggests is feasible and appropriate for a single case study research (e.g., [52]).

The unit of analysis of this study is the organization, but we enlarged the scope of investigation with the intent to capture diverse elements that enabled organizational speed (i.e., at the individual, organizational, and partner levels). Methodologically, this was done by triangulating the data at distinct levels in the organization, as explicitly suggested by Ritala et al. [50], to cope with level multiplicity in innovation management research.

For this purpose, both authors carried out eight interviews with the CEO and two scientists from Polo GGB following the research protocol (available upon request), which were recorded and transcribed to increase reliability [53]. After a preliminary interview, the CEO was selected as the main key informant due to her deep knowledge and involvement in all ongoing initiatives in the organization regarding both strategic and operational aspects, as frequently happens in small organizations. Furthermore, we interviewed two researchers with a senior position and responsibility for projects who were selected in different areas of research of Polo GGB to get complementary feedback from employees at a different level in the organization, as anticipated. Using multiple sources of evidence and having different researchers involved in data collection are techniques that contribute to constructing validity and reliability, respectively. Data were collected on different occasions, approximately one per week, to have the possibility, on the one hand, to follow the course of events as they happened and, on the other hand, to understand well the relationship between events and connection with the theory. This allowed us to go into deeper details of the different facts interview by interview and sharpen interview protocols according to the theoretical background. During these interviews, we collected information on several aspects: the organization activities and structures, the new projects launched within the COVID-19 domain, the recognition of opportunities, the decision-making process, the execution of projects, and the factors that enabled the organization to be so rapid in taking advantage of opportunities offered after the COVID-19 outbreak. We also checked secondary sources such as reports and YouTube videos to increase the reliability of the data.

Considered the fast-paced nature of the study, immediately after each interview, one of the researchers wrote extensive minutes that carefully documented the questions asked, the responses given, and other relevant observations. Each researcher analyzed data collected independently, inferring some understandings that were then discussed to converge towards a shared interpretation.

To ensure the quality of data, we triangulated information derived from primary data with secondary sources, as archival analysis. To further increase construct validity, we asked the key informants to review the case study draft.

Data analysis involved identifying categories or data themes, as well as sub-categories, naming them, and finding relationships among these categories and data themes, as recommended by the literature (e.g., [47,49]). This step reflects the standard coding process, which, in this study, was partially "theory-driven" [54] since most categories were not unknown to the literature. The technique was to code the transcript of interviews, grouping together quotes by means of a thematic criterion, so that they pointed to the same construct and had a connection with organizational speed. Figure 1 exhibits an example of the coding process.

We made extensive use of tables to organize and present data [55,56], thus making evidence emerge from the analysis, which was the result of a reiterated process of examining the data, going over the theory, and going back to the data. The comparison of evidence with the literature during data analysis contributes to increasing external validity [53].

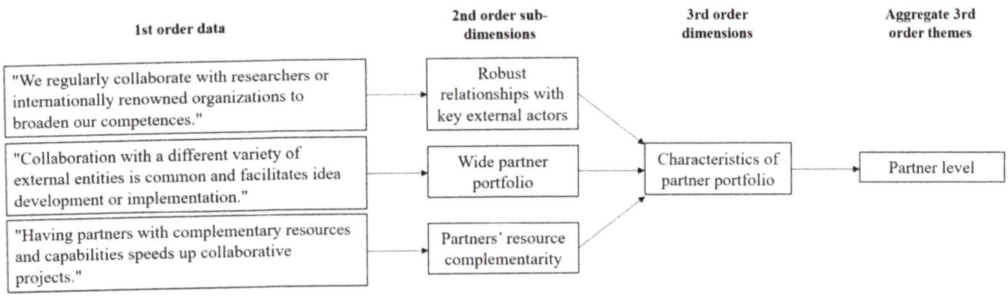

Figure 1. Example of coding—main dimensions and themes that emerged during the interviews.

5. Results

5.1. Case Description

Polo GGB is a small R&D organization and a service facility employing 19 people and founded in 2011 mainly by private companies, with the minor participation of one university. This center of excellence provides operational capacity for innovation and research infrastructure for projects involving genetics, genomics, biology, microbiology, human diagnostics, agro-food, and environment. In particular, it provides researchers with leading-edge facilities holding some of the most advanced instruments in three different laboratories: Genomics & Bioinformatic Laboratory, Ecology and Genetics Research Center, and the Immunology Laboratory. The Genomic and Computational Facility of Polo GGB, located in a bio-incubator inside the Tuscany Life Science and Pharma Valley district in central Italy, has two main objectives: driving innovation in genetics and genomics and promoting technology transfer with industry, with a strong competence in sequencing applications in all fields of biology. The Ecology & Genetics Research Centre is involved in state-of-the-art research projects aimed at developing genetic measures for the control of vector-borne diseases (malaria). The Immunology Laboratory is specialized in the production of monoclonal antibodies. These two laboratories are located in Terni within the Medicine Campus of the University of Perugia, still in the center of Italy but in an adjoining region.

The pool of competences this R&D organization possesses makes this case particularly relevant; indeed, the CEO explains: "Genomics, molecular biology, and immunodiagnostics serve the purpose of COVID-19-related projects; they are a sort of tools to develop projects, and we have all of them inside our organization". This feature is not common in such small organizations, which, combined with the strength of speed, repeatedly stressed by interviewees ("Our strength is speed of action and reaction to changes"), makes this case exemplary to inform academics and practitioners regarding best practices to be prepared to face disruptive changes in the environment, such as the COVID-19 pandemic. Finally, yet importantly, Polo GGB, being involved in a number of projects that have a high scientific value, has been able to combine speed with quality of projects, overcoming the common tension between speed or efficiency and quality or accuracy (e.g., [57]), which further contributes to making it a case of excellence.

5.2. How to Take Advantage of Opportunities Deriving from the COVID-19 Outbreak Rapidly

In this paragraph, we describe how Polo GGB was able to react quickly to the COVID-19 outbreak, showing a high level of organizational speed, in all of its three components, namely recognition, decision, and execution.

5.2.1. Recognition Speed

The COVID-19 pandemic has posed many challenges in different fields, such as economics, improvement of healthcare systems, safety, and security. In the healthcare domain, the widely recognized opportunities have revolved around the development of a drug against COVID-19 and of swabs and serologic tests to detect the presence of the virus and antibodies, respectively.

In this scenario, the CEO of Polo GGB, who holds the responsibility for launching and managing projects for the organization, immediately senses the opportunity to use the competences of the Polo along two main lines: "*We are able to help and we are able to do business in this situation*", she asserts.

Different idea generation sessions, organized very rapidly with the purpose of understanding what they can do practically, are organized. The first outcome is the recognition of the type of competences they have and how they could be used fruitfully, as the CEO outlines: "*We have the competences in molecular biology, and we have an authorization to do diagnostics; let's exploit them and work immediately to validate the test to search for COVID-19 antibodies in real time*". In other words, the small R&D laboratory does not put routinized procedures in place for scanning the external environment, but mainly decides to analyze the scientific and multidisciplinary internal competences they possess to promptly recognize the opportunities they could pursue. This allows them to identify potentially interesting ideas even though they are not in their core business, as the CEO highlighted: "*We have never realized a real-time test with a diagnostic purpose, but we can do it*". This means they are able to re-think their way of doing business by combining the emergent opportunities with alternative use of their competence with respect to their standard way of doing business.

Furthermore, the role of an outstanding Scientific Director of the Polo, based at the University of Padova, must also be stressed: as the CEO outlines "*This new idea is an epiphany of our Scientific Director*", which means that the intuition of scientists can make the difference in recognizing an opportunity in a timely manner.

In parallel with internal actions to identify opportunities, external ideas are also evaluated: "*We always welcome ideas coming from outside our company and never discard them a priori*". On such a basis, despite the intense activity within the company, the CEO and staff accept discussion of proposals from other firms or institutions without delaying them as explicitly declared: "*A firm contacted us thanks to a collaboration I had years ago with a business angel that is now part of the advisory board of that firm, and I started listening and discussing with them straightaway*".

To summarize, the recognition phase has proceeded internally and externally without excluding any idea and trying to catch the necessary information as soon as possible to check the feasibility and interest for Polo GGB.

5.2.2. Decision Speed

Thanks to the degrees of freedom the CEO has in terms of decisions on which projects to launch, she quickly shares the ideas and opportunities with the Scientific Director of Polo GGB to verify their scientific value and feasibility: "*Polo GGB has a very quick decision process and the possibility to set aside funds for research: it can use them without the intervention of the Board if the investments are below 40,000 euros, as in the case of COVID-19-related projects*", the CEO explains. "*Moreover*", she adds, "*I have daily contact with the Scientific Director to check the feasibility and quality of the projects*". Overall, among the different opportunities that emerged, four new projects are launched in a very short time lapse, as the CEO confirms: "*We, the Scientific Director and I, took the decision to proceed with these projects in a few hours*". She also explains that this is not the standard way of progressing: "*Normally, we make a business plan […], but in this case we had to be fast in deciding to start with the project, so we gave priority to reaching the aim instead of doing the best financial choices; for example, we considered technological feasibility and quality of the output first instead of preparing a detailed budget*". This means that preliminary activities involving planning and budgeting, which normally take place before deciding whether to launch a project or not, are given less importance in favor of more operative tasks associated with whether the project could be carried out in practice.

The decision falls on four main projects, among which two are carried out within the laboratories of Polo GGB, and the other two in collaboration with a set of external partners.

The first internal project has an explorative nature, ambitiously aiming to produce a monoclonal antibody (i.e., a potential drug) against SARS-CoV-2; furthermore, they realize they could use these immunology-related competences to identify SARS-CoV-2 to also develop a serologic test (i.e., direct test to detect whether a person has developed antibodies against SARS-CoV-2). For the second internal project, which is more exploitative, they could use internal competences to execute the real-time analysis of a swab to detect infection by SARS-CoV-2, to be offered as a new service to clients.

As far as project ideas coming from external partners, the first project concerns explorative research to develop a platform for contagion risk management based on an idea developed by the research center MATEMA and the University of Padova. Indeed, the core idea is to identify clusters of employees that are more or less susceptible to SARS-CoV-2 based on social, work, and environmental conditions, determining a risk score that can improve decision-making on COVID-19 test execution. This project can contribute to active surveillance of COVID-19 spread and, thus, safety in workplaces. Polo GGB is contacted for its strong ability to analyze swabs and manage the logistics of swabs, considering the relevance of a timely diagnosis of the presence of SARS-CoV-2. For this purpose, on-site campers in remote zones are also set up thanks to the collaboration of a large Italian company.

The second proposal, whose idea comes from the National Centre of Research (CNR) of Italy, has the purpose of validating a method to test for the presence of SARS-CoV-2 in PM10 particles from filters made available by regional agencies for environmental safeguarding. The CEO understands that Polo GGB is called to bridge the lack of instruments required to analyze filters safely, as well as of the possibility to move them safely. The last explorative project departs from an idea of the CEO, who contacted another laboratory in the north of Italy with the aim of defining an algorithm meant to identify conditions increasing the probability of getting infected by COVID-19, which could reveal useful insights for the other ongoing projects.

5.2.3. Execution Speed

A few days after the decision to proceed, Polo GGB starts being operative on the four new projects, which are set as priorities. Concretizing these ideas into ongoing projects passes through the rapid re-allocation of different resources to the different new and existing projects, which makes the need for new resource acquisition emerge. As the CEO outlines, "*We immediately realized we needed more technicians due to the increase in workload in the laboratories after the COVID-19-related projects, so we hired four new employees straightaway, which, for an organization such as Polo GBB, is a lot*". Normally, new employees go through a training period when they are involved in some of the activities of ongoing projects; instead, in this case, "*We needed new technicians to be operative immediately; therefore, we integrated them into all project activities, making senior technicians and scientists take responsibility for them*", as the CEO and scientists explained. Indeed, in this case and contrarily from what normally happens at Polo GGB, mid-level project leaders are defined and, in general, senior technicians and scientists are given more autonomy and responsibility regarding the new projects. Obviously, this does not exclude the supervision of the CEO: "*I have multiple daily meetings, even extemporaneous, with staff to check work in progress, identify problems, and decide how to solve them, and frequent interactions with external partners*".

Interviewees admitted they have to face some difficulties in organizing the work and some resistance against some activities, especially at the beginning, but then results are surprising. As the CEO recognizes, "*Our team was able to quickly switch from the routinized activities that were mainly related to the genomic domain to molecular biology and diagnostics with a willingness to perform these activities even unexpectedly [...]; they want to be active and help in this emergency situation, and the result is that they react in a proactive manner*".

One last relevant point raised during the interviews is the IT infrastructure: "*We put resources and efforts into creating an appropriate IT infrastructure to support the new activities that need to be done rapidly. Normally, we didn't need it, or at least we were used to a less developed infrastructure, but now that we have it, it is another opportunity*". The high quantity of tests to be processed requires strengthening the IT infrastructure to proceed more rapidly.

What is interesting for this study, beyond the scientific value of the projects that may have a significant impact on society, is the richness and variety of initiatives that a small R&D organization has been able to launch in such a short time span since the COVID-19 pandemic. Indeed, Polo GGB has gone through the recognition–decision–execution process in less than three months. Using the words of the CEO: "*Catching and developing new ideas and implementing them quickly is key for our organization; I am active in doing it*".

5.3. Antecedents of Organizational Speed

The data analysis shows that antecedents of organizational speed can be found at three levels, namely the individual, organizational, and partner levels, as described in the following paragraphs.

5.3.1. Antecedents at the Individual Level

As far as antecedents at the individual level are concerned, they refer to characteristics of the CEO and of the employees. The CEO clearly explains that human capital in an R&D organization is at the basis of the ability to react quickly, because if scientists are highly educated, they have a more flexible mindset, which is required to face challenges and solve problems. As the CEO outlines, a company like Polo GGB "*needs scientists and technicians with high skills to take advantage of challenging opportunities from the market*", and she further stresses, "*this profile (where 80% of the staff have a PhD) is necessary to have flexible resources*". People are encouraged "*to learn from others in several ways to enlarge their competences and be more flexible*".

On the other hand, scientists emphasize the fundamental role of the CEO in terms of proactivity and ability to be a leader while also being part of the team: "*(She) is involved in all issues and involves us researchers in the different projects*", explains one of the researchers. Moreover, the CEO describes herself as a person who "*welcomes challenges; it is in my nature*": her strong attitude towards taking risks emerges clearly from the passion in describing the potential relevance, although uncertain, of the new projects. In addition, she explained that she was rapidly able to recognize and evaluate new challenges both at the technical and managerial levels, which reflects her dual competence.

5.3.2. Antecedents at the Organizational Level

At the organizational level, the flat organizational **structure** with a low degree of hierarchy further allows the CEO to make decisions autonomously: "*There are no other middle managers between top management and other employees, which speeds up all processes*". The organizational context, described by the interviewees as characterized by employees "*who normally carry out explorative and exploitative projects*" and where "*people are used to it and it helps switching from one to another easily*", seems to be a fruitful environment to catch the challenges posed by the COVID-19 emergency. Put in other words, the ability to start both explorative and exploitative projects rapidly is obtained through a contextual form of ambidexterity where scientists are used to easily moving between different tasks with a different nature within the same laboratory.

A bundle of broad financial, technological, and knowledge **resources** complements the flexibility allowed by the scientists and the structure, because it facilitates the number and variety of new projects Polo GGB is able to implement. As the employees declare, "*We have a wide variety of technologies available in house, which, together with a wide spectrum of competences, allows us to start new projects easily and rapidly*". Moreover, the CEO adds, "*financial resources are not a problem for us, and we have visibility on resource availability in the medium term, which poses no problems in carrying out new projects*".

The **culture** plays a fundamental role in this R&D organization, and the CEO emphasizes this aspect a lot during the interviews: *"We have put a lot of effort into creating a culture of continuous change to be ready to learn new things and adapt to the changing environment rapidly"*. Indeed, shared values seem to be at the core of their mission and permeate all levels of the organization. The CEO has worked on fostering and stimulating a culture of mutual learning and inclination towards change that looks at the success of the organization instead of at the success of individuals. This culture is reflected by the rewards system, since the company *"does not reward individuals, but the team, which is pushed towards accomplishing the team goal"*. Furthermore, scientists and employees have always been used to working with external partners, because *"for a small R&D organization such as Polo GGB, being open to collaboration is essential to being more ready and reactive; of course you are able to collaborate if you are an excellence in your field, so our employees are always pushed to maintain a high reputation"*, the CEO underlines.

Finally, internal **operative systems and processes** reflect the climate where everybody is motivated to work and interact frequently with colleagues sharing different backgrounds, and communication is facilitated not only horizontally but also vertically. The speediness is also guaranteed by the fact that employees *"frequently check the work in progress through meetings, so that if there is a problem, it emerges immediately"*; in addition, *"communication is frequent and open; we use different channels to be always updated, almost in real time"*, highlights the CEO. This open line of communication facilitates the monitoring of projects, problem solving, and control of the scientists' work, while maintaining a climate of collaboration. Generally, the employees *"are always open to accepting, evaluating, and discussing solutions proposed by third parties to find the most effective ones"*. This is backed by the fact that scientists and technicians working in teams have different competences, as the CEO explains: *"People work in teams, and in each team there are different competences; problem solving is more rapid in this way"*.

5.3.3. Antecedents at the Partner Level

The same open culture has made Polo GGB establish a wide portfolio of relationships with external partners, most of which are based on a complementarity principle. This factor further increases the potential of the organization to recognize new opportunities and implement projects more rapidly due to their previous experience of collaboration and lack of competition. In the words of the CEO, *"collaborations with a wide variety of external entities are common and facilitate idea development or implementation"*; complementarity among partners seems to play a positive role, as the CEO confirms: *"Having partners with complementary resources and capabilities speeds up collaborative projects"*.

Figure 2 presents an overarching picture of how organizational speed is obtained and the antecedents of organizational speed, which is the main result of this study.

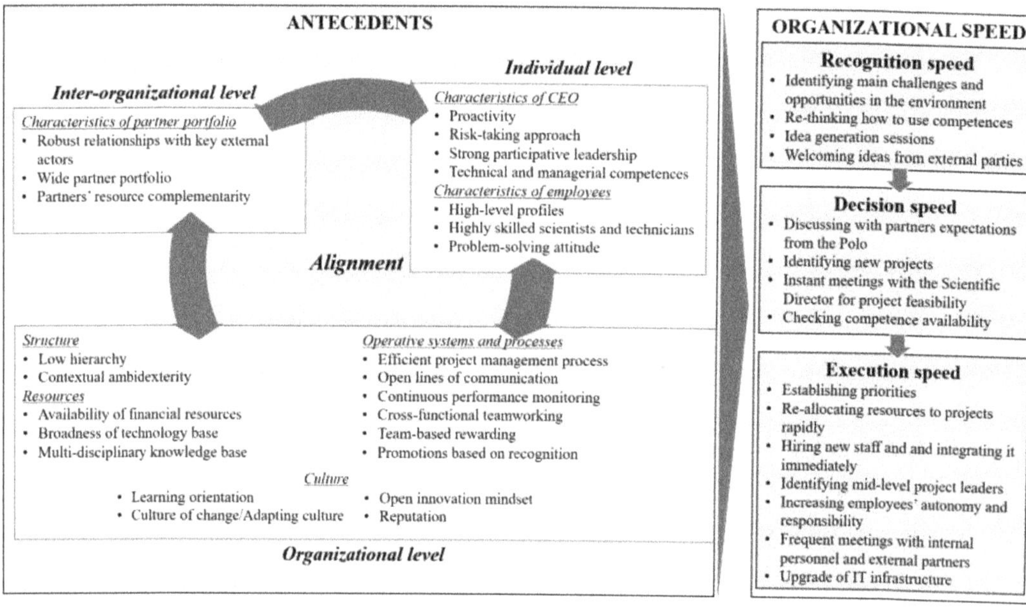

Figure 2. Overview of the antecedents of organizational speed.

6. Discussion and Conclusions

6.1. Discussion of Findings

The purpose of this study was to understand how an R&D organization can rapidly take advantage of the opportunities offered by the COVID-19 outbreak and to unveil the factors that enable the organizational speed of an R&D organization in the face of the COVID-19 outbreak.

Our case shows that Polo GGB is able to revise its strategy, integrating both explorative and exploitative projects, balancing the exploitation of the current capabilities with developing new ones.

In the face of the dramatic external circumstances, the R&D organization displays a great ability to adapt its strategic direction to seize the newly emerged opportunities fully; the whole organization is able to respond by adding a new set of projects to the current portfolio, thus keeping pace with the challenges posed by the external environment. Organizational speed is achieved through a quick process that involves fast awareness and recognition, decision, and action. Overall, it seems that the internal scientific competences, jointly with the outstanding profile of the Scientific Director of the Polo, are at the basis of the first two steps. Firstly, Polo GGB recognizes they have a strong and differentiated pool of competences they can use for emerging opportunities, identified through idea generation sessions, despite not being in line with their core business. Secondly, in the decision phase, the attention is focused more on whether the Polo has all the competences required for the project and whether the project is feasible from a technological point of view, rather than on detailed planning and budgeting. During the execution phase, these new projects become operative very rapidly, revealing the strong potential of the small R&D organization that is the object of study. Indeed, through a rapid reallocation of resources and the hiring of new employees who become operative straightaway thanks to the supervision of senior technicians and researchers who are granted enhanced autonomy and responsibility, Polo GGB has reached the first results in a record time.

Results show there is a set of factors that enable the CEO to successfully recognize and act promptly when it is time to boost changes and to quickly commit the resources to new courses of actions in response to such changes. According to the few previous contributions on dynamic capabilities in R&D organizations, these factors are spread out over three different levels, namely the partner, organizational, and individual levels. For organizational speed to be increased, the combination of the specific identified elements at the three levels enables a quick recognition, decision, and execution process: what we add to the previous debate is an overarching and integrated picture of these antecedents that jointly allow these small R&D organizations to react promptly.

At the partner level, the open innovation attitude combined with the company's reputation, also boosted by the fact of having two laboratories located in an incubator and another one in a university, enables the proliferation of interesting external opportunities and partnerships and facilitates the identification and subsequent execution of some of the new projects. Furthermore, the partnerships already in place facilitate the scanning and sensing of new opportunities and projects as well as their subsequent execution. At the organizational level, the flat organization with a low hierarchy favors a climate of effective open communication and collaboration throughout the organization, where silos are broken up; as a result, employees are used to working in a team, where their multidisciplinary competences are enhanced and a process of mutual learning and cooperation is encouraged. The incentive system and the values of the organization promote cooperation instead of competition, nurtured by the fact that the evaluation is made on the team instead of on the individuals. At the individual level, employees are fully embodied in the culture and systems of the organization, handling exploitative and exploratory activities simultaneously. Furthermore, the strong knowledge in several different scientific fields and the commitment devoted to its development allow the company to be flexible in accepting new projects. In other words, commitment and flexibility coexist and contribute to organizational speed; this flexibility is rooted in the deep knowledge and experience of the organization in the three main scientific fields, representing the core competences, developed over time, which permits the generation of options for future explorative and exploitative projects.

Overall, the organization's attitude towards teamwork, the multidisciplinary collaboration among departments, the availability of financial resources, the partnership ecosystem, and the creativity and skills of the employees combined with their strong attitude in problem solving are all fundamental ingredients of the formula that leads the company to show this high level of organizational speed. These factors, embracing the partner, organizational, and individual levels, are inter-related, aligned, and mutually reinforcing. The principles and values of the organization permeate the systems, the practices, and the behavior of employees, fully nurtured by the strong participative leadership style of the CEO who is able to mentor and at the same time coach, orchestrate, and bring together diverse employees to create an effective team [57–59]. Furthermore, the existing contextual ambidextrous structure makes it possible to realize the strategic goals of the company on several fronts, also thanks to the highly skilled personnel of the laboratories. It seems that the organization, at a time just as the new projects are introduced, is well-aligned along its central building blocks (structure, strategy, system, share value, style, skills, staff), often mentioned by the literature [60] as fundamental for firm competitive advantage. Indeed, maintaining the alignment along these elements keeps on being a key priority for the CEO: the implementation of new projects is consistent with the organizational structure of the laboratories and the systems in place; the new staff share the systems and values of the organization; cross-functional teams are maintained as much as possible, thanks to the support of virtual technologies. This allows the organization and the employees to reconfigure their activities to embrace these new opportunities in their daily work and strive to work together to achieve clear and consistent goals, while avoiding tensions and conflicts that could have emerged. While previous contributions on DC in R&D organizations have mainly investigated some specific factors, this study emphasizes the importance of having a set of different factors at different levels aligned and working in a synergic way to reach high levels of organizational speed.

6.2. Theoretical and Practical Implications and Limitations

This study offers some interesting theoretical and practical implications.

From an academic standpoint, the present article extends the literature on organizational speed and dynamic capabilities from the context of manufacturing firms to the much less investigated context of R&D organizations. Furthermore, it offers an overarching picture of how a combination of different and aligned elements allows an R&D organization to face the challenges posed by the COVID-19 outbreak promptly. Other studies can further enrich this literature by deeply investigating the interplay among the different enablers and the role of the CEO, as well as by shedding light on the process that leads to organizational speed, maybe enlarging the studies to other R&D organizations of different sizes and industries. Considering the increasingly mutable social, economic, and environmental scenario, R&D theorists are likely to be particularly interested in a better understanding of how R&D organizations or departments can face disruptive changes efficiently, as well as how they can prepare to be ready to do so. This study, aiming to elaborate on organizational speed, goes in that direction by providing an overarching picture on how to unfold the recognition–decision–execution process rapidly and identifying antecedents at the individual, organizational, and partner levels, which represented a gap in the literature. In addition, it contributes to the recent stream connecting the triple bottom-line dimensions of environmental, social, and economic sustainability with the COVID-19 pandemic (e.g., [3,61]), with particular attention on economic sustainability.

From a managerial viewpoint, the case shows that the antecedents of organizational speed are located at different levels of the organization, in the mind and knowledge of employees, in the values, systems, and processes in which they are embodied, and in the ecosystem of external relationships. In this context, CEOs must play the role of orchestrators who keep the communication channels open, interacting consistently, frequently, and effectively with employees and other stakeholders about strategic goals, priorities, and values. Due to the small size of the organization, the CEO becomes the point of reference for all employees and can easily expand the decision portfolio, adding variations to the initiatives already in place [62]. Managers of small R&D organizations should interestingly notice that an upgrade of the IT infrastructure is required to speed up processes, as widely recommended by the literature (e.g., [63]), which suggests investments in this area are useful even in those contexts where a deep use of IT is not widespread. At the same time, investments in improving the infrastructure, or in more general processes, may offer further opportunities to change the standard way of doing things, leading to processes that are more efficient. This example suggests that the first move towards change may be the most difficult one, but then it can open new scenarios for development in small contexts.

With particular reference to the global practicalities of curtailing the COVID-19 pandemic, this study can have an impact by showing how small R&D organizations can react to the disruption this outbreak has brought about and proving the fundamental role they can play in this situation thanks to their ability to react promptly. Small R&D organizations that have the right intuition and a structure able to support innovative ideas are making the difference in such a challenging situation.

Of course, we recognize the limitations related to the analysis of a single case, which prevents us from making our analysis generalizable to different contexts. For this purpose, future qualitative research could contribute to replicating or enriching our proposed framework across multiple settings, and subsequent quantitative studies could statistically test relationships among variables.

Author Contributions: Conceptualization, L.A. and A.N.; methodology, L.A. and A.N.; formal analysis, L.A. and A.N.; writing—original draft preparation, L.A. and A.N.; writing—review and editing, L.A. and A.N.; project administration, L.A. and A.N. All authors have read and agreed to the published version of the manuscript.

Funding: This research received no external funding.

Institutional Review Board Statement: Not applicable.

Informed Consent Statement: Informed consent was obtained from all subjects involved in the study.

Conflicts of Interest: The authors declare no conflict of interest.

References

1. Surico, P.; Galeotti, A. The Economics of a Pandemic: The Case of COVID-19, London Business School. 2020. Available online: https://www.dropbox.com/s/wm521646rszpl90/slides_COVID19_final.pdf?dl=0 (accessed on 1 December 2020).
2. Rai, S.S.; Rai, S.; Singh, N.K. Organizational resilience and social-economic sustainability: COVID-19 perspective. *Environ. Dev. Sustain.* **2021**, *23*, 1–18. [CrossRef] [PubMed]
3. Jiang, J.; Park, E.-M.; Park, S.-T. The Impact of the COVID-19 on Economic Sustainability—A Case Study of Fluctuation in Stock Prices for China and South Korea. *Sustainability* **2021**, *13*, 6642. [CrossRef]
4. Sarkis, J. Supply chain sustainability: Learning from the COVID-19 pandemic. *Int. J. Oper. Prod. Manag.* **2020**, *41*, 63–73. [CrossRef]
5. Birkinshaw, J.; Zimmermann, A.; Raisch, S. How do firms adapt to discontinuous change? Bridging the dynamic capabilities and ambidexterity perspectives. *Calif. Manag. Rev.* **2016**, *58*, 36–58. [CrossRef]
6. Sanchez, R. Strategic flexibility in product competition. *Strat. Manag. J.* **1995**, *16*, 135–159. [CrossRef]
7. Li, Y.; Su, Z.; Liu, Y.; Li, M. Fast adaptation, strategic flexibility and entrepreneurial roles. *Chin. Manag. Stud.* **2011**, *5*, 256–271. [CrossRef]
8. Dykes, B.J.; Hughes-Morgan, M.; Kolev, K.D.; Ferrier, W.J. Organizational speed as a dynamic capability: Toward a holistic perspective. *Strat. Organ.* **2018**, *17*, 266–278. [CrossRef]
9. Horst, S.O.; Järventie-Thesleff, R.; Baumann, S. The practice of shared inquiry: How actors manage for strategy emergence. *J. Media Bus. Stud.* **2019**, *16*, 202–229. [CrossRef]
10. Schwarz, G.M.; Stensaker, I. Researching a Pandemic: Letting COVID-19 Drive Our Research. *J. Appl. Behav. Sci.* **2020**, *56*, 261–265. [CrossRef]
11. Amis, J.M.; Janz, B.D. Leading Change in Response to COVID-19. *J. Appl. Behav. Sci.* **2020**, *56*, 272–278. [CrossRef]
12. Arora, P.; Suri, D. Redefining, relooking, redesigning, and reincorporating HRD in the post Covid 19 context and thereafter. *Hum. Resour. Dev. Int.* **2020**, *23*, 438–451. [CrossRef]
13. Gerald, E.; Obianuju, A.; Chukwunonso, N. Strategic agility and performance of small and medium enterprises in the phase of Covid-19 pandemic. *Int. J. Financ. Account. Manag.* **2020**, *2*, 41–50. [CrossRef]
14. Akhigbe, E.A.; Onuoha, B.C. Strategic Agility and Organizational Resilience of Food and Beverage Firms in Rivers State, Nigeria. *Int. J. Bus. Syst. Econ.* **2019**, *12*, 80–93.
15. Chesbrough, H. To recover faster from Covid-19, open up: Managerial implications from an open innovation perspective. *Ind. Mark. Manag.* **2020**, *88*, 410–413. [CrossRef]
16. Hirt, M.; Laczkowski, K.; Mysore, M. Bubbles Pop, Downturns Stop. McKinsey Quarterly. 2019. Available online: https://www.mckinsey.com/business-functions/strategy-and-corporate-finance/our-insights/bubbles-pop-downturns-stop (accessed on 1 February 2021).
17. Verma, S.; Gustafsson, A. Investigating the emerging COVID-19 research trends in the field of business and management: A bibliometric analysis approach. *J. Bus. Res.* **2020**, *118*, 253–261. [CrossRef] [PubMed]
18. Liu, Y.; Lee, J.M.; Lee, C. The challenges and opportunities of a global health crisis: The management and business implications of COVID-19 from an Asian perspective. *Asian Bus. Manag.* **2020**, *19*, 277–297. [CrossRef]
19. Shani, A.B.; Coghlan, D.; Alexander, B.N. Rediscovering abductive reasoning in organization development and change research. *J. Appl. Behav. Sci.* **2020**, *56*, 60–72. [CrossRef]
20. Yi, Y.; He, X.; Ndofor, H.; Wei, Z. Dynamic Capabilities and the Speed of Strategic Change: Evidence from China. *IEEE Trans. Eng. Manag.* **2014**, *62*, 18–28. [CrossRef]
21. Kraatz, M.S.; Zajac, E.J. How organizational resources affect strategic change and performance in turbulent environments: Theory and evidence. *Organ. Sci.* **2001**, *12*, 632–657. [CrossRef]
22. Abdelilah, B.; El Korchi, A.; Balambo, M.A. Flexibility and agility: Evolution and relationship. *J. Manuf. Technol. Manag.* **2018**, *29*, 1138–1162. [CrossRef]
23. Nafei, W.A. Organizational agility: The key to organizational success. *Int. J. Bus. Manag.* **2016**, *11*, 296–309.
24. Singh, J.; Sharma, G.; Hill, J.; Schnackenberg, A. Organizational agility: What it is, what it is not, and why it matters. In *Academy of Management Proceedings*; Academy of Management: Briarcliff Manor, NY, USA, 2013; Volume 1, pp. 1–40.
25. Baškarada, S.; Koronios, A. The 5S organizational agility framework: A dynamic capa-bilities perspective. *Int. J. Organ. Anal.* **2018**, *26*, 331–342. [CrossRef]
26. Walter, A.T. Organizational agility: Ill-defined and somewhat confusing? A systematic literature review and conceptualization. *Manag. Rev. Q.* **2020**, 1–49. [CrossRef]
27. Kor, Y.Y.; Mahoney, J.T. How dynamics, management, and governance of resource deployments influence firm-level performance. *Strat. Manag. J.* **2005**, *26*, 489–496. [CrossRef]
28. Biedenbach, T.; Müller, R. Absorptive, innovative and adaptive capabilities and their impact on project and project portfolio performance. *Int. J. Proj. Manag.* **2012**, *30*, 621–635. [CrossRef]

29. Thal, A.E., Jr.; Shahady, D.E. Innovation in a military research laboratory: An initial exploratory study. *Technol. Anal. Strateg. Manag.* **2010**, *22*, 171–187. [CrossRef]
30. Löfsten, H. Organisational capabilities and the long-term survival of new technology-based firms. *Eur. Bus. Rev.* **2016**, *28*, 312–332. [CrossRef]
31. Do Nascimento Welter, C.V.; Sausen, J.O.; Rossetto, C.R. The development of innovative capacity as a strategic resource in technology-based incubation activities. *Rev. Gestão* **2020**, *27*, 169–188. [CrossRef]
32. Deligianni, I.; Voudouris, I.; Spanos, Y.; Lioukas, S. Non-linear effects of technological competence on product innovation in new technology-based firms: Resource orchestration and the role of the entrepreneur's political competence and prior start-up experience. *Technovation* **2019**, *88*, 102076. [CrossRef]
33. Salehi, F.; Zolkiewski, J.; Perks, H.; Bahreini, M.A. Exploration of capability and role development in an emerging technology network. *J. Bus. Ind. Mark.* **2018**, *33*, 931–944. [CrossRef]
34. Deakins, D.; Bensemann, J. Entrepreneurial learning and innovation: Qualitative evidence from agri-business technology-based small firms in New Zealand. *Int. J. Innov. Learn.* **2018**, *23*, 318–338. [CrossRef]
35. Jensen, A.; Clausen, T.H. Origins and emergence of exploration and exploitation capabilities in new technology-based firms. *Technol. Forecast. Soc. Change.* **2017**, *120*, 163–175. [CrossRef]
36. Cho, C.; Son, J.K.; Lee, S.; Park, S. Comparative Analysis of R&D-Based Innovation Capabilities in SMEs to Design Innovation Policy. *Sci. Public Policy* **2016**, *44*, 403–416.
37. Hutterer, P.; Gattringer, R.; Irschik, H.; Strehl, F. Managing Contradictory Stakeholder Demands of a Publicly Funded Research Center. *Int. J. Innov. Technol. Manag.* **2015**, *12*, 1550002. [CrossRef]
38. Ortín-Ángel, P.; Vendrell-Herrero, F. University spin-offs vs. other NTBFs: Total factor productivity differences at outset and evolution. *Technovation* **2014**, *34*, 101–112. [CrossRef]
39. Brinckmann, J.; Hoegl, M. Effects of initial teamwork capability and initial relational capability on the development of new technology-based firms. *Strat. Entrep. J.* **2011**, *5*, 37–57. [CrossRef]
40. Strehle, F.; Katzy, B.R.; Davila, T. Learning capabilities and the growth of technology-based new ventures. *Int. J. Technol. Manag.* **2010**, *52*, 26–45. [CrossRef]
41. Wu, L.-Y.; Wang, C.-J. Transforming resources to improve performance of technology-based firms: A Taiwanese Empirical Study. *J. Eng. Technol. Manag.* **2007**, *24*, 251–261. [CrossRef]
42. Andries, P.; De Backere, K. Adaptation in new technology-based ventures: Insights at the company level. *Int. J. Manag. Rev.* **2006**, *8*, 91–112. [CrossRef]
43. Atuahene-Gima, K.; Li, H.; De Luca, L.M. The contingent value of marketing strategy innovativeness for product development performance in Chinese new technology ventures. *Ind. Mark. Manag.* **2006**, *35*, 359–372. [CrossRef]
44. Garud, R.; Tuertscher, P.; Van de Ven, A.H. Perspectives on innovation processes. *Acad. Manag. Ann.* **2013**, *7*, 775–819. [CrossRef]
45. Elsahn, Z.; Callagher, L.; Husted, K.; Korber, S.; Siedlok, F. Are rigor and transparency enough? Review and future directions for case studies in technology and innovation Management. *R&D Manag.* **2020**, *50*, 309–328.
46. Van de Ven, A.H.; Poole, M.S. Alternative approaches for studying organizational change. *Organ. Stud.* **2005**, *26*, 1377–1404. [CrossRef]
47. Yin, R.K. *Case Study Research: Design and Methods*, 4th ed.; Sage: Thousand Oaks, CA, USA, 2009.
48. Voss, C.; Tsikriktsis, N.; Frohlich, M. Case research in operations management. *Int. J. Oper. Prod. Manag.* **2002**, *22*, 195–219. [CrossRef]
49. Gaya, H.J.; Smith, E.E. Developing a qualitative single case study in the strategic management realm: An appropriate research design. *Int. J. Bus. Manag. Econ. Res.* **2016**, *7*, 529–538.
50. Ritala, P.; Schneider, S.; Michailova, S. Innovation management research methods: Embracing rigor and diversity. *R&D Manag.* **2020**, *50*, 297–308.
51. Pettigrew, A.M. Longitudinal Field Research on Change: Theory and Practice. *Organ. Sci.* **1990**, *1*, 267–292. [CrossRef]
52. Mukhija, V. N of One plus Some: An Alternative Strategy for Conducting Single Case Research. *J. Plan. Educ. Res.* **2010**, *29*, 416–426. [CrossRef]
53. Riege, A.M. Validity and reliability tests in case study research: A literature review with "hands-on" applications for each research phase. *Qual. Mark. Res. Int. J.* **2003**, *6*, 75–86. [CrossRef]
54. Ryan, G.W.; Bernard, H.R. Data management and analysis methods. In *Handbook of Qualitative Research*, 2nd ed.; Denzin, N., Lincoln, Y., Eds.; Sage: Thousand Oaks, CA, USA, 2000; pp. 769–802.
55. Miles, M.B.; Huberman, A.M. *Qualitative Data Analysis: An Expanded Sourcebook*; Sage: Thousand Oaks, CA, USA, 1994.
56. Eisenhardt, K.M.; Graebner, M.E. Theory building from cases: Opportunities and challenges. *Acad. Manag. J.* **2007**, *50*, 25–32. [CrossRef]
57. Johnson, L.A.; Hayashi, H. Synthesis Efforts in Disaster Recovery Research. *Int. J. Mass Emergencies Disasters* **2012**, *30*, 212–238.
58. Keller, R.T. A longitudinal study of the individual characteristics of effective R&D project team leaders. *R&D Manag.* **2017**, *47*, 741–754.
59. Kim, C.C.; Hays, J.M. The Australian National University Renaissance Leaders Global Trends and Emerging Forms of Leadership. *J. Manag.* **2010**, *4*, 1–27.
60. Feurer, R.; Chaharbaghi, K. Strategy development: Past, present and future. *Manag. Decis.* **1995**, *33*, 11–21. [CrossRef]

61. Bartle, J.; Lutte, R.; Leuenberger, D. Sustainability and Air Freight Transportation: Lessons from the Global Pandemic. *Sustainability* **2021**, *13*, 3738. [CrossRef]
62. Jacobs, P. Five steps to thriving in times of uncertainty. *Harv. Manag. Update* **2005**, *10*, 1.
63. Cai, Z.; Liu, H.; Huang, Q.; Liang, L. Developing organizational agility in product innovation: The roles of IT capability, KM capability, and innovative climate. *R&D Manag.* **2019**, *49*, 421–438.

sustainability

MDPI

Article

Impact of COVID-19 on the Educational Process in Saudi Arabia: A Technology–Organization–Environment Framework

Khloud Alshaikh , Shaikhah Maasher, Afnan Bayazed, Farrukh Saleem *, Sahar Badri and Bahjat Fakieh

Information System Department, Faculty of Computing and Information Technology, King Abdulaziz University, Jeddah 21589, Saudi Arabia; khassanalshaikh@stu.kau.edu.sa (K.A.); salimaasher@stu.kau.edu.sa (S.M.); aahmedbayazed@stu.kau.edu.sa (A.B.); skbadri@kau.edu.sa (S.B.); bfakieh@kau.edu.sa (B.F.)
* Correspondence: fsaleem@kau.edu.sa

Abstract: The lockdown of universities and educational institutions during the COVID-19 pandemic has negatively impacted the educational process. Saudi Arabia became a forerunner during COVID-19 by taking initial precautions of curfews and total restrictions. However, these restrictions had a disruptive effect on various sectors, specifically the educational sector. The Ministry of Education strived to cope with the consequences of these changes swiftly by shifting to online education. This paper aims to study the impact of COVID-19 on the educational process through a comparative study of the responses collected from different cases, and the challenges that are faced throughout the educational process. The study conducted a cross-sectional, self-administered online questionnaire during the outbreak and distance learning, which was designed based on the Technology–Organization–Environment (TOE) framework of students. Most questions used a five-point Likert scale. The responses were randomly collected from 150 undergraduate and postgraduate students who were studying in Saudi Arabian universities, to study the overall performance of education institutions during COVID-19. The collected data were analyzed and compared to the results in the literature. The main factors impacted by this transformation are addressed. These factors are based on research and observations and aim to overcome the encountered limitations and to present their level of impact on distance education. The research framework can be useful for higher educational authorities aiming to overcome the issues highlighted and discussed in this study.

Keywords: COVID-19; pandemic; educational process; higher education; online education

Citation: Alshaikh, K.; Maasher, S.; Bayazed, A.; Saleem, F.; Badri, S.; Fakieh, B. Impact of COVID-19 on the Educational Process in Saudi Arabia: A Technology–Organization–Environment Framework. *Sustainability* **2021**, *13*, 7103. https://doi.org/10.3390/su13137103

Academic Editors: Miltiadis D. Lytras, Abdulrahman Housawi and Basim Alsaywid

Received: 22 May 2021
Accepted: 17 June 2021
Published: 24 June 2021

1. Introduction

When COVID-19 spread, authorities and organizations around the world found it difficult to respond quickly to the consequences of the pandemic. Consequently, many emergency changes were implemented worldwide in different domains, including government [1–3], public services [4,5], healthcare [6,7], education [8], transnational labor [9], geopolitical issues [10–12], the global energy market [13], spatial dimensions [14], and the economy [15–17]. These changes imposed serious challenges on the global higher education sector in terms of coping with its high impact on the educational process. In March 2020, at least 100 countries announced national school closures to combat virus transmission. The closures affected roughly half of the global student population [18]. Reviews also cited the negative economic impact of educational institutions' closures on society, which lost (a) a significant number of workers, who were forced to leave work to take care of their children's education; and (b) the privilege of attending classes, especially for students with poor understanding and educational attainment [18]. This issue prompted many universities and educational institutions to conduct significant transformations, review essential elements of their business processes, and adopt technology to continue their operations according to new guidelines and procedures [19]. Given the speed of change necessitated by social distancing, they had to confront challenges that arose with the emergence of the online delivery of teaching content, specifically the rapid digitization of the curriculum [20].

In Portugal, for example, after identifying the first case of COVID-19 on 7 March 2020, the Presidential Council announced the suspension of all activities, including education. Accordingly, face-to-face teaching was suspended due to the closure of schools and universities, and online teaching and learning was promoted by the continuity of the activities through teacher–student interaction via digital tools, such as Blackboard [21,22]. Another precautionary measure was taken by the United Arab Emirates (UAE) to contain the spread of the virus. The authorities in the UAE applied several safety measures by closing universities and schools, suspending any outside entry to the country, canceling public events and gatherings, and moving to online education [20,23,24]. Higher education institutions in the UAE also worked on engaging students in interactive discussions, whether synchronously through web meeting tools or in asynchronous ways via discussion boards and other tools. As a result of this situation, many universities in the UAE adopted new digital delivery methods [20]. The smart university of Hamdan Bin Mohammed has a rich experience in delivering online content, as it is considered to be the first e-University in the UAE [25]. Heriot Watt University in Dubai is working on "Vision", a virtual learning delivery tool, and the universities of Sharjah and the UAE have adopted the Blackboard system [20]. Saudi Arabia is another example that faced the same scenario. After the World Health Organization (WHO) pandemic announcement, the Saudi Arabian government applied a lockdown in most sectors. On 9 March 2020, the Ministry of Education (MOE) in Saudi Arabia immediately started to deliver online education in all government and private educational institutions [26,27].

Given this history, this paper aims to study the impact of COVID-19 on higher education by highlighting the educational processes that were followed during COVID-19, and how they could be improved from a student's perspective. In addition, this research mainly focuses on studying the impact of those rapid changes on student satisfaction in Saudi Arabian universities. It was crucial to identify the factors that play important roles in facilitating and improving the higher education responses to the pandemic. This research would benefit educational institutions located in Saudi Arabia in understanding the current status of learning and education in universities. Furthermore, it is important to understand how to improve the explored status based on the recommendations discussed in this study. The identified factors can also be useful and applicable for other countries, to enhance the performance of their educational system.

The rest of the paper is organized as follows: in the next section, a literature review and related work is presented to give more insights regarding the context of this paper. This is followed by highlighting the methodology used during the study. After that, data analysis and results are discussed side by side. The discussion section highlights major developments and achievements of research objectives. Finally, the paper concludes with a summary of this study and ideas for future research.

2. Literature Review

The lockdown of universities and educational institutions around the world during the COVID-19 pandemic showed a rapid transformation from face-to-face learning to online learning, which comes in two forms of interactions: synchronous and asynchronous [28]. Synchronous education encourages human interaction between an instructor and a student to assist the student's engagement and interactivity. Asynchronous education relies on technologies that do not require real-time interaction between instructors and participants, such as email, recorded video, discussion forums, etc. Most online education provides an asynchronous learning environment that encloses the flexible practical educational process [28].

In this section, previous work was reviewed to understand the degree of the educational process change in universities and other educational institutions globally, regionally, and locally. It aims to find out how the COVID-19 pandemic has changed the educational process and to what extent this emergency situation has created a considerable impact on the performance of academic staff and students. The subsequent sections describe the educational situation after COVID-19 in different regions.

2.1. Education during COVID-19 in Europe

In Portugal, telework was implemented in terms of teaching and administration. Later on, it was announced that the education process would be conducted remotely for the second semester [21,29]. The second semester extended to 27 June 2020, to ensure the completion of curriculum activities more flexibly. The working scenario was applied as follows: first, teachers started to coordinate with program supervisors and educational councils to adjust teaching and learning strategies considering the distance learning context, with educational resources available to students at different times to work on activities and assignments. Second, teachers provided students with regular feedback on their educational progress by recording all activities in a summary form for evaluation and adding related notes. Third, teachers should also have shared the changes made in the teaching process with students and agreements on the assessment methodologies that were to be adopted. Last, in order to support teachers in this transformation, the Center for Innovation and Teaching and Learning Development developed a number of procedures for posting advice and suggestions throughout the blog and sharing ideas through webinars [21,29]. This training was focused on how to teach online, how to manage online activities, etc. Despite this, there were limitations to investigate, such as the way faculty members use the platforms and develop teaching methods to meet the program's institutional requirements [21]. This means, despite the effectiveness of the platforms in the continuity of the education process, there is no potential to ascertain the methods of faculty members and their development in using the platforms to meet the institutional requirements of the programs.

Similarly, COVID-19 created visible impact over educational institutes located in the United Kingdom (UK). A research study conducted a survey by taking opinions from the final-year medical students in the UK. The purpose was to illustrate the impact of COVID-19 on final examination and placement processes. The research managed to collect responses from 33 medical institutes across the UK. The collected results identified that 38.4% of participants' examinations were cancelled due to the pandemic situation [30]. Moreover, the digital higher education status in Germany was analyzed in research to investigate the impact of COVID-19 on learning processes. The research discussed the positive side, as due to this pandemic situation, the success of the digital education and innovation process widely increased [31]. Another significant issue is outdoor education for university students (or for younger people), which has been presented in different ways related to transitional justice [32,33].

One more study was conducted in Spain to analyze the effect of the COVID-19 quarantine on the learning performance of students in higher education [34]. A field experiment was applied to 458 students from three subjects at Universidad Autónoma de Madrid (Spain). The researcher studied the differences in assessments by splitting the students into two groups. The first group was of the academic years 2017/2018 and 2018/2019 while the second group was of the academic year 2019/2020, which is the group that was disrupted by the pandemic and shifted to online education. The results showed a positive effect of COVID-19 quarantine on the student's performance and the continuity of the activities that did not change after the confinement. It even led to an increase in activity assessments and subjects. Additionally, the results showed that the first group did not study continuously, while the second one did. This means that COVID-19 confinement changed the study strategy to a continuous habit that improves student efficiency in the studies. Based on that, better student assessment scores are expected due to COVID-19 confinement, which would lead to the improvement in their educational performance [34]. The research highlighted that despite the global negative impact of COVID-19, it may have had a positive effect on students (with additional autonomous activities), which leads to spending more time understanding and mastering all lessons and activities.

2.2. Education during COVID-19 in Middle East

In the Middle East, the situation is different, where many developing countries in the region, such as Egypt, have limited access to formal learning management systems (LMS) for distance learning and academic communication. [35]. This situation forced many institutions to look for alternatives in free communication software, e.g., Zoom, Microsoft Teams, and Google Classroom, or through social media platforms, e.g., WhatsApp, Facebook, and YouTube. A study was conducted to focus on investigating the extent to which faculty members and students rely on informal communication platforms and social media to sustain academic communication in developing countries [35]. The study conducted surveys and in-depth interviews to gather opinions from faculty members and students from nine educational institutes. The results showed that the proper use of such platforms could promote a new era of social e-learning and social media can be used effectively to create a positive learning experience.

Further research was conducted in Jordan to examine distance e-learning amongst medical students during the COVID-19 pandemic [36]. The study highlights possible challenges and limitations in medical education to deliver educational material and clinical training using synchronous live streaming sessions supported by advanced communication technologies. The research showed that advanced technologies and social media platforms represent a new approach to the teaching methods and could provide the ideal solution to preserve the educational process in exceptional and emergency situations such as the COVID-19 pandemic. However, presenting educational materials sessions using synchronous live streaming technologies could represent the biggest challenge for students, as nearly 69% of students in the research stated that this was the main obstacle for them due to the internet quality and coverage [36]. To conclude this section, Table 1 represents the summary of some previous studies that covered the educational process during the COVID-19 situation.

Table 1. Summary of the related literature outside of Saudi Arabia.

Methodology	Theory	Purpose	Country	Year	Ref.
Analysis of the initiatives and responses from the authors' university	Reflective Practice Theory	To provide the Portuguese context of the initiatives and responses to COVID-19 on education.	Portugal	July 2020	[21]
Linear test—computer adaptive test (CAT)	Item Response Theory	To identify the effect of COVID-19 confinement on students' performance.	Spain	October 2020	[34]
Surveys and in-depth interviews among faculty members and students	Mixed-methods sequential explanatory design—theory of practice	To investigate the usage of informal communication platforms and social media to sustain academic communication in the developing countries.	Egypt	August 2020	[35]
Students questionnaire	N/A	To study the challenges and limitations to deliver medical material and clinical training using synchronous live streaming sessions	Jordan	October 2020	[36]

2.3. Education during COVID-19 in Saudi Arabia

To keep the education process uninterrupted in Saudi Arabia, the MOE developed a temporary emergency policy to control and to facilitate remote education for both schools and university education. For public education, the MOE fully supervised the educational process by employing online education platforms for supporting virtual classrooms and enriched digital materials, such as (Vschool.sa), which is the formal online school education platform in Saudi Arabia [37]. Additionally, all the lessons for all the student's levels were asynchronously available on the Ain channel run on Arabsat and YouTube [37]. On the other side, higher education institutions and universities were given further freedom for managing educational programs. Nevertheless, the MOE established several strategies and procedures for safeguarding student's learning, their future opportunities, and their cumulative average [27]. Currently, all Saudi universities utilize e-learning platforms by one of the most commonly used systems, which is Blackboard. This practice of education made the experience more flexible by utilizing reliable communication tools. Subsequently, universities had overcome several obstacles regarding the communication between instructors and students, and empowered all stakeholders to reach each other [38,39]. A review on some published material related to academic education and the learning process investigated in Saudi Arabia (during the COVID-19 period) is illustrated in Table 2, and further explained in this section.

Table 2. Summary of related work in Saudi Arabia.

Methodology	Theory	Purpose	Country	Year	Ref.
Questionnaire for the academic staff	A Model of Information and Communication Technology	To assess academic staff satisfaction with distance learning and its affect on the administrative work.	King Khalid University	July 2020	[37]
Surveys among students and faculty members	N/A	To analyze the impact of the COVID-19 pandemic on online education at the College of Medicine.	Al Faisal University	July 2020	[40]
Focus group	SWOT Model	To analyze the processes management followed by Health Sciences Colleges for delivering online education.	King Saud Bin Abdulaziz University	September 2020	[26]
Focus group and interviews	N/A	A study on the effectiveness of simultaneous online learning on medical students.	Qassim University	August 2020	[28]
Survey among students and faculty members	Formulated hypothesis method	To examine the effectiveness of the fourth industrial revolution technologies that was the emergence of digital educational transformation, in relieving the impact of COVID-19 on higher education.	Many Universities	June 2020	[41]

In the College of Medicine and Medical Sciences in Qassim University, a group of researchers conducted a quantitative study on the effectiveness of live online streaming learning sessions [28]. The result showed high level of acceptance and approval from medical students. All participants agreed that online sessions save time, and thus their performance has improved due to extra time they have for studies. However, students indicated some challenges, including methodological challenges, content perception, technical and behavioral challenges during the online sessions, and exams. The participating students emphasized the rigorous and regular evaluation of the principles of the online learning model and learning outcomes to monitor its effectiveness [28].

Another study was conducted in Pharmacy College at King Khalid University (KKU), which aimed to evaluate the satisfaction of the academic staff with shifting into virtual education, and additionally, how suspension has effected the academic work [37]. The online questionnaire completed by the entire pharmacy's academic staff reported that the educational environment in KKU was already prepared for the emergency digital transformation. In addition, the Pharmacy College's staff noticed that virtual education is more flexible than the traditional model. However, the successful shift into virtual learning environment required some online training for instructors provided by the Electronic Learning Deanship, providing the virtual classrooms, recording the lectures, and facilitating the online discussion. Furthermore, the results showed that the shift into e-learning encountered issues regarding the lack of face-to-face human interaction. A majority of instructors have faced challenges of directly engaging the students besides evaluating and judging the participants with integrity. In addition to the previous limitations, covering all the content of the syllabus, a lacking attendance in practice sessions were some other issues highlighted in this study [37].

Similar research was sponsored by Al Faisal University in Riyadh [40]. The research studied the impact of conquering the COVID-19 pandemic during the shift from traditional to online education at the College of Medicine (COM). The developed questionnaire was emailed to targeted faculty members and students, which revealed 41% of the responders had a low level or no online education experience (and some students had no experience of online learning). Therefore, the issues highlighted in this study were associated with some relevant tools and academic processes including communication, the assessment of students, the use of technologies, time management, online experience, and technophobia. Nevertheless, in general, there is a largely affirmative influence of online education at the COM, Al Faisal University, where the confidence of the responders in the effectiveness of virtual medical education increased during the first few weeks of COVID-19 [40].

In addition, a study highlighted the processes management that was followed by health science colleges at King Saud Bin Abdulaziz University (KSAU-HS) to respond to the changes by delivering online education during the COVID-19 crisis [26]. The focus groups were constructed to develop emergency planning using the SWOT (strengths, weaknesses, opportunities, and threats); an analysis model that focused on four main areas, (i) faculty abilities development, (ii) curriculum management, (iii) assessment policies, and (iv) technology infrastructure and support. The authors proposed a model established by KSAU-HS as a framework that could be taken as the guidelines for delivering online education in health science colleges [26]. Similarly, another study prepared and distributed a survey among two samples: faculty members and students in some Saudi universities [41]. The questionnaire was designed to examine the formulated hypothesis, to discuss the effectiveness of the fourth industrial revolution technologies (4IR) that was the emergence of digital educational transformation, in relieving the impact of COVID-19 on higher education at Saudi universities [41].

Furthermore, this literature review revealed the changes related to the shift of the educational process to the virtual form in universities and educational institutions globally, regionally, and locally. Some previous studies were focused on studying the impact of COVID-19 on the medical education from different perspectives, such as faculty members, students, and administrative staff. Table 3 shows a summary of the main factors and areas investigated by previous studies in the local universities.

Table 3. Summary of main factors investigated by Saudi universities.

Ref.	[28]	[37]	[40]	[26]	[41]
University	Qassim University	Khalid University	Alfaisal University	King Saud Bin Abdulaziz University	Some Saudi universities
Faculty	College of Medicine	Pharmacy College	College of Medicine	Health Science	All
Objective	To study the effectiveness of synchronized digital learning	To evaluate the satisfaction of academic staff	To study the impact of switching to online education	To highlight process management that was taken by KSAU-HS	To ighlight effectiveness of digital educational transformation
			Measuring Factors		
Time Management	√		√		√
Assessment		√	√	√	
Communication and Engagement		√	√		
Course Material		√		√	
Technology Infrastructure		√	√	√	
			Focus on		
Faculty Members		√	√	√	√
Students	√		√	√	√

In this research, a study was conducted to examine the effectiveness of distance learning education during the COVID-19 pandemic. To fulfill this main objective, the study was designed based on the Technology-Organization-Environment (TOE) Framework. The TOE examined the adoption of new technology based on three substantial perspectives contexts, which are: (i) the technology context, (ii) the organization context, and (iii) the environment context [42,43]. Hence, in this paper, the authors employed TOE to investigate the effectiveness of virtual distance education during this pandemic by identifying three main aspects that may impact the educational process. Therefore, the main research question of this study is: "Is the distance learning impact effectively on the learning curve of higher education during the COVID-19 pandemic?", which further breaks down into the three following research questions discussed in the next sections:

- How has distance learning in high education impacted students during COVID-19?
- What are the perceptions that students have about the effectiveness of distance learning processes during COVID-19?
- What are the difficulties and challenges which students are facing with distance learning in terms of educational processes and tools?

3. The TOE Framework

The study builds on the TOE framework to investigate the major factors that impact the effectiveness of the educational process on students during the COVID-19 pandemic in Saudi Arabian universities. The TOE theoretical framework was first introduced by Tornatzky and Fleischer [44]. TOE framework has been exceedingly used in technology adoption relevant studies. The theory is based on three contexts: the environmental context, the organizational context, and the technological context, which is shown in Figure 1 and further discussed in previous research [44,45]. Therefore, three aspects of the TOE are introduced to suggest influential factors that could be beneficial for the universities and the higher educational institutions.

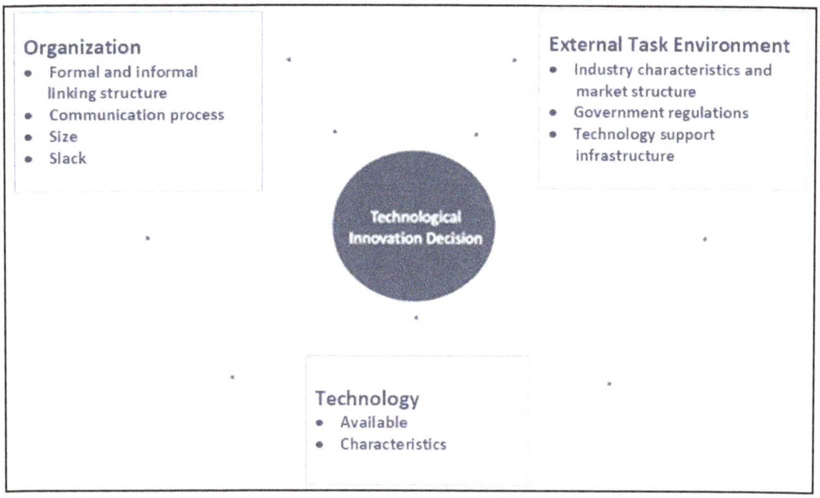

Figure 1. The technology–organization–environment framework [44].

3.1. Environmental Context (EC)

The environmental context consists of some environmental characteristics surrounding an organization. These characteristics include factors from stakeholders, such as supporters, suppliers, customers, the government, society, and competitive pressures, which can influence the organization's structure, decisions, and performance. The stakeholders may either support or prevent technology adoption [45,46]. Government regulations can mandate the allocation of resources for compliance. The Kenyan government, for example, has applied some policy and strategy documents to guide the integration of Information and Communications Technology (ICT) into education through its ministries of Education, Science, Technology, and ICT [43]. This application has created awareness of the place of ICT in education in all its forms, especially in distance education, which represents a large segment of education at the present time [47].

Saudi Arabia's Ministry of Education implemented some regulations to determine the evaluation mechanism in final exams, which was applied to all the educational institutions nationwide [48]. These regulations were circulated for the interest of students in light of the exceptional circumstances of education during COVID-19. The regulations included the mechanisms followed for the continuation of the educational process, the distribution of students' grades, the activation of analyzing the student's score for the past semesters by the admission and registration deanships and information technology in universities, when exams will be conducted, how to calculate the student's GPA, and how to deal with preparatory year programs for the fresh students in universities [48].

3.2. Organizational Context (OC)

The organizational context embraces multiple aspects, such as regulatory background, organizational culture, the quality of human resources, management structure, the degree of centralization, top management support, and the organization's size. From the perspective of new technology adoption, the organization context refers to conditions of an institution, such as employee competence and management support that impact this adoption [49]. Furthermore, the OC explores the extent of the leadership quality at the level of management along with the departments to achieve the institution's goal [43].

In the case of education institutions, the educational culture comprises the vision, plans, standards, and values shared by the top management, staff, and students [43]. The support of top management is demonstrated by providing supportive regulations and decisions. The higher administration has an impact on implementing proper processes and technologies by professors and administrative staff through its influence on participation in decision making, promoting professional developments and reducing feelings of uncertainty [45]. The educational institution's culture plays a mediating role in adopting technologies, which measures their readiness to adopt ICT in the educational process. Therefore, studying the educational institution's responses to the unexpected changes that led to rapid decisions and the adoption of new technologies to support distance education is important to examine its impact on the learning curve from the student's perspective during the pandemic.

3.3. Technological Context (TC)

The technological context points out the technology's boundaries, including external and internal technologies correlated with the institution. Hence, the technology extends to the capabilities of the institution by adapting existing available technologies in the industrial market. Further, the TC refers to the internal variables regarding technologies' performances that were taken into account in the pre-implementation phase. This perspective aimed to gain the benefits of external and internal technologies that may assist in increasing the efficiency of the institution's processes and outcomes besides speeding up the response to the market changes [49,50]. For applied TOE in the higher learning institution of Kenya, the cost and technology competence were the main factors investigated from the technology perspective. The technical competence of institutions is a fundamental infrastructure enabling the adoption of information technology (IT) as the basic form in such institutions' resources are built. The technology competence in the case of Kenya reflects the performance of technical infrastructure and resources in addition to the internet availability and bandwidth [43]. The TC includes the following main factors: compatibility, complexity (usability and learning curve), expectancy, performance, and reliability [51].

Hence, based on the literature review, the paper focused on two main factors: performance and complexity of the internal and external technologies that were employed in distance education during the COVID-19 pandemic to examine their impact on the learning curve from the student perspective.

4. Research Framework and Methodology

It is worth recalling that the main objective of this paper is to examine the effectiveness of distance learning on higher education during the COVID-19 pandemic. This section discusses the research framework and major research methodological steps taken in this study to answer the research questions. This study used a cross-sectional, self-administered online questionnaire directed to the fresh, undergraduate, and postgraduate students who were studying in Saudi Arabian Universities during the first semester of the 2020–2021 academic year during the COVID-19 outbreak. Most of the data collected were qualified for a quantitative analysis except two open-ended questions. The questionnaire was prepared based on the enhanced TOE framework that was modified with the help of the literature review and initial investigation, as shown in Figure 2. The questionnaire was prepared using each factor presented in the figure and further associated with multiple attributes categorized under each factor.

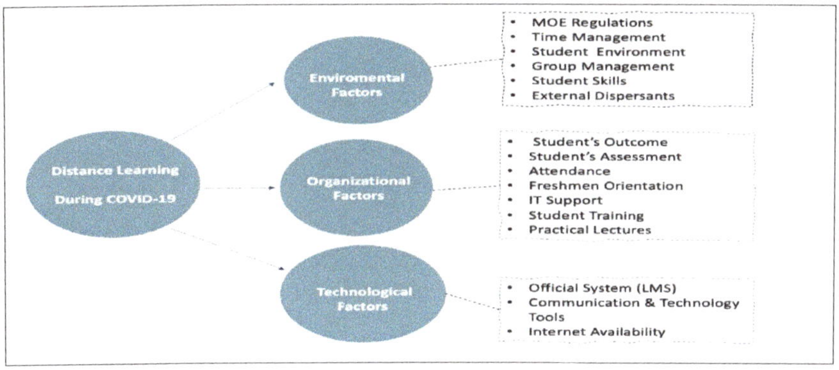

Figure 2. The conceptual research framework. (Adapted from [44]).

Due to the COVID-19 precautionary measures, the survey was shared over the WhatsApp social media platform and directed to the university students. For their ease and to understand the research context, the purpose of the study also provided using an online link. Overall, the online survey was designed using the Google Forms online survey platform. All responses were anonymous and confidential. In this survey, the total number of participants contacted was 580, where we communicated the survey link using different universities' WhatsApp groups. The received and correct responses (i.e., error free and answered completely) were 150, which provided feedback on 27 compulsory questions. In addition, there were 5 optional questions were asked from the participants, in which, 2 linear-scale questions received 116 and 115 responses, 1 check box question received 138 responses, and 2 open-ended questions received 84 and 59 responses, respectively. All of the received responses were used for data analysis, for which the Google Forms spreadsheet tool and Microsoft Excel were used, which is discussed in the Results and Discussion Section.

5. Data Collection

The structured questionnaire was designed based on the TOE theory of students during the outbreak and distance learning. The questionnaire was prepared in Arabic, consisted of three sections, with demographic information included in the introduction section. Most questions used a five-point Likert scale ranging from "strongly agree" to "strongly disagree". Section 2 consisted of (8) Likert scale questions to assess the sudden shift towards distance education, students' communication with colleagues and professors, their ability to focus and comprehend during the distance lectures, group work in projects, barriers of practical applications, and time management in the distance education.

Section 3 consisted of (10) Likert scale questions to assess the advantage of recorded lectures and their impact on increasing the understanding of educational content, as well as students' assessment in distance education and the challenges compared to the traditional way. In addition, the official system (LMS) effectiveness and the technical support provided to students was assessed, as well as the alternative solutions for practical lectures provided by the university and their applications. The difficulties faced by freshmen students with LMS were also assessed. The fourth section consisted of (7) questions to assess the technical tools used in distance education in terms of reliability, usability, and effectiveness, the availability of the necessary tools for the distance education process, such as a computer, printer, headset, and the quality of internet connection to attend lectures, submitting assignments, and exams. The survey consisted of one checklist question to determine which technical tools/apps were used in addition to the official LMS system. At the end of the survey, two open-ended questions were provided to assess students' experience in distance education compared to traditional education and the challenges they faced throughout the process.

To assess the understandability and readability of the asked questions, the questionnaires were reviewed with the help of two professors and one student, who were representative of the study population. This process helped to determine the clarity of the language and the structure of the questionnaire. The questionnaire then was modified based on the feedback received in the pilot test.

6. Result and Analysis

As a result of the distributed questionnaire, one hundred and fifty student's responses were received. Some questions required a cross-analysis, and some provide direct answers by the respondents. The rest of the section will present the result of each questionnaire section.

6.1. The Demographical View

The presented data in Table 4 show that the majority of the participants were (82.7%) females, (71.3%) between 20–30 years old, (70.7%) and studying in the bachelor programs. A total of 34% were freshmen students at the university and 71.3% already used technical tools in education, such as the Blackboard system, before the COVID-19 pandemic.

Table 4. The result of the demographics section in the questionnaire. (N = 150).

Items	Number	Percentage
Gender		
Female	124	82.7%
Male	26	17.3%
Age		
Less than 20 years old	33	22%
20–30	107	71.3%
31–40	8	5.3%
Over 40 years old	2	1.3%
Educational level		
BSC	106	70.7%
Masters	43	28.7%
PhD	1	0.7%
Are you a freshman at the university?		
Yes	51	34%
No	99	66%
Have you ever used technical tools in education, such as the Blackboard system before the corona pandemic?		
Yes	107	71.3%
No	43	28.7%

6.2. The Environmental View

The environmental factors are evaluated based on the responses collected through university students using a five-point Likert scale, as presented in Table 5. The transformation from traditional to distance education was not carried out very smoothly from the perspective of freshmen, where only 31.4% of them agreed or strongly agreed about that. The possible reasons could be that a freshman has no idea about the university education as that was the first experiment and of course, psychological factors may play critical roles in a freshman's opinion. For the student who had a good experience with the educational tools and system, the results show that 43.7% of them agreed or strongly agreed with the idea that it was a smooth shift.

Table 5. Survey results of the environmental factors.

Items			1 Strongly Disagree	2 Disagree	3 Neutral	4 Agree	5 Strongly Agree
The sudden shift towards distance education was smooth	Without Cross-Analysis		16 (10.7%)	25 (16.7%)	45 (31.3%)	35 (23.3%)	27 (18%)
	Are you a freshman?	Yes	7 (13.7%)	10 (19.6%)	18 (35.3%)	8 (15.7%)	8 (15.7%)
		No	9 (9.3%)	15 (15.5%)	27 (27.8%)	27 (27.8%)	19 (19.6%)
	Have you ever used educational tools before COVID-19 pandemic?	Yes	9 (8.6%)	18 (17%)	32 (30.4%)	27 (25.7%)	19 (18%)
		No	7 (16.3%)	7 (16.3%)	13 (30.2%)	8 (18.6%)	8 (18.6%)
Communication with colleagues and professors was easy during distance education, during the Corona pandemic	Without cross-analysis		13 (8.7%)	35 (23.3%)	36 (24.7%)	34 (23.3%)	30 (20%)
	Are you a freshman?	Yes	3 (5.9%)	11 (21.7%)	16 (31.4%)	11 (21.7%)	10 (19.6%)
		No	10 (10.3%)	24 (24.7%)	20 (20.6%)	23 (23.7%)	20 (20.6%)
	Have you ever used educational tools before COVID-19 pandemic?	Yes	10 (9.5%)	26 (24.8%)	24 (22.8%)	24 (22.8%)	21 (20%)
		No	3 (7%)	9 (21%)	12 (28%)	10 (23.3%)	9 (21%)
Compared to university attendance, the ability to focus and comprehend was greater during the distance lectures			49 (32.7%)	29 (19.3%)	33 (22%)	15 (10%)	24 (16%)
I have a greater opportunity to use the time between lectures to complete the required assignments			30 (20%)	13 (8.7%)	23 (15%)	23 (15.3%)	61 (40.7%)
Presenting presentations from a distance has negatively affected the development of my presentation and speaking skills			48 (32%)	26 (17.3%)	42 (28%)	13 (8.7%)	21 (14%)
Geographical spacing has negatively affected the performance and communication of the members in the group projects	Without cross-analysis		36 (24%)	20 (13.3%)	29 (20.7%)	25 (16.7%)	38 (25.3%)
	Are you a freshman?	Yes	15 (29.4%)	8 (15.6%)	6 (11.8%)	8 (15.7%)	14 (27.5%)
		No	21 (21.6%)	12 (12.4%)	23 (23.7%)	17 (17.5%)	24 (24.7%)
	Have you ever used educational tools before COVID-19 pandemic?	Yes	27 (25.7%)	14 (13.3%)	23 (21.9%)	18 (17.1%)	23 (21.9%)
		No	9 (20.9%)	6 (14%)	6 (14%)	7 (16%)	15 (34.9%)
Distance education form an obstacle to practical training, in subjects that include practical hours	Without cross-analysis		14 (9.3%)	12 (8%)	33 (23.3%)	28 (18.7%)	61 (40.7%)
	Have you ever used educational tools before COVID-19 pandemic?	Yes	12 (11.4%)	8 (7.6%)	27 (25.7%)	21 (20%)	37 (35.2%)
		No	2 (4.7%)	4 (9.3%)	6 (13.9%)	7 (16.2%)	24 (55.8%)
Distance education has contributed to making use of the time wasted in commuting between home and university			16 (10.7%)	6 (4%)	20 (13.3%)	13 (8.7%)	95 (63.3%)

About the communication process with colleagues and professors, 41.3% of freshmen and 42.8% of the students who used educational tools before agreed and strongly agreed that it was easy. This is due to the availability and diversity of social media applications that have contributed to overcome the difficulties that may face them in this periode specially. More than half of the students (52%) disagreed or strongly disagreed that the ability to focus and comprehend was greater during the distance lecturers compared with in-campus studies while 56% agreed and strongly agreed that they had a greater opportunity to use the time between lectures to complete the required assignments. At the same time, only 22.7% of the students agreed or strongly agreed that giving presentations from a distance created a negative impact on preparing presentations and speaking skills.

Another question was asked to take their opinion on geographical distance and its impact on group work. A total of 43.2% of the freshmen agreed or strongly agreed that it negatively affected group work, whereas only 39% of the students who had experiences with the educational tools before the COVID-19 pandemic agreed or strongly agreed on its negative impact. Besides, 55.2% of the students who used educational tools before agreed or strongly agreed that distance education formed an obstacle for practical training, especially in the subjects that include practical hours. The majority of the students (72%) agreed or strongly agreed that distance education contributed to saving time traveling from home to university.

6.3. The Organizational View

By focusing on the TOE theory organizational factors, university students assessed the factors by using a five-point Likert scale, as shown in Table 6. The majority of the students (79.3%) agreed or strongly agreed that the possibility of returning to the recorded lectures helped them to increase their understanding and realization of the scientific content, whereas more than half (56.6%) agreed or strongly agreed that remote tests characterized and depended on perception and understanding of the scientific content more than memorization and narration.

Meanwhile, 46% agreed or strongly agreed that for online examination, there is no unified mechanism in terms of the duration and method of the exam. The majority of the students (71.6%) agreed or strongly agreed that the challenges they faced while taking the exam remotely were the exam submission via the Blackboard system, in addition to the possibility of losing the network connection, while only 24% agreed or strongly agreed that they could easily reach technical support or educational affairs when they needed help.

Nearly half of the freshmen (47.1%) agreed or strongly agreed that they received help in installing and setting up the software that required to use in distance learning, while (41.9%) of the students who used educational tools before the COVID-19 pandemic agreed or strongly agreed with that. Nearly half of the students (47%) also agreed or strongly agreed that they were trained in the applications and systems they needed to use in the courses. Moreover, about half (48.6%) of the students who used educational tools before the COVID-19 pandemic agreed or strongly agreed with that, while only 31.4% of the students agreed or strongly agreed that the university provided an alternative solution to apply the subjects that include practical hours. The number of freshmen students (24.1%) agreed or strongly agreed that they encountered problems and difficulties in dealing with the systems approved by the university and only (25.2%) of them agreed or strongly agreed that they received the necessary training to deal with the university's systems.

Table 6. Survey results of the organizational factors.

Items			1 Strongly Disagree	2 Disagree	3 Neutral	4 Agree	5 Strongly Agree
The possibility of returning to the recorded lectures helped to increase the understanding and realizing of the scientific content			4 (2.7%)	5 (3.3%)	22 (14.7%)	35 (23.3%)	84 (56%)
Remote tests were characterized by their dependence on perception and understanding of the scientific content more than memorization and narration			19 (12.7%)	17 (11.3%)	29 (19.3%)	20 (13.3%)	65 (43.3%)
When the course exam is remote, there is no unified mechanism in terms of the duration and method of the exam			13 (8.7%)	18 (12%)	48 (32%)	31 (20.7%)	38 (25.3%)
One of the challenges I faced while taking the exam remotely was the submission of the exam via the blackboard system and the possibility of losing the network connection			7 (4.7%)	14 (9.5%)	21 (14.2%)	34 (23%)	72 (48.6%)
I can easily reach technical support or educational affairs when I need help			25 (16.7%)	24 (16%)	65 (43.3%)	21 (14%)	15 (10%)
I received help in installing and setting up the software that required to use	Without cross-analysis		31 (20.7%)	11 (7.3%)	44 (30.7%)	20 (13.3%)	42 (28%)
	Are you a freshman?	Yes	11 (21.6%)	2 (3.9%)	14 (27.5%)	8 (15.7%)	16 (31.4%)
		No	20 (20.6%)	9 (9.3%)	30 (30.1%)	12 (12.4%)	26 (26.8%)
	Have you ever used educational tools before COVID-19 pandemic?	Yes	18 (17.1%)	10 (9.5%)	33 (31.4%)	14 (13.3%)	30 (28.6%)
		No	13 (30.2%)	1 (2.3%)	11 (25.6%)	6 (14%)	12 (27.9%)
I have been trained in the applications and systems I need to use in the course materials	Without cross-analysis		19 (12.7%)	12 (8%)	44 (30.7%)	36 (24%)	37 (24.7%)
	Are you a freshman?	Yes	9 (17.6%)	4 (7.8%)	14 (27.5%)	12 (23.5%)	12 (23.5%)
		No	10 (10.3%)	8 (8.2%)	30 (30.9%)	24 (24.7%)	25 (25.8%)
	Have you ever used educational tools before COVID-19 pandemic?	Yes	12 (11.4%)	10 (9.5%)	29 (27.6%)	24 (23%)	30 (25.6%)
		No	7 (16.3%)	2 (4.7%)	15 (34.9%)	12 (27.9%)	7 (16.3%)
The university provided an alternative solution to apply the subjects that include practical hours			16 (10.7%)	18 (12%)	69 (46%)	25 (16.7%)	22 (14.7%)
As a freshman at the university, I encountered problems and difficulties in dealing with the systems approved by the university			28 (24.1%)	18 (15.5%)	42 (36.2%)	14 (12.1%)	14 (12.1%)
As a freshman at the university, I received the necessary training to deal with the university's systems			22 (19.1%)	22 (19.1%)	42 (36.5%)	13 (11.3%)	16 (13.9%)

6.4. The Technological View

Like other factors, the technological factors were also evaluated using a five-point Likert scale, as shown in Table 7. Most of the students were satisfied with the LMS tools used in distance education, as more than 67% agreed or strongly agreed. To support that reason, another question was asked regarding the interaction in the virtual classroom during the lectures, where more than 65% agreed or strongly agreed that the instant chat available in the virtual classroom positively increased the classroom interactions.

Table 7. Survey results of the technological factors.

Items	1 Strongly Disagree	2 Disagree	3 Neutral	4 Agree	5 Strongly Agree
I am satisfied with the use of technical tools such as: the blackboard system in a distance education environment	18 (12%)	11 (7.3%)	19 (12.7%)	49 (32.7%)	53 (35.3%)
In the virtual classroom, instant chat increased interaction and possibility to ask questions during lectures	14 (9.3%)	20 (13.3%)	17 (11.3%)	38 (25.3%)	61 (40.7%)
Technical tools such as: the blackboard system covers all my needs as a student during the distance education process	22 (14.7%)	18 (12%)	24 (16%)	38 (25.3%)	48 (32%)
Other technical tools were adopted in addition to the university approved system	11 (7.3%)	15 (10%)	46 (30.7%)	44 (29.3%)	34 (22.7%)
I have the tools I need for the distance education process, such as: a computer, a printer, a headset	6 (4%)	12 (8%)	25 (16.7%)	33 (22%)	74 (49.3%)
I have a good internet connection that helps me: attend lectures, deliver assignments, give presentations and take exams	5 (3.3%)	29 (19.3%)	28 (18.7%)	42 (28%)	46 (30.7%)

More than half of the students (57%) agreed or strongly agreed that technical tools such as the Blackboard system covered all their educational needs during distance education, while 52% agreed or strongly agreed that there were other technical tools used during the pandemic in addition to the official LMS system. In addition, more than 71% stated that they used WhatsApp, and 43% used Telegram in distance education communication during the COVID-19 pandemic. Other tools were used for the verbal communication and classroom presentations, where around 49% of students claimed that they used Zoom, 30% used Google Meet, and 25% used MS Teams and Skype.

Technical equipment availability was not an issue for most students, as about 71% claimed that they had all the distance education tools, such as a computer device, printer, and a headset for the virtual classroom. The same applies for the internet connection, where around 59% of students stated that they had a good internet connection, enabling them to attend the virtual classroom, deliver their assignments and projects, and to appear in the exams.

A number of challenges were encountered by the students through open-ended questions centered on two main areas, whether in terms of technological limitation or regarding the education process. Internet availability and instability were significant obstacles that students worried about, especially during online examinations. Another issue was the increase in the number of homework assignments that were caused by non-uniform established assessment methods. For examinations, there were some obstacles, such as less time, the difficulty of questions, and dealing with tools. Some students faced obstacles in communication with the professors due to a lack of defined office hours.

Many answers were in favor of distance education on the condition that future enhancements to the education process would be implemented. Suggestions were mostly focused on the quality of the teaching methods in the virtual classrooms, where some claimed that the issue is with creating an interactive environment to compensate for face-to-face communication, facial expressions, and body language that was easily filled in the traditional education. There were opinions that focused on the assignments, where some claimed that it was challenging to complete the assignments that were designed for traditional education. Other opinions were related to the psychological side of the student, as isolation and learning from home had an effect on their learning experience in general.

7. Discussion

The results of this study showed that the sudden shift towards distance education was smooth, and the existing educational tools were ready for the sudden shift towards online education. The respondents found that communication with colleagues and professors was easy, and they did not face any difficulties in communication with each other. This was due to the availability and diversity of social media applications that contributed to overcome the difficulties that one may have faced, especially in this period. Meanwhile, the ability to focus and comprehend was less than the traditional face-to-face education, which probably depends on the student's environment, social status, and family circumstances.

On the other hand, distance learning provides the students with a greater opportunity to use the time between lectures to complete the required assignments. As for the course projects, presentations continued even in distance education and did not adversely affect the presentation and delivery skills, while geographical distance negatively affected the performance and communication between members in project groups. In subjects that include practical hours, there were obstacles in the practical training application, and thus the benefits of practical training were not achieved successfully. Finally, distance education has the advantage of saving time wasted in traveling between home and university. It provides the student more time to review subjects and work on the projects and assignments required, which helps to improve the student's academic performance; the research in [34] also found similar results. The same advantage was also identified in previous research that online education saved time in different ways, which could be further utilized for different purposes such as extra reading and exam preparation [28].

In this research, most of the participants showed their experience of using online tools before this pandemic situation. Actually, the prior experience provided an extra advantage to them while using online tool during lectures. Previously, we found that during a survey in a university, it took some time for students and faculty members to understand the online tools [40]. The strategy of recording lecturers assists in raising the understanding and realization of the scientific content, since the students can access a lecture more than once. Thus, it is easy to take notes and repeat a particular lesson in the case of a complex concept or a hard lesson. This feature makes distance education more flexible and efficient than traditional education [28]. Besides, the online examination focused on evaluating the realization and understanding of the student by asking analytical and discussion questions instead of memorization and narration of the information. The nature of these questions solidifies the information in the student's mind with the minimum exhaustion of the student required to memorize the information which is quickly forgotten by the end of the exam.

Despite the established regulation and planning regarding the examination process by universities and the efforts made by the administrative and academic staff, there are still some obstacles encountered by students. One of the issues is there is no unified mechanism in terms of the duration and way of the exams that resulted in confusing the student. Therefore, the online exam process needs more clarification and standard procedures to be followed by higher educational institutions. On the other hand, proper guidelines will also be helpful for the students to understand the process and improve their performance in the exam. Based on the findings in this study, network problems, time allocation, the way of examination, and assessment procedures are some of the common factors that require equal importance in standardizing the overall mechanism of online examination.

Furthermore, the submission of the exam via the Blackboard system and the possibility of losing the network connection were significant challenges for most of the students. The results of this study indicated that the universities provided sufficient training for students for using the Blackboard system to join the virtual classroom, submit the assignment, and other useful features. However, the majority of the fresh students did not receive adequate training in the approved system, probably due to difficulties in contacting technical support, where some of the students faced a few issues in reaching technical support. Online education is still inappropriate to deliver practical sessions, as was reported by [36,37], where they suggested employing a hybrid model with a small group of the students present in a face-to-face session while other students attend the simulation-based online session.

The variety of options, capabilities, and the capability of customization that is offered by LMS provides universities with a powerful mechanism to represent a comprehensive educational content. Most of the educational resources and activities in LMS can be provided via a unified platform that is used by students as well as professors and laboratory technicians. However, this study found that other mediums are used in an influential way, as there is a huge percentage who stated that they have used many social media platforms during the distance educational experience. This study shows that many social media platforms have been used to support communications from students with their academic instructors and among students themselves. Social media platforms support the idea of speeding up the response, alongside ease and mobility being some extra features. It provides the space for practitioners to use it in official forums, including distance education, which is the same idea discussed in previous research [35]. Considering the integration between the official LMS and social media platforms could be proposed as one of the future enhancements of the distance educational experience as this may encourage the convergence between students and professors as well as their fellow colleagues.

8. Conclusions

This study aimed to evaluate the effectiveness of distance education during the COVID-19 pandemic in higher education in Saudi Arabia. In addition, this study reported the main considerable difficulties and challenges encountered by students during the pandemic. The authors examined students' opinions by conducting an online survey based on the modified TOE framework. The results of the survey indicate the shift into online education was positive, with a learning curve to consider for future enhancement to the overall distance education experience. Meanwhile, group projects, student outcomes, freshmen orientation, and practical lectures have been strongly impacted by the university's lockdown. Furthermore, according to the study, universities need to establish a unified mechanism for the examination process, including regulations regarding convenient questions with exam duration and methods. This study introduced the main factors and their impact on the educational process for the decision makers of universities to overcome the encountered limitations and adopt new solutions in order to enhance the overall experience. The identified factors and obstacles discussed in this study can be beneficial for the educational institutes in Saudi Arabia, to overcome problems highlighted in this study. The results of this study can also be useful to improve different processes, such as how to conduct online lectures, a list of preferable online assessments, and most importantly, preparing the standard guidelines for the exams. The main research limitation was the time constraints; due to this, a limited number of participants were contacted in this study. Future research could focus on measuring the faculty members' satisfaction with online teaching tools and students' academic performance in the distance learning environment.

Author Contributions: Conceptualization, K.A., S.M. and A.B.; methodology, K.A., S.M. and A.B.; Literature Review, K.A., S.M. and A.B.; data collection, K.A., S.M. and A.B.; writing—original draft preparation, K.A., S.M. and A.B.; review and editing, F.S., S.B. and B.F.; supervision, F.S., S.B. and B.F.; project administration, F.S., S.B. and B.F.; revision, F.S.; rechecking, B.F. and S.B. All authors have read and agreed to the published version of the manuscript.

Funding: This research received no external funding.

Institutional Review Board Statement: Not applicable.

Informed Consent Statement: Not applicable.

Data Availability Statement: Not applicable.

Conflicts of Interest: The authors declare no conflict of interest.

References

1. Shah, A.U.M.; Safri, S.N.A.; Thevadas, R.; Noordin, N.K.; Rahman, A.A.; Sekawi, Z.; Ideris, A.; Sultan, M.T.H. COVID-19 outbreak in Malaysia: Actions taken by the Malaysian government. *Int. J. Infect. Dis.* **2020**, *97*, 108–116. [CrossRef]
2. Tolu, L.B.; Ezeh, A.; Feyissa, G.T. How prepared is Africa for the COVID-19 pandemic response? The case of Ethiopia. *Risk Manag. Healthc. Policy* **2020**, *13*, 771–776. [CrossRef]
3. Shafi, M.; Liu, J.; Ren, W. Impact of COVID-19 pandemic on micro, small, and medium-sized Enterprises operating in Pakistan. *Res. Glob.* **2020**, *2*, 100018. [CrossRef]
4. Johnson, S.U.; Ebrahimi, O.V.; Hoffart, A. PTSD symptoms among health workers and public service providers during the COVID-19 outbreak. *PLoS ONE* **2020**, *15*, e0241032. [CrossRef]
5. Chen, X. Spaces of care and resistance in China: Public engagement during the COVID-19 outbreak. *Eurasian Geogr. Econ.* **2020**, *61*, 435–447. [CrossRef]
6. Dewey, C.; Hingle, S.; Goelz, E.; Linzer, M. Supporting Clinicians During the COVID-19 Pandemic. *Ann. Intern. Med.* **2020**, *172*, 752–753. [CrossRef] [PubMed]
7. Mayer, J.D.; Lewis, N.D. An inevitable pandemic: Geographic insights into the COVID-19 global health emergency. *Eurasian Geogr. Econ.* **2020**, *61*, 404–422. [CrossRef]
8. Mishra, L.; Gupta, T.; Shree, A. Online teaching-learning in higher education during lockdown period of COVID-19 pandemic. *Int. J. Educ. Res. Open* **2020**, *1*, 100012. [CrossRef]
9. Creţan, R.; Light, D. COVID-19 in Romania: Transnational labour, geopolitics, and the Roma outsiders. *Eurasian Geogr. Econ.* **2020**, *61*, 559–572. [CrossRef]
10. Moisio, S. State power and the COVID-19 pandemic: The case of Finland. *Eurasian Geogr. Econ.* **2020**, *61*, 598–605. [CrossRef]
11. Mionel, V.; Neguţ, S.; Mionel, O. Pandemopolitics. How a public health problem become a geopolitical and geoeconomic issue. *Eurasian Geogr. Econ.* **2020**, *61*, 389–403. [CrossRef]
12. Chan, K.W.; Gentile, M.; Kinossian, N.; Oakes, T.; Young, C. "More-than-viral" Eurasian geographies of the covid-19 pandemic: Interconnections, inequalities, and geopolitics. *Eurasian Geogr. Econ.* **2020**, *61*, 343–361. [CrossRef]
13. Connolly, R.; Hanson, P.; Bradshaw, M. It's déjà vu all over again: COVID-19, the global energy market, and the Russian economy. *Eurasian Geogr. Econ.* **2020**, *61*, 511–531. [CrossRef]
14. Krzysztofik, R.; Kantor-Pietraga, I.; Spórna, T. Spatial and functional dimensions of the COVID-19 epidemic in Poland. *Eurasian Geogr. Econ.* **2020**, *61*, 573–586. [CrossRef]
15. Abu Hammour, M.; Abuhammour, W.; Alfalah, S. *Agility and COVID-19 Pandemic Success and Failure*; Elsevier: Amsterdam, The Netherlands, 2020.
16. Kumar, S.; Maheshwari, V.; Prabhu, J.; Prasanna, M.; Jayalakshmi, P.; Suganya, P.; Malar, B.A.; Jothikumar, R. Social economic impact of COVID-19 outbreak in India. *Int. J. Pervasive Comput. Commun.* **2020**, *16*, 309–319.
17. Alsolami, F.J.; Alghamdi, A.S.A.-M.; Khan, A.I.; Abushark, Y.B.; Almalawi, A.; Saleem, F.; Agrawal, A.; Kumar, R.; Khan, R.A. Impact Assessment of COVID-19 Pandemic Through Machine Learning Models. *Comput. Mater. Contin.* **2021**, *68*, 2895–2912. [CrossRef]
18. Viner, R.M.; Russell, S.J.; Croker, H.; Packer, J.; Ward, J.; Stansfield, C.; Mytton, O.; Bonell, C.; Booy, R. School closure and management practices during coronavirus outbreaks including COVID-19: A rapid systematic review. *Lancet Child Adolesc. Health* **2020**, *4*, 397–404 [CrossRef]
19. Dwivedi, Y.K.; Hughes, D.L.; Coombs, C.; Constantiou, I.; Duan, Y.; Edwards, J.S.; Gupta, B.; Lal, B.; Misra, S.; Prashant, P.; et al. Impact of COVID-19 pandemic on information management research and practice: Transforming education, work and life. *Int. J. Inf. Manag.* **2020**, *55*, 102211. [CrossRef]
20. Crawford, J.; Butler-Henderson, K.; Rudolph, J.; Malkawi, B.; Glowatz, M.; Burton, R.; Magni, P.A.; Lam, S. COVID-19: 20 countries' higher education intra-period digital pedagogy responses. *Int. Perspect. Interact. Educ.* **2020**, *3*, 1–20.
21. Flores, M.A.; Gago, M. Teacher education in times of COVID-19 pandemic in Portugal: National, institutional and pedagogical responses. *J. Educ. Teach.* **2020**, *46*, 507–516. [CrossRef]
22. Office of Technology and Higher Education. Clarification Note from the Office of the Minister for Science, Technology and Higher Education. 2020. Available online: https://www.sec-geral.mec.pt/noticia/nota-de-esclarecimento-do-gabinete-do-ministro-da-ciencia-tecnologia-e-ensino-superior (accessed on 1 January 2021).
23. The Media Line Staff, Schools, Public Events Canceled in Gulf as Coronavirus Spreads. 2020. Available online: https://themedialine.org/headlines/schools-public-events-canceled-in-gulf-as-coronavirus-spreads/ (accessed on 24 December 2020).
24. Abigail, N.G. UAE Takes Steps to Curb the Coronavirus Spread Including Shutting Schools for Four Weeks. 2020. Available online: https://www.cnbc.com/2020/03/04/coronavirus-uae-cancels-events-announces-school-closures.html (accessed on 24 December 2020).
25. Ministry of Education and Hamdan Bin Mohammed Smart University Qualifies more than 42,000 Teachers in UAE & the Region. 2020. Available online: https://www.zawya.com/mena/en/press-releases/story/Ministry_of_Education_and_Hamdan_Bin_Mohammed_Smart_University_qualifies_more_than_42000_teachers_in_UAE__the_region-ZAWYA20200315143149/ (accessed on 24 December 2020).
26. Al-Kadri, H.M.; Al Moamary, M.; Al Knawy, B. Framework for curriculum delivery during COVID-19 pandemic in a health sciences university. *Ann. Thorac. Med.* **2020**, *15*, 185–189. [CrossRef]

27. Tanveer, M.; Bhaumik, A.; Hassan, S.; Haq, I.U. Covid-19 Pandemic, Outbreak Educational Sector and Students Online Learning in Saudi Arabia. *J. Entrep. Educ.* **2020**, *23*, 1–14.
28. Khalil, R.; Mansour, A.E.; Fadda, W.A.; Almisnid, K.; Aldamegh, M.; Al-Nafeesah, A.; Alkhalifah, A.; Al-Wutayd, O. The sudden transition to synchronized online learning during the COVID-19 pandemic in Saudi Arabia: A qualitative study exploring medical students' perspectives. *BMC Med. Educ.* **2020**, *20*, 1–10. [CrossRef] [PubMed]
29. University of Minho. Dispatches and Circulars RT-33. 2020. Available online: https://www.uminho.pt/PT/viver/COVID-19/Paginas/DespachosCirculares.aspx (accessed on 1 January 2021).
30. Choi, B.; Jegatheeswaran, L.; Minocha, A.; AlHilani, M.; Nakhoul, M.; Mutengesa, E. The impact of the COVID-19 pandemic on final year medical students in the United Kingdom: A national survey. *BMC Med. Educ.* **2020**, *20*, 1–11. [CrossRef] [PubMed]
31. Zawacki-Richter, O. The current state and impact of Covid-19 on digital higher education in Germany. *Hum. Behav. Emerg. Technol.* **2021**, *3*, 218–226. [CrossRef]
32. Light, D.; Creţan, R.; Dunca, A.-M. Museums and Transitional Justice: Assessing the Impact of a Memorial Museum on Young People in Post-Communist Romania. *Societies* **2021**, *11*, 43. [CrossRef]
33. Light, D.; Creţan, R.; Dunca, A.-M. Education and post-communist transitional justice: Negotiating the communist past in a memorial museum. *Southeast Eur. Black Sea Stud.* **2019**, *19*, 565–584. [CrossRef]
34. Gonzalez, T.; De La Rubia, M.A.; Hincz, K.P.; Comas-Lopez, M.; Subirats, L.; Fort, S.; Sacha, G.M. Influence of COVID-19 confinement on students' performance in higher education. *PLoS ONE* **2020**, *15*, e0239490. [CrossRef]
35. Sobaih, A.; Hasanein, A.; Abu Elnasr, A. Responses to COVID-19 in higher education: Social media usage for sustaining formal academic communication in developing countries. *Sustainability* **2020**, *12*, 6520. [CrossRef]
36. Al-Balas, M.; Al-Balas, H.I.; Jaber, H.M.; Obeidat, K.; Al-Balas, H.; Aborajooh, E.A.; Al-Taher, R.; Al-Balas, B. Distance learning in clinical medical education amid COVID-19 pandemic in Jordan: Current situation, challenges, and perspectives. *BMC Med. Educ.* **2020**, *20*, 1–7.
37. Almaghaslah, D.; Alsayari, A. The effects of the 2019 Novel Coronavirus Disease (COVID-19) outbreak on academic staff members: A case study of a pharmacy school in Saudi Arabia. *Risk Manag. Healthc. Policy* **2020**, *13*, 795–802. [CrossRef] [PubMed]
38. El Zawaidy, H.A. Using Blackboard in online learning at Saudi universities: Faculty member's perceptions and existing obstacles. *Int. Interdiscip. J. Educ.* **2014**, *1*, 1–9. [CrossRef]
39. Daniel, S.J. Education and the COVID-19 pandemic. *Prospects* **2020**, *49*, 91–96. [CrossRef]
40. Rajab, M.H.; Gazal, A.M.; AlKattan, K. Challenges to Online Medical Education During the COVID-19 Pandemic. *Cureus* **2020**, *12*, e8966. [CrossRef] [PubMed]
41. Abdulrahim, H.; Mabrouk, F. COVID-19 and the Digital Transformation of Saudi Higher Education. *Asian J. Distance Educ.* **2020**, *15*, 291–306.
42. Yeh, C.-H.; Lee, G.-G.; Pai, J.-C. Using a technology-organization-environment framework to investigate the factors influencing e-business information technology capabilities. *Inf. Dev.* **2014**, *31*, 435–450. [CrossRef]
43. Micheni, E.M. Using the Technology Organization Environment Framework for Adoption and Implementation of Cloud Computing in Institutions of Higher Learning in Kenya. 2015. Available online: http://41.89.56.62:8080/handle/123456789/1451 (accessed on 24 June 2021).
44. Tornatzky, L.G.; Fleischer, M.; Chakrabarti, A.K. *Processes of Technological Innovation*; Lexington Books: Lanham, MD, USA, 1990.
45. Pudjianto, B.; Zo, H.; Ciganek, A.P.; Rho, J.J. Determinants of e-government assimilation in Indonesia: An empirical investigation using a TOE framework. *Asia Pacific J. Inf. Syst.* **2011**, *21*, 49–80.
46. Angeles, R. Using the Technology-Organization-Environment framework and Zuboff's concepts for understanding environmental sustainability and RFID: Two case studies. *Int. J. Soc. Educ. Econ. Manag. Eng.* **2013**, *7*, 1599–1608.
47. Kashorda, M.; Waema, T.; Omosa, M.; Kyalo, V. *E-Readiness Survey of Higher Education Institutions in Kenya: A Study Funded by Partnership for Higher Education in Africa*; Education Network Report: Nairobi, Kenya, 2007.
48. The Ministry of Education Sets a Mechanism for Evaluating Final Exams to Ensure Fair Implementation in the Interest of the Student. Ministry of Education: Saudi Arabia. 2020. Available online: https://www.moe.gov.sa/ar/news/pages/un-2020-547.aspx (accessed on 25 September 2020).
49. Eze, S.C.; Chinedu-Eze, V.C.A.; Okike, C.K.; Bello, A.O. Factors influencing the use of e-learning facilities by students in a private Higher Education Institution (HEI) in a developing economy. *Humanit. Soc. Sci. Commun.* **2020**, *7*, 1–15. [CrossRef]
50. Alone, K. Adoption of e-learning technologies in education institutions/organizations: A literature review. *Asian J. Educ. Res.* **2017**, *5*, 63–71.
51. Namisiko, P.; Munialo, C.; Nyongesa, S. Towards an optimization framework for e-learning in developing countries: A case of private universities in Kenya. *J. Comput. Sci. Inf. Technol.* **2014**, *2*, 131–148.

sustainability

MDPI

Article

Factors Influencing the Adoption of COVID-19 Preventive Behaviors in Chile

Nicolás C. Bronfman [1,2,*], Paula B. Repetto [2,3], Pamela C. Cisternas [2,4] and Javiera V. Castañeda [3]

1 Engineering Sciences Department, Universidad Andres Bello, Santiago 7500973, Chile
2 Research Center for Integrated Disaster Risk Management ANID/FONDAP/15110017, Santiago 7820436, Chile; prepetto@uc.cl (P.B.R.); pccisternas@uc.cl (P.C.C.)
3 Department of Psychology, Pontificia Universidad Católica de Chile, Santiago 7820436, Chile; jvcastaneda@uc.cl
4 Industrial and Systems Engineering Department, Pontificia Universidad Católica de Chile, Santiago 7820436, Chile
* Correspondence: nbronfman@unab.cl

Abstract: The COVID-19 pandemic forced people worldwide to implement a series of preventive hygiene and distancing measures that have significantly altered their way of life. This study examined an adapted version of the Theory of Planned Behavior (TPB) on adopting preventive behavior against COVID-19. Data was collected using a web survey completed by 1004 college students a few weeks after the first wave of infections in Chile. Our findings show that the subjective norm was the strongest predictor of adopting preventive behaviors, followed by the knowledge level and perceived behavioral control. Contrary to expectations, we found no evidence that an attitude towards preventive action predicted actual adoption of preventive behavior against COVID-19. However, knowledge and social norms play a significant role. We discuss implications for effective risk communication.

Keywords: preventive behaviors; theory of planned behavior; subjective norms; COVID-19

Citation: Bronfman, N.C.; Repetto, P.B.; Cisternas, P.C.; Castañeda, J.V. Factors Influencing the Adoption of COVID-19 Preventive Behaviors in Chile. *Sustainability* **2021**, *13*, 5331. https://doi.org/10.3390/su13105331

Academic Editor: John Rennie Short

Received: 12 March 2021
Accepted: 7 May 2021
Published: 11 May 2021

1. Introduction

The health crisis sparked in 2020 by the SARS-CoV-2 virus causing the COVID-19 pandemic has significantly altered people's lives worldwide and has caused hundreds of millions of confirmed infections and millions of deaths [1]. The pandemic has had a significant economic impact not seen since World War II [2].

Because of the high rate of contagion of the virus, health authorities imposed a series of sanitary measures and restrictions to reduce its spread, which forced the suspension of economic and social activities in many places worldwide. The implementation of these measures revealed that one of the most significant challenges posed by this crisis was encouraging the adoption of preventive behaviors in the population [3–5], particularly among younger adults [6]. Youths are an important group for compliance with preventive behaviors, though their situation involving a low risk of contracting serious symptoms as a result of COVID-19 [7] and the high mental health impacts they could face due to the interruption of their social and academic life [8–10] discourages their adherence to sanitary measures. Consequently, they are an important vector of contagion [11,12].

Due to the particular aforementioned situation for young people during the pandemic, it is essential to know the psychosocial factors that influence the adoption of preventive measures for this group and identify the differences with the factors that motivate the adoption of preventive measures in the general population. Consequently, this study aimed to investigate the influence of several predictive variables on the adoption of preventive behavior by university students against COVID-19 in Chile. To contribute to knowledge about preventive behaviors against COVID-19, this study assessed the effect of attitudes, subjective norms, and perceived behavioral control on adopting preventive behaviors. In addition, knowledge about COVID-19 was incorporated as a behavior predictor since, as a new hazard, it can be a determining factor in people's behavior [13–15].

1.1. *Theory of Planned Behavior (TPB)*

Several theoretical models have been used to understand the adoption of preventive behaviors in response to health problems. The Theory of Planned Behavior (TPB) is widely used to explain these behaviors [16–25]. TPB is an extension of the Theory of Reasoned Action (TRA) [26] and postulates that the direct predictor of behavior is the behavioral intention, which in turn is explained by three components: attitudes, subjective norms, and perceived behavioral control [27]. In this model, attitudes refer to each person's evaluation of a behavior based on their beliefs about the particular behavior's characteristics, probable outcomes associated with it, and whether it is favorable or unfavorable. Subjective norms correspond to each individual's beliefs about whether their reference groups approve or disapprove of the behavior and whether they are motivated to comply with their reference group's norms [27,28]. Perceived behavioral control refers to the person's perceived power to perform a behavior, considering the beliefs about his or her ability to perform it [28].

In this model, the authors propose that a favorable attitude, combined with the approval of the behavior by others close to them and a positive perception of control, strengthens the intention to perform that particular behavior [26]. Theoretically, the three components of the theory have independent effects on behavioral intention [26,27]: attitudes and subjective norms directly affect behavioral intention. In contrast, perceived control affects behavioral intention and preventive behavior [29,30]. While the results of studies using the TPB are diverse, attitude is generally recognized as the best predictor of behavioral intention, as is the perception of control. Subjective norms appear to have a more limited effect on intention [16,31]. Nevertheless, other studies have noted that the strength of the effect of attitude, subjective norms, and perception of control varies according to the behavior to which they are applied [17,26,29].

1.2. *TPB and Preventive Behaviors in Pandemic*

The study by Cheng and Ng [19] was one of the few to use the TPB to examine the factors influencing preventive behavior during an epidemic, particularly the Severe Acute Respiratory Syndrome (SARS) epidemic between 2002 and 2004. They concluded that the subjective norm was the strongest predictor of the intention and adoption of preventive behaviors. The knowledge level about the virus also had a significant effect on behavioral intention [19].

More recently, Barile et al. [32] integrated the Health Belief Model (HBM) and the TPB to explore both barriers and facilitators for public use of face coverings to prevent the contagion and spread of the SARS-CoV-2 virus. They found that perceived subjective norms, self-efficacy, and attitudes were positively associated with an intention to wear a face-covering in public [32]. In another study which aimed to determine the main predictors of actual social distancing behavior to prevent the spread of the SARS-CoV-2 virus, Das et al. [33] found that perceived behavioral control was the most robust predictor on both intentions and actual social distancing behavior in Bangladesh. They also found that subjective norms were also significant. Similarly, in a study to investigate the influence of the several predictive factors on the intention to adopt preventive behaviors against COVID-19 in China, Ahmad et al. [13] developed a behavioral framework composed of TRA and TPB, and incorporated a set of additional factors (risk perception, epidemic knowledge, and risk aversion, among others). They found that governments' guidelines on epidemic prevention, risk perception, epidemic knowledge, perceived behavioral control, and subjective norms were the most important and influential factors on individuals' intention. The attitudes towards prevention exhibited a lower influence [13]. These findings suggest that other variables play a relevant role and should be considered to understand these behaviors, particularly among groups that appear to adhere less to these behaviors.

Overview of the Present Study

In this study, we used an adapted version of TPB to characterize the adoption of preventive behaviors and determine the factors with the most significant predictive power. We focused on behaviors that were already present and prescribed for the population. Consequently, we did not measure future intention. Previous studies have made similar adjustments that have faced the same difficulty [34,35]. Thus, we hypothesized that attitudes (H1), subjective norms (H2), and perceived behavioral control (H3) have a significant and positive effect on COVID-19 preventive behavior. Also, we evaluated knowledge about COVID-19 and we hypothesized that the knowledge level about COVID-19 has a significant and positive effect on adopting preventive behaviors (H4).

2. Materials and Methods

2.1. Materials

Data were obtained from a web-based survey aimed to assess preventive behaviors against the COVID-19 pandemic developed by the research team. The survey was organized into 10 sections. We used five for the purposes of this study. Section 1 measured preventive behaviors in three dimensions: social distancing, hygiene, and information seeking. Section 2 contained questions associated with attitudes to preventive measures. Section 3 measured subjective norms and perceived behavioral control. Section 4 addressed knowledge questions related to myths and realities about COVID-19. Finally, the last one included sociodemographic questions to characterize the respondents. Each of the TPB latent variables was constructed based on a set of items used in previous studies [16,17,19,29,36]. The items used to build the latent variables of the model are described in detail below.

Attitude. Measured through two concepts: perceived benefits (3 items) and barriers (3 items). Participants had to respond to the six items on a five-point Likert scale, from (1) "very unlikely" to (5) "very likely" (see Table 1).

Table 1. Mean values and standard deviation (*SD*) for each item measuring attitude, subjective norms, and perceived behavioral control.

	Item Description	**All Sample (n = 1004)**	
	Attitude [1]	**Mean**	**(*SD*)**
	Benefits (α = 0.84)		
Be1	If I adopt these measures, how likely am I to contract the coronavirus (COVID-19)?	2.25	(*0.93*)
Be2	If I adopt these measures, how likely am I to infect others with the coronavirus (COVID-19)?	2.09	(*0.91*)
Be3•	If I adopt these measures, I will be less anxious about contracting the coronavirus (COVID-19).	2.64	(*1.35*)
	Barriers (α = 0.67)		
Ba1	If I adopt these measures, they will cause me considerable inconvenience.	1.96	(*1.12*)
Ba2	If I adopt these measures, they will greatly change my daily routines.	2.92	(*1.39*)
Ba3	If I adopt these measures, I will take longer to perform some activities (e.g., preparing food, studying, etc.).	2.78	(*1.41*)
	Subjective Norms [2]		
	Normative beliefs (α = 0.81)		
NB1	My family thinks I should adopt the measures to prevent the coronavirus (COVID-19).	4.68	(*0.70*)
NB2	My best friends think I should adopt the measures to prevent the coronavirus (COVID-19).	4.36	(*0.85*)

Table 1. *Cont.*

	Item Description	All Sample (*n* = 1004)	
NB3	My colleagues think I should adopt the measures to prevent the coronavirus (COVID-19).	4.31	*(0.87)*
NB4	My teacher(s) thinks I should adopt the measures to prevent the coronavirus (COVID-19).	4.58	*(0.74)*
	Motivation to comply (α = 0.89)		
MC1	What my family thinks motivates me to adopt coronavirus prevention measures (COVID-19).	4.13	*(1.09)*
MC2	What my best friends think motivates me to adopt coronavirus prevention measures (COVID-19).	3.75	*(1.16)*
MC3	What my colleagues think motivates me to adopt coronavirus prevention measures (COVID-19).	3.57	*(1.12)*
MC4	What my teacher(s) thinks motivates me to adopt coronavirus prevention measures (COVID-19).	3.87	*(1.14)*
	Perceived Behavioral Control [2] (α = 0.52)		
PBC1	If I want to, I can adopt coronavirus prevention measures (COVID-19).	4.51	*(0.98)*
PBC2	I am confident that I am able to implement coronavirus prevention measures (COVID-19).	4.67	*(0.66)*
PBC3	I find it difficult to implement coronavirus prevention measures (COVID-19).	4.05	*(1.14)*

Mean values and standard deviations (*SD*) for each item. Cronbach's Alpha of each predictor variable included in the model (α) was incorporated. [1] Items were rated on a 5-point Likert scale, from [1] very unlikely to [2] very likely. [2] Items were assessed on a 5-point Likert scale, from [1] strongly disagree to [2] strongly agree. • item removed from the model.

Subjective norm. This was measured through four normative belief items and four compliance motivation items (see Table 1). The following significant groups were included: family, best friends, peers, and teachers. Higher education teachers were incorporated as a reference group, since college students in previous interviews mentioned them. Participants had to respond on a five-point Likert scale, from (1) "strongly disagree" to (5) "strongly agree".

Perceived behavioral control. We used three items to which participants had to respond through a five-point Likert scale, from (1) "strongly disagree" to (5) "strongly agree" (see Table 1), and that correspond to the three type of behaviors included: social distancing, hygiene, and information seeking.

Knowledge of COVID-19. To assess knowledge, we developed 11 items based on the study by Cheng and Ng [19]. Five items were assertions related to myths associated with COVID-19 (items K3, K6, K7, K9, and K10 in Table 2), and six were assertions of fundamental knowledge of COVID-19 (items K1, K2, K4, K5, K8, and K11 in Table 2). Respondents had to answer whether each statement was (1) "true", (2) "false", or (3) "I do not know". The items associated with myths were recoded to (0) true, and I do not know, (1) false, while items related to fundamental knowledge were recoded to (1) true, (0) false, and I do not know.

Table 2. Percentage of correct answers for each item of knowledge about COVID-19.

	Item Description	All Sample (n = 1004)
	Knowledge (α = 0.54)	**(%)**
K1•	Fever is one of the main symptoms of the coronavirus.	83.1
K2•	Young people like me cannot get sick from the coronavirus (COVID-19).	98.5
K3	Drinking alcohol prevents the spread of the coronavirus (COVID-19).	93.8
K4	There is still no vaccine to prevent the coronavirus.	78.0
K5•	Asymptomatic people infected with the coronavirus (COVID-19) cannot infect others.	97.5
K6	Vitamin C prevents the transmission of the coronavirus (COVID-19).	67.8
K7	Exposure to the sun or to temperatures above 25 °C prevents the spread of the coronavirus (COVID-19).	74.7
K8•	The incubation period of the virus can be 14 days.	86.3
K9	The coronavirus can be transmitted through mosquito bites.	61.5
K10•	The exchange of banknotes and coins CANNOT transmit COVID-19.	81.9
K11	The best protection against the coronavirus is physical distancing.	85.4

Note: Cronbach's Alpha of the predictor variable (α). • item removed from the model.

Preventive behaviors. Fourteen items were developed based on three behavioral dimensions: social distancing behavior (4 items), hygiene (7 items), and information seeking (3 items) (see Table 3). Participants had to respond on a frequency scale: (1) "never", (2) "rarely", (3) "sometimes", (4) "very often", (5) "always", and (6) "I do not know".

Table 3. Mean values and standard deviation (*SD*) for each item measuring preventive behaviors against COVID-19.

	Item Description	All sample (n = 1004)	
	Social distancing Behavior (α = 0.67)	**Mean**	***SD***
LB1	During this period, I avoided leaving my home and being in contact with other people.	4.63	*(0.78)*
LB2	During this period, I avoided going out to shop if I could do it online.	4.41	*(0.98)*
LB3	When I left home, I did so using the individual temporary permits for essential activities (requested at www.comisariavirtual.cl).	4.75	*(0.69)*
LB4	I avoided visits and holding or participating in social events such as parties and gatherings.	4.84	*(0.57)*
	Hygiene Behavior (α = 0.69)		
HB1	I wore a face mask on the street and in closed places (supermarket, pharmacy, etc.).	4.96	*(0.27)*
HB2	I disinfected purchased products with bleach or disinfectant.	4.11	*(1.20)*
HB3	I washed my hands when I got home with an alcohol-based hand sanitizer or soap and water.	4.86	*(0.48)*
HB4	I changed clothes immediately upon arriving home after going out.	3.48	*(1.36)*
HB5	I washed my hands with an alcohol-based hand sanitizer after sneezing, coughing, or wiping my nose.	4.13	*(1.10)*
HB6	When I left my home, I avoided hugging, shaking hands, or kissing on the cheek when greeting another person.	4.71	*(0.63)*
HB7	When I left my home, I tried to keep at least one meter of distance from other people.	4.71	*(0.58)*

Table 3. *Cont.*

	Item Description	All sample (n = 1004)	
	Information-seeking Behavior (α = 0.72)		
ISB1	I was aware of the sanitary or isolation restrictions of the territory where I live (mandatory quarantine, *cordon sanitaire*, curfew, etc.).	4.70	*(0.67)*
ISB2	I was regularly informed of the national evolution of the pandemic.	4.04	*(1.04)*
ISB3	I looked for updated information on the figures provided by the Ministry of Health.	3.47	*(1.29)*

Mean values and standard deviations (*SD*) for each item. Cronbach's Alpha of each behavior variable included in the model (α) was incorporated. Items were rated on the following frequency scale: [1] never, [2] rarely, [3] sometimes, [4] very often, and [5] always.

2.2. Procedure and Participants

The questionnaire was initially validated by the research team and the opinion of experts, to later be evaluated and validated in a focus group with 9 college students (five women) between 18 and 23 years old from various universities located in Santiago, Chile. The information obtained allowed the items to be adjusted and validated.

In Chile, the first confirmed case of COVID-19 was reported on 3 March 2020, almost two months after the WHO declared a health emergency. Due to the rapid spread of the virus, two weeks after the first confirmed case, the Chilean government declared a state of emergency, and closed the borders. The peak of confirmed infections during the first wave in Chile occurred between May and June 2020.

We studied preventive behaviors after the first contagion peak, using a web survey administered to undergraduate and graduate students at Universidad Andrés Bello between 24 August and 8 November 2020. One thousand and four students successfully completed the survey: 67.3% female, mean age 25.7 years old (SD = 7 years; range 18 to 50 years old), and 82.1% single. The average response time was 20 min.

At the beginning of the survey, all respondents were informed about their participation conditions and voluntarily agreed to participate in the study by signing an informed consent form. The Ethics Committee approved all procedures of the Universidad Andrés Bello.

2.3. Data Analysis

First, we conducted descriptive analyses to examine all items and measures included. Then, we used Cronbach's alpha coefficient (α-Cronbach) to evaluate the internal reliability of the scales, Kline [37] suggests values above 0.7 for highly consistent scales, and that values close to 0.6 are acceptable [38]. However, in analyses with latent variables, values lower than 0.7 are acceptable if the sample size is large enough to estimate all the model parameters [30,37]. Finally, a structural equations model was used to test the hypotheses of the study.

Structural Equation Model. We used the software IBM SPSS AMOS 26.0 to analyze the proposed structural equation model. We replaced missing values for each variable with the mean (the non-stochastic imputation method). The model fit was evaluated through the chi-squared (χ^2) indices, robustness of mean square error approximation (RMSEA), and comparative fix index (CFI) [39]. Finally, the proportion of variance explained by the model was measured using the Squared Multiple Correlation (R^2_{SMC}).

Model variables were entered as follows (see Figure 1). Attitude comprised the average of the benefits items and the average of the barrier items. The subjective norm was constructed using the average of the four normative belief items and the four items of motivation to comply. The three items that comprised the perceived behavioral control were entered as observed variables. The knowledge level about COVID-19 was included in the model using the score for each of the calculated factors. Finally, the adoption of preventive behaviors consisted of the average of each behavioral dimension: social distancing behavior, hygiene, and information seeking.

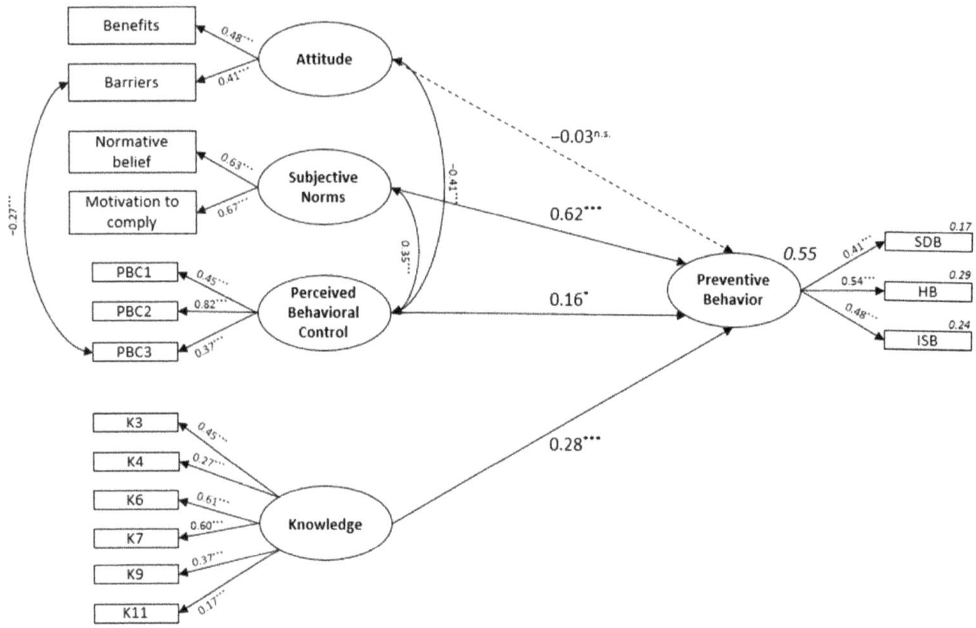

Figure 1. General model of COVID-19 preventive behavior. Two-headed arrows indicate the covariances suggested by the analysis to improve the fit of the model. Arrows depicted the direct relationships between constructs, and the values above them are the standardized regression coefficients that represent their strength. Dotted lines indicate non-significant coefficients. The values in cursive represent the proportion of explained variance of COVID-19 preventive behavior. SDB: Social distancing behavior, HB: Hygiene behavior, and ISB: Information-seeking behavior. *** $p < 0.001$; ** $p < 0.05$; * $p < 0.1$; n.s. non-significant at p-value ($p > 0.05$).

3. Results

3.1. Internal Consistency Analysis

An internal consistency analysis through Cronbach's alpha for each latent variable was developed.

Attitude. Before calculating internal reliability, the Be3 benefits item (see Table 1) was inverted, so the score was in the same direction as the remaining two items. The three benefit items' internal reliability (Be1, Be2, and Be3) was $\alpha = 0.45$. After the inverted item Be3 was eliminated, the reliability of the scale rose to $\alpha = 0.84$. The reliability of the barriers scale (items Ba1, Ba2, and Ba3) was $\alpha = 0.67$.

Subjective norm. The internal reliability of the four normative belief items (NB1, NB2, NB3, and NB4) was $\alpha = 0.81$, while for the four compliance motivation items (MC1, MC2, MC3, and MC4), it was $\alpha = 0.89$ (see Table 1).

Perceived behavioral control. Item PBC3 was inverted so that it went in the same direction as the remaining items. Considering the three items (PBC1, PBC2, and PBC3), the internal reliability of this scale was $\alpha = 0.52$ (see Table 1).

Knowledge. The internal reliability of the 11 knowledge items was $\alpha = 0.46$. We eliminated the five items with the lowest item-to-total correlation to improve the scale's reliability, increasing it to $\alpha = 0.54$ (see Table 2).

Preventive behavior. The internal reliability of the social distancing, hygiene, and information-seeking scales was $\alpha = 0.67$, $\alpha = 0.69$, and $\alpha = 0.72$, respectively (see Table 3).

3.2. TPB Model

The initial general model did not have a good fit (see Table 4). Three covariances were suggested by the modification indices, these were: between perceived behavioral control and subjective norm, between perceived behavioral control and attitude, and finally between the average of the perceived barriers (attitude) and the PBC3 item of perceived behavioral control. The incorporation of covariance between the TPB latent variables is reasonable because, although the variables are conceptually independent, empirically, they are free to correlate with each other [30]. Incorporating the third covariance is explained by the fact that the items are related to perceived barriers to performing preventive behaviors regarding COVID-19. Once we performed the adjustments to the model, the overall fit improved and reached the fit criteria' satisfactory values (see Table 4). The results of the Final General Model are shown in Figure 1.

Table 4. Fit Indexes for the Structural Models.

	χ^2	*df*	CFI	RMSEA	CI90%$_{RMSEA}$	Δdf
Initial General Model (n = 1004)	411.52	100	0.80	0.056	[0.050–0.061]	
Final General Model (n = 1004)	233.94	97	0.91	0.038	[0.031–0.044]	3

Fit criteria: Accepted values for the adjustment indices are based on Hair et al. (2010). The RMSEA should be less than 0.05 for a good fit, but values close to 0.1 are acceptable, as they indicate a moderate fit of the model to the data. For the CFI, values greater than 0.95 indicate an excellent fit, although values of up to 0.8 are acceptable. Δdf difference on degrees of freedom regarding the initial general model.

The proposed model explains 55% of the variance of preventive behavior adoption against COVID-19. Only the subjective norm, perceived behavioral control, and knowledge predictors obtained statistically significant coefficients, confirming H2, H3, and H4 and rejecting H1. Based on the value of the standardized coefficients (see Figure 1), the subjective norm was the strongest predictor of the behaviors studied. Finally, based on the mean values of the items that comprise the subjective norm reported in Table 1, family members and teachers would be the greatest motivators for adopting preventive measures.

4. Discussion

This study's main objective was to examine the influence of several predictor variables on adopting COVID-19 preventive behaviors in Chile, based on a solid theoretical framework such as TPB. Our findings indicate that the subjective norm is the strongest predictor of adopting preventive behaviors, followed by the knowledge level and the perceived behavioral control. Contrary to expectations, attitudes had no statistically significant influence on the adoption of preventive behaviors.

4.1. Subjective Norm

In line with previous studies [13,19,32,33], the subjective norm was positively and significantly related to adopting preventive behaviors. Compared to the other predictors included in the study, subjective norms had the highest predictive power on adopting hygiene, social distancing, and information-seeking behaviors to prevent the transmission and spread of the SARS-CoV-2 virus. Subjective norms reflect the individual's general observation of whether important reference groups (family, friends, teachers) accept, encourage, and perform the preventive behaviors during the crisis [33]. Then, in the context of the COVID-19 situation, social pressures from significant others on adopting preventive behaviors would be the primary motivator for adopting preventive hygiene and social distancing behaviors in Chile. These results are consistent with previous research that emphasizes the role of social norms in the adoption of these behaviors [40–43]. It is important to understand that interaction between different groups is not only essential for the adoption of preventive behaviors but also for strengthening community resilience [5].

Given our sample's characteristics (undergraduate and graduate students), our results suggest that higher education professors emerge as one of the reference groups that can stimulate students to adopt COVID-19 preventive behaviors. This is likely because social distancing restrictions and voluntary or mandatory isolation has been imposed since March 2020 by Chile's health authority, and most higher education students had to remain at home during the first months of the pandemic, carrying out their learning activities online. Thus, university students' most significant interactions in the early months of the pandemic were with their family members and professors.

4.2. Knowledge

The knowledge level is not commonly included as a predictive variable in TPB. However, since COVID-19 poses a new hazard to people's health, we believe that the knowledge level about COVID-19 should significantly influence preventive behavior adoption and our results supported this hypothesis. The knowledge level about COVID-19 was positively and significantly related to adopting preventive behaviors: the higher the knowledge level, the higher the actual adoption of preventive actions against COVID-19. This result confirms reports in previous studies indicating that knowledge is relevant in contexts of high uncertainty such as a pandemic [13–15,19]. In light of this result, we conclude that risk communication strategies implemented to counter COVID-19 require sharing knowledge and addressing the gaps they may find to be more effective. In fact, the TPB proposes that individuals assess each behavior. This assessment may be influenced by the knowledge they have about how useful or not this behavior might be, and therefore, this knowledge may influence attitudes. These findings are consistent with what West and colleagues [44,45] have argued about the need to implement interventions that recognize the need for people to understand how behaviors reduce the risk of being infected and that are associated with how the virus is transmitted.

4.3. Perceived Behavioral Control

The more people feel that they have the resources and skills necessary to perform and maintain preventive behavior, the more likely they are to adopt the behavior [27]. Our results show that perceived behavioral control had a positive and significant impact on the actual adoption of COVID-19 preventive behaviors, suggesting that study participants believe they can adopt measures to avoid infection and prevent the spread of the virus. This result agrees with that reported in previous studies, which reinforce the importance of perceived behavioral control over the final adoption of prevent behaviors [13,32,33,41]. Consequently, it is necessary that communication strategies reinforce the idea that people have the necessary skills to adopt preventive measures that protect them and their families against COVID-19.

4.4. Attitude

In general terms, studies based on TPB show that the predictive power of attitude on the adoption of preventive behaviors varies according to the behavior being studied [17,26,29]. Our research found no evidence that attitudes toward the preventive behaviors were associated with the actual adoption of these behaviors. This result is supported by recent studies related to preventive behaviors against COVID-19, in which attitudes had the least predictive power over the adoption of preventive behaviors [13,33,41]. It is likely that since this is a new hazard, there is considerable ambiguity about the real pros and cons of adopting preventive behaviors; therefore, other predictor variables such as subjective norms and knowledge of a pandemic or epidemic acquire greater significance because they facilitate individual decisions to adopt preventive behaviors. In line with our assumption, Cheng and Ng [19] also found that attitude had the least predictive power on the adoption of preventive behaviors in the SARS epidemic (2002–2004 epidemic), and argue that subjective norms are more relevant when the context or the behavior under study is ambiguous [19].

4.5. Implications

The results reported in our study have important implications for those responsible for designing programs, plans, and strategies to encourage the adoption of preventive behaviors in the face of an epidemic such as COVID-19. In light of the significant predictive power of the subjective norm on the adoption of preventive behaviors in front of COVID-19, the formulators of preventive programs (local governments, health institutions, educational establishments, etc.) should strategically consider the participation of reference groups as the main driving and motivating agents for the adoption of preventive behaviors. It is particularly worth noting the role that higher education teachers can play in implementing programs and strategies to encourage preventive behaviors in young adults in the face of a new epidemic outbreak. Higher education teachers are potential motivators for their students to adopt preventive behaviors and objectively represent a practical, direct, and permanent communication channel with young people.

The effectiveness of the communication strategies implemented by local governments and other preventive program formulators is a key element in combating the COVID-19 pandemic [46–49]. As this is a new hazard, people need accurate, reliable, and timely information to adopt effective preventive behaviors [46]. Our results are consistent with this premise in that the knowledge level about COVID-19 was positively and significantly related to the actual adoption of preventive behaviors. In light of this result, the more influential the future communication strategies associated with COVID-19, the more active and effective the adoption of preventive behaviors will be in the population.

Finally, perceived behavioral control was positively and significantly related to the adoption of preventive behaviors. Also, our results show that participants report having the necessary resources and skills to perform and maintain the preventive behaviors associated with hygiene and social distancing. Under this favorable scenario, formulators of preventive programs should focus on strengthening and preserving people's heightened sense of behavioral control, for example, by reinforcing the benefits of performing the behavior.

5. Conclusions

The study on the adoption of preventive behaviors to cope with the COVID-19 epidemic is new in the health sector, and to date, there are more questions than answers. Based on a solid theoretical framework such as TPB, we conclude that subjective norms represent the main predictor of the adoption of preventive behaviors in the face of COVID-19, followed by the knowledge level and perceived behavioral control. We also conclude that, since this is a new hazard where there is considerable ambiguity about the real pros and cons of adopting preventive behaviors, the attitude toward preventive behaviors decreases its predictive power on adopting preventive behaviors, probably when knowledge is limited.

Certain limitations in this study are worth mentioning. First, our study used cross-sectional data, which prevented us from generalizing our results and the inference causality. Future research should use longitudinal data for more robust results. Second, given that our sample consisted of undergraduate and graduate university students, we should be cautious about extending these results to other specific groups that are unlikely to possess the same level of knowledge. The model used could better predict the adoption of preventive behaviors and provide more comprehensive results if future research could include other population groups and obtain reliability coefficients over the 0.7 value for all the latent variables of the model. Finally, despite the studies' self-reporting nature, participants overall reported high compliance about these behaviors, while findings from other studies, such as the COVID-19 Social Study, also indicate that overall compliance with behaviors is high [50].

Author Contributions: Conceptualization, N.C.B., P.B.R., P.C.C. and J.V.C.; Formal analysis, N.C.B. and P.C.C.; Funding acquisition, N.C.B. and P.B.R.; Investigation, N.C.B., P.B.R., P.C.C. and J.V.C.; Methodology, N.C.B., P.B.R., P.C.C. and J.V.C.; Project administration, N.C.B.; Resources, N.C.B.; Software, P.C.C.; Supervision, N.C.B.; Validation, N.C.B., P.B.R. and P.C.C.; Visualization, P.C.C.; Writing—original draft, P.C.C. and J.V.C.; Writing—review & editing, N.C.B., P.B.R., P.C.C. and J.V.C. All authors have read and agreed to the published version of the manuscript.

Funding: This research was partially funded by the National Agency of Research and Development of Chile (ANID) of the Ministry of Science, Technology, Knowledge and Innovation of Chile through the National Fund for Scientific and Technological Research (grant no. ID1180996/FONDECYT/ANID) and by the Research Center for Integrated Disaster Risk Management (CIGIDEN), (grant no. ANID/FONDAP/15110017).

Institutional Review Board Statement: The study was approved by the Institutional Ethics Committee of Universidad Andres Bello (023 12 August 2020).

Informed Consent Statement: Informed consent was obtained from all subjects involved in the study.

Data Availability Statement: The data presented in this study are available upon request from the corresponding author. Data is not publicly available due to confidentiality agreement with participants.

Conflicts of Interest: The authors declare no conflict of interest.

References

1. World Health Organization. Coronavirus Disease (COVID-19) Dashboard. Available online: https://covid19.who.int/ (accessed on 15 February 2021).
2. World Bank. COVID-19 to Plunge Global Economy into Worst Recession since World War II. Available online: https://www.worldbank.org/en/news/press-release/2020/06/08/covid-19-to-plunge-global-economy-into-worst-recession-since-world-war-ii (accessed on 30 March 2021).
3. Betsch, C. How behavioural science data helps mitigate the COVID-19 crisis. *Nat. Hum. Behav.* **2020**, *4*, 438. [CrossRef]
4. Van Bavel, J.J.; Baicker, K.; Boggio, P.S.; Capraro, V.; Cichocka, A.; Cikara, M.; Crockett, M.J.; Crum, A.J.; Douglas, K.M.; Druckman, J.N. Using social and behavioural science to support COVID-19 pandemic response. *Nat. Hum. Behav.* **2020**, *4*, 460–471. [CrossRef]
5. Drury, J.; Carter, H.; Ntontis, E.; Guven, S.T. Public behaviour in response to the COVID-19 pandemic: Understanding the role of group processes. *BJPsych Open* **2021**, *7*, e11. [CrossRef]
6. Nivette, A.; Ribeaud, D.; Murray, A.; Steinhoff, A.; Bechtiger, L.; Hepp, U.; Shanahan, L.; Eisner, M. Non-compliance with COVID-19-related public health measures among young adults in Switzerland: Insights from a longitudinal cohort study. *Soc. Sci. Med.* **2021**, *268*, 113370. [CrossRef] [PubMed]
7. Snape, M.D.; Viner, R.M. COVID-19 in children and young people. *Science* **2020**, *370*, 286–288. [CrossRef]
8. Capone, V.; Caso, D.; Donizzetti, A.R.; Procentese, F. University student mental well-being during COVID-19 outbreak: What are the relationships between information seeking, perceived risk and personal resources related to the academic context? *Sustainability* **2020**, *12*, 7039. [CrossRef]
9. O'Reilly, A.; Tibbs, M.; Booth, A.; Doyle, E.; McKeague, B.; Moore, J. A rapid review investigating the potential impact of a pandemic on the mental health of young people aged 12–25 years. *Ir. J. Psychol. Med.* **2020**, 1–16. [CrossRef] [PubMed]
10. Xiong, J.; Lipsitz, O.; Nasri, F.; Lui, L.M.; Gill, H.; Phan, L.; Chen-Li, D.; Iacobucci, M.; Ho, R.; Majeed, A. Impact of COVID-19 pandemic on mental health in the general population: A systematic review. *J. Affect. Disord.* **2020**, *277*, 55–64. [CrossRef]
11. Conway, L.G., III; Woodard, S.R.; Zubrod, A. Social Psychological Measurements of COVID-19: Coronavirus Perceived Threat, Government Response, Impacts, and Experiences Questionnaires. 2020. Available online: https://psyarxiv.com/z2x9a/download?format=pdf (accessed on 15 February 2021).
12. Oosterhoff, B.; Palmer, C.A.; Wilson, J.; Shook, N. Adolescents' motivations to engage in social distancing during the COVID-19 pandemic: Associations with mental and social health. *J. Adolesc. Health* **2020**, *67*, 179–185. [CrossRef]
13. Ahmad, M.; Iram, K.; Jabeen, G. Perception-based influence factors of intention to adopt COVID-19 epidemic prevention in China. *Environ. Res.* **2020**, *190*, 109995. [CrossRef]
14. Chen, X.; Chen, H. Differences in preventive behaviors of COVID-19 between urban and rural residents: Lessons learned from a cross-sectional study in China. *Int. J. Environ. Res. Public Health* **2020**, *17*, 4437. [CrossRef] [PubMed]
15. Siegrist, M.; Luchsinger, L.; Bearth, A. The Impact of Trust and Risk Perception on the Acceptance of Measures to Reduce COVID-19 Cases. *Risk Anal.* **2021**, *0*, 1–14.
16. Montaño, D.E.; Kasprzyk, D. Theory of reasoned action, theory of planned behavior, and the integrated behavioral model. *Health Behav. Theory Res. Pract.* **2015**, *70*, 231.

17. Armitage, C.J.; Conner, M. Efficacy of the theory of planned behaviour: A meta-analytic review. *Br. J. Soc. Psychol.* **2001**, *40*, 471–499. [CrossRef]
18. Bish, A.; Michie, S. Demographic and attitudinal determinants of protective behaviours during a pandemic: A review. *Br. J. Health Psychol.* **2010**, *15*, 797–824. [CrossRef] [PubMed]
19. Cheng, C.; Ng, A.K. Psychosocial factors predicting SARS-preventive behaviors in four major SARS-affected regions. *J. Appl. Soc. Psychol.* **2006**, *36*, 222–247. [CrossRef]
20. Kassem, N.O.; Lee, J.W. Understanding soft drink consumption among male adolescents using the theory of planned behavior. *J. Behav. Med.* **2004**, *27*, 273–296. [CrossRef]
21. Kassem, N.O.; Lee, J.W.; Modeste, N.N.; Johnston, P.K. Understanding soft drink consumption among female adolescents using the Theory of Planned Behavior. *Health Educ. Res.* **2003**, *18*, 278–291. [CrossRef]
22. Kirk, T.N.; Haegele, J.A. Theory of planned behavior in research examining physical activity factors among individuals with disabilities: A review. *Adapt. Phys. Act. Q.* **2019**, *36*, 164–182. [CrossRef] [PubMed]
23. Montanaro, E.A.; Kershaw, T.S.; Bryan, A.D. Dismantling the theory of planned behavior: Evaluating the relative effectiveness of attempts to uniquely change attitudes, norms, and perceived behavioral control. *J. Behav. Med.* **2018**, *41*, 757–770. [CrossRef]
24. Shi, J.; Kim, H.K. Integrating risk perception attitude framework and the theory of planned behavior to predict mental health promotion behaviors among young adults. *Health Commun.* **2019**, *35*, 597–606. [CrossRef]
25. Zhang, N.; Campo, S.; Yang, J.; Janz, K.F.; Snetselaar, L.G.; Eckler, P. Effects of social support about physical activity on social networking sites: Applying the theory of planned behavior. *Health Commun.* **2015**, *30*, 1277–1285. [CrossRef] [PubMed]
26. Ajzen, I.; Et Fisbein, M. *Understanding Attitudes and Predicting Social Behavior*; Prentice Hall: Englewood Cliffs, NY, USA, 1980.
27. Ajzen, I. The theory of planned behavior. *Organ. Behav. Hum. Decis. Process.* **1991**, *50*, 179–211. [CrossRef]
28. Ajzen, I. Perceived behavioral control, self-efficacy, locus of control, and the theory of planned behavior 1. *J. Appl. Soc. Psychol.* **2002**, *32*, 665–683. [CrossRef]
29. Ajzen, I. The theory of planned behaviour: Reactions and reflections. *Psychol. Health* **2011**, *26*, 1113–1127. [CrossRef] [PubMed]
30. Ajzen, I. The theory of planned behavior: Frequently asked questions. *Hum. Behav. Emerg. Technol.* **2020**, *2*, 314–324. [CrossRef]
31. McEachan, R.R.C.; Conner, M.; Taylor, N.J.; Lawton, R.J. Prospective prediction of health-related behaviours with the theory of planned behaviour: A meta-analysis. *Health Psychol. Rev.* **2011**, *5*, 97–144. [CrossRef]
32. Barile, J.P.; Guerin, R.J.; Fisher, K.A.; Tian, L.H.; Okun, A.H.; Vanden Esschert, K.L.; Jeffers, A.; Gurbaxani, B.M.; Thompson, W.W.; Prue, C.E. Theory-based Behavioral Predictors of Self-reported Use of Face Coverings in Public Settings during the COVID-19 Pandemic in the United States. *Ann. Behav. Med.* **2021**, *55*, 82–88. [CrossRef] [PubMed]
33. Das, A.K.; Abdul Kader Jilani, M.M.; Uddin, M.S.; Uddin, M.A.; Ghosh, A.K. Fighting ahead: Adoption of social distancing in COVID-19 outbreak through the lens of theory of planned behavior. *J. Hum. Behav. Soc. Environ.* **2020**, *31*, 373–393. [CrossRef]
34. Ashraf, M.A.; Joarder, M.H.R.; Ratan, S.R.A. Consumers' anti-consumption behavior toward organic food purchase: An analysis using SEM. *Br. Food J.* **2019**, *121*, 104–122. [CrossRef]
35. George, J.F. Influences on the intent to make Internet purchases. *Internet Res.* **2002**, *12*, 165–180. [CrossRef]
36. Bae, S.Y.; Chang, P.-J. The effect of coronavirus disease-19 (COVID-19) risk perception on behavioural intention towards 'untact'tourism in South Korea during the first wave of the pandemic (March 2020). *Curr. Issues Tour.* **2021**, *24*, 1017–1035. [CrossRef]
37. Kline, R.B. *Principles and Practice of Structural Equation Modeling*; Guilford publications: New York, NY, USA, 2015.
38. Ursachi, G.; Horodnic, I.A.; Zait, A. How reliable are measurement scales? External factors with indirect influence on reliability estimators. *Procedia Econ. Financ.* **2015**, *20*, 679–686. [CrossRef]
39. Hair, J.F. *Multivariate Data Analysis*; Prentice Hall: Upper Saddle River, NJ, USA, 2009.
40. Latkin, C.A.; Dayton, L.; Yi, G.; Konstantopoulos, A.; Boodram, B. Trust in a COVID-19 vaccine in the US: A social-ecological perspective. *Soc. Sci. Med.* **2021**, *270*, 113684. [CrossRef]
41. Lee, Y.; Li, J.Y.Q. The role of communication transparency and organizational trust in publics' perceptions, attitudes and social distancing behaviour: A case study of the COVID-19 outbreak. *J. Contingencies Crisis Manag.* **2021**, 1–17. [CrossRef]
42. Masser, B.M.; Hyde, M.K.; Ferguson, E. Exploring predictors of Australian community members' blood donation intentions and blood donation–related behavior during the COVID-19 pandemic. *Transfusion* **2020**, *60*, 2907–2917. [CrossRef]
43. Kim, H.K.; Tandoc, E.C., Jr. Wear or Not to Wear a Mask? Recommendation Inconsistency, Government Trust and the Adoption of Protection Behaviors in Cross-Lagged TPB Models. *Health Commun.* **2021**, 1–9. [CrossRef]
44. West, R.; Michie, S.; Rubin, G.J.; Amlôt, R. Applying principles of behaviour change to reduce SARS-CoV-2 transmission. *Nat. Hum. Behav.* **2020**, *4*, 451–459. [CrossRef]
45. Hameleers, M.; van der Meer, T.G.; Brosius, A. Feeling "disinformed" lowers compliance with COVID-19 guidelines: Evidence from the US, UK, Netherlands and Germany. *Harv. Kennedy School Misinformation Rev.* **2020**, *1*. [CrossRef]
46. World Health Organization. *Risk Communication and Community Engagement Readiness and Response to Coronavirus Disease (COVID-19): Interim Guidance, 19 March 2020*; World Health Organization: Geneva, Switzerland, 2020.
47. Dryhurst, S.; Schneider, C.R.; Kerr, J.; Freeman, A.L.; Recchia, G.; Van Der Bles, A.M.; Spiegelhalter, D.; van der Linden, S. Risk perceptions of COVID-19 around the world. *J. Risk Res.* **2020**, *23*, 994–1006. [CrossRef]
48. Malecki, K.; Keating, J.A.; Safdar, N. Crisis communication and public perception of COVID-19 risk in the era of social media. *Clin. Infect. Dis.* **2021**, *72*, 697–702. [CrossRef] [PubMed]

49. Zhang, L.; Li, H.; Chen, K. Effective risk communication for public health emergency: Reflection on the COVID-19 (2019-nCoV) outbreak in Wuhan, China. *Healthcare* **2020**, *8*, 64. [CrossRef] [PubMed]
50. University College London. Understanding the Psychological and Social Impact of the Pandemic. Available online: https://www.covidsocialstudy.org/ (accessed on 8 March 2021).

sustainability

MDPI

Article

The Effectiveness of Online Education during Covid 19 Pandemic—A Comparative Analysis between the Perceptions of Academic Students and High School Students from Romania

Gina Ionela Butnaru [1], Valentin Niță [1], Alexandru Anichiti [2,*] and Geanina Brînză [2]

[1] Department of Management, Marketing and Business Administration, Faculty of Economics and Business Administration, Alexandru Ioan Cuza University of Iași, 700505 Iași, Romania; gina.butnaru@uaic.ro (G.I.B.); valnit@uaic.ro (V.N.)

[2] Department of Business Administration, Ștefan cel Mare University of Suceava, 13 University Street, 720229 Suceava, Romania; brinza.geanina@gmail.com

* Correspondence: alexandru.anichiti@usm.ro

Abstract: The COVID-19 pandemic has disrupted normal activities such as going to school, moving education online. Based on data gathered through a survey (N = 784), this paper analyses students' perceptions regarding the effectiveness of online education in a period when this type of education is the only available option. Although several studies suggest that online education can be as effective as traditional education which requires attendance, few studies have focused on learner satisfaction with online instruction, particularly in the transition to online learning from traditional approaches. The results indicate that students react differently to online education, and their reaction is based on their proficiency in using online tools, their ability to technically access online courses, and the instructors' manner in conducting learning activities.

Keywords: the COVID-19 pandemic; online education; online courses; the satisfaction of students; higher education

check for updates

Citation: Butnaru, G.I.; Niță, V.; Anichiti, A.; Brînză, G. The Effectiveness of Online Education during Covid 19 Pandemic—A Comparative Analysis between the Perceptions of Academic Students and High School Students from Romania. *Sustainability* **2021**, *13*, 5311. https://doi.org/10.3390/su13095311

Academic Editors: Miltiadis D. Lytras, Abdulrahman Housawi and Basim Alsaywid

Received: 9 April 2021
Accepted: 4 May 2021
Published: 10 May 2021

1. Introduction

In order to prevent the spread of the SARS-COV2 virus, Romania developed a strategy to prevent and control the situation, it includes: the cancellation or closure of flights, the closure of restaurants and hotels, and mandatory quarantine for infected people or people who are suspected to be infected. Thus, institutions including schools, high schools and universities have temporarily suspended their traditional courses, migrating to a model of online education.

The SARS-CoV2 virus has threatened the activity of high schools, colleges and universities, which is why the issue of whether or not to continue teaching and learning has been raised, provided that the health of students and of education staff is protected. Educational institutions preferred to close traditional (face-to-face) teaching activities, including laboratories, and transferred teaching activities to the online environment in order to prevent the spread of the virus. Forcibly moving educational activity to the online environment allows for flexibility in teaching and learning because courses are easily accessed. Despite the crisis, courses were transferred online at an astonishing and unprecedented speed [1–3].

As a distinctive feature, this study aims to compare the perceptions of university students (both undergraduate and postgraduate) and of high school students regarding the effectiveness of online courses. Although the study of university students'/high school students' perceptions regarding the effectiveness of online courses has been an intensely studied topic by authors such as Bentley [4], Bali and Liu [5], and Platt, Raille and Yu [6], their studies used a scenario in which there was the possibility of attending both online courses and courses in the study rooms (face to face with university students/high school students). Therefore, this research analyses the perceptions of undergraduate and postgraduate students and of high school students in a time when there is only the possibility of online courses, in a state of emergency caused by the COVID-19 pandemic.

In order to study the perceptions of undergraduate and postgraduate students and of high school students regarding the effectiveness of online courses, we used a survey applied to undergraduate and postgraduate students from Alexandru Ioan Cuza University of Iași and students from Virgil Madgearu Economic Technological High School of Iași, between 25 March and 30 April 2020. The survey included questions with Likert-type scale responses, with control questions regarding age, gender, education level, access to the use of technologies, etc. A total of 784 responses were received in this survey.

Therefore, this paper is structured in the following sections: introduction, literature analysis and development of research hypotheses, research methodology, results, discussions, and conclusions.

2. Literature Analysis

2.1. Advantages vs. Disadvantages of Online Education

The impact of the COVID-19 pandemic has led to the temporary interruption of educational activities in the classroom. University students and high school students in their final years are in an unprecedented situation, which does not allow for a clear perspective of the future. The length of the pandemic and its effects on daily life, costs and other financial issues can directly affect the continuation of the education of university students and high school students. The vulnerability caused by the disturbances in the academic space is worrying. The situation of both undergraduate and postgraduate students has generated unfavorable conditions, such as the necessity to drop out of education. A feeling of exclusion was created by the pandemic situation, outlining an image of inequity in the academic education system [7–9].

Online education has the potential to transform the education system by expanding educational opportunities, transforming student populations and encouraging the development of new pedagogical methods [6,10,11], making the learning process more reliable, more efficient, and less stressful for both instructors and students. Although there are studies suggesting that online and traditional education are comparable in terms of learning outcomes, it is also admitted that online learning is perceived as lacking in interactivity compared with classroom learning [5,12]. However, in a study including 156 students, Fortune, Spielman and Pangelinan [13] concluded that there was no statistically significant difference in learning preferences between students attending online courses and students attending courses in person.

The effectiveness of online education has shown a number of advantages due to increased flexibility and learning opportunities: easy access to experts, exposure to educational environments, a wide range of types of courses, and joining student communities. There are also several disadvantages of online education, such as: internet browsing issues, computer compatibility, or technical issues [14,15].

At the beginning of the COVID-19 pandemic, students had to reorganize their daily schedule to adapt to a situation of isolation. Those studying abroad had to go back home, but, at the same time, many of them were blocked due to the closure of airports and borders. The lack of socialization affected students and their socio-emotional balance suffered, especially in young people with pre-existing problems of this nature. Students claimed that main the effects of isolation were anxiety and depression [16].

Therefore, this study proposes to test the following hypotheses:

Hypothesis 1 (H1). *The perception of the effectiveness of face-to-face courses will influence students' desire for physical class attendance.*

Hypothesis 2 (H2). *The perception of the effectiveness of face-to-face courses will negatively influence the perception of the effectiveness of online courses.*

2.2. Premises of a Good Experience with Online Education

A significant relationship has been reported between the degree to which students feel comfortable using the internet and their general feeling of satisfaction with the online experience [17].

Research has proven that advances regarding information technology and the evolution of computers positively affect the approach of university/high school students on learning in contemporary educational environments [18,19]. Mobile devices and easy internet access have changed the paradigm of communication in the context of the education system. Social networks used in everyday life can bring about changes in teaching methods and generate improvements by stimulating and involving students in the learning process [20,21].

In the context of the COVID-19 pandemic, the sudden and forced transition from traditional education to online education has had negative effects on the preparation of university students and high school students [22,23]; these effects are generated by problems with access to technology and internet networks, and by the diminished quality of teaching [24].

In Romania, during the period of isolation, parents had to ensure students' access and participation in online learning activities organized by their educational institutions. The technology required to participate in online courses is as follows: laptop/smart phone/tablet and internet access. This technology may belong to the students' families or educational institutions. The damages caused to loaned technical and communication equipment falls entirely on the parents or legal guardians. The access to platforms, digital learning resources and virtual libraries is free [24,25].

According to Kebritchi et al. [26], in order to support students in online education, prior training in technical skills is necessary for the efficient use of computers and the internet. Other factors are important as well, such as: the students' perception and attitude towards the internet, their level of knowledge of English, and how the students manage their time. The success of online education is given by the dimensions suggested by Hung et al. [27]: learning motivation, self-directed learning, computer and internet quality, effectiveness of online communication, and student control.

Therefore, this study proposes to test the following hypotheses:

Hypothesis 3 (H3). *The students who do not possess the technology needed for easy access to online courses will perceive them as ineffective compared with the students who possess the technology.*

Hypothesis 4 (H4). *The students who can browse the internet easily will perceive the effectiveness of online courses differently compared with the other students.*

2.3. Adapting Educational Institutions to Online Education vs. the Effect on the Trust of University Students and High School Students

Educational institutions have had to adapt to the restrictions created by the COVID-19 pandemic [28,29]. A series of applications, platforms and educational resources have been launched to support high school and university students, teachers and professors, parents, educational institutions and school managers, with the aim of facilitating the study process of high school/university students, providing social support and facilitating interaction during the closure of educational institutions [3,30,31].

Online courses were initially designed in the format of a teacher/professor teaching material in a classroom, transmitted and stored with the help of technology, so that the information could be accessed later and repeatedly by students. Students seem to appreciate this format because they have the possibility of easy access to the information. However, undergraduate students tend to be more conservative, which is why they do not appreciate the format of online courses, compared with postgraduate students who are more open to modern work methods in the educational environment, and prefer participatory methods and direct interaction within the university [32,33]. Online courses require more discipline from the students. Higher success rates for online courses can be seen in postgraduate students due to their maturity, while undergraduate students are more disinterested in this format, and high school students are much less inclined to online courses [10]. High school and university students feel attached to the teachers/professors and/or schoolmates with whom they work. Interaction and social relationships give individuals a sense of security, trust and love, and these aspects can generate significant success for the educational process [34,35]. High aspirations, care and positive attitude towards learning, motivation and encouragement of group members, professors for university students, and teachers for high school students, can contribute to positive academic results. University students and high school students who do not receive significant moral support from their family compensate with the moral and social support provided by teaching staff, which can have significant positive effects on school results and general success [36,37].

The students who appreciate participation in traditional classroom courses consider it is important to interact with academia for face-to-face meetings with professors and schoolmates, and for immediate feedback generated by the stated circumstances [38,39]. High school students' perception in the pre-academic environment is similar. The challenge of transferring contents from a traditional teaching format to an online format generates certain problems, and there are differences in perception [26]. Thus, students consider online courses ineffective due to the effects generated by distance learning: limited interaction with the school environment and schoolmates, and increased interaction with technology [40,41]. The teaching staff faces problems regarding content delivery and lower student interest due to the absence of face-to-face meetings, lower control over student groups, and a blockage in the ability to correct these situations [26,42].

Therefore, this study proposes to test the following hypotheses:

Hypothesis 5 (H5). *Students' desire for physical class attendance will generate the perception of ineffectiveness of online courses.*

Hypothesis 6 (H6). *The capacity of educational institutions to manage online courses will have a positive effect on the students' perception of the effectiveness of online courses.*

2.4. Training of the Teaching Staff Regarding Online Education vs. the Effect on the Trust of University Students and High School Students

The COVID-19 pandemic has caused unprecedented damage to the education system, especially in educational institutions, and in particular to teaching staff—the most critical resource of any educational institution, which have to face financial, physical, and mental challenges [43,44].

Teaching staff use teaching methods virtually, which are not affected by the additional demands generated by the context of the coronavirus pandemic. Every academic or pre-academic educational institution have the necessary tools for the formation of a virtual class, as an extension of the real class. However, the ability of the teaching staff to adapt to the new requirements is greatly affected by the formation of the working skills with new technologies. The teaching staff who had access to distance learning courses and digital resources have a lower risk of maladaptation to the difficulties generated by online courses and have the confidence to continue teaching according to the new conditions. In the context generated by the COVID-19 pandemic, the simplest option of the teaching staff is to teach their classes by video, either live or digitally transmitted [8,30,31].

It is essential to ensure the continuity of teaching in the online environment, so the teaching staff must build their own skills to be able to work in virtual environments that are of high quality and technological complexity, for which special training is not necessarily required [16,45].

Traditional teaching used body language, facial expressions, and the teacher's voice as important learning tools. The transition to online teaching courses created restrictions on the above and highlighted the importance of emphasis in teachers' voice and reducing the amount of speech presented to high school students on key knowledge, using a smaller number of words [46].

Teaching staff must take into account three elements when designing the content for online courses: theoretical materials, pedagogy, and technology. If these elements are perfectly combined, they ensure efficient and high-quality teaching [11,47]. Transmission of content with the help of technology is much more efficient if the teaching staff create collaborative, reflective activities and draw clear criteria regarding evaluation—in online education, evaluation is carried out with the help of digital technologies [48]. According to Kebritchi et al. [26] and Duncan [49], the transition from traditional courses to online courses requires content redesign strategies to integrate technology. Redesigning courses lowers costs and can increase students' memorization speed. Course redesign strategies involve interactive learning in the online environment with the support of multimedia tools.

Cross [50] and Song, Kim and Luo [51] consider that the teacher-student relationship is based on a series of experiences and expectations. Students appreciate teachers who provide quick, permanent and open feedback, who provide prompt responses, who know the curricula content and the teaching methods in detail, who constantly assist the students in the preparation stages, and who have a positive behavior in the instructional/educational act. Students want their teachers to have initiative, to take the first step in the educational activities, to prepare the students for the online classes, and to have a positive behavior, as the support and modelling offered by the teachers contribute to their subsequent success. Mather and Sarkans [38] also identified factors that can improve the learning process and student satisfaction, such as: promptness of teachers'/professors' responses, clarity of tasks, and easy access to the content. In online learning activities, students appreciate flexibility and convenience. Cross [50] concludes that all the teaching staff, regardless of the level of education (academic or pre-academic), should pay more attention to students' perceptions and consider them in instructional/educational activities.

Educational institutions have adopted educational platforms that have the advantage of collecting digital teaching resources, whose availability and quality are particularly important; these are alternatives to online courses conducted by videoconference [8,52].

Teaching staff were trained within projects aiming to capitalize on new technologies in teaching-learning activities. The development of the digital skills of the teaching staff is a priority for the education system. Training through courses or projects for online education aims to use online tools in distance learning activities with university students and/or high school students [53,54].

Isolation has made the situation of education in Romania uncertain and uncomfortable, the lack of clear, coherent and specific measures to continue instructive/educational activities has created stress and anxiety for most teaching staff, high school students and parents/legal guardians [43,55,56].Teachers were forced to identify new methods to offer all high school students the opportunity to participate directly/indirectly in online courses, which caused them to be more lenient and more understanding of students' personal situations. The materials sent by the teachers through the learning platforms, the work tasks sent to the high school students in a less favorable situation, the exclusively online communication, and the lower quality of the pedagogical act, in addition to the students' lack of interest, lowered the teachers' expectations regarding the quality of the topics and contents to be prepared for classes [43,57].

Therefore, this study proposes to test the following hypotheses:

Hypothesis 7 (H7). *The capacity of the teaching staff to be more lenient will have a positive effect on the perception of the effectiveness of online courses.*

Hypothesis 8 (H8). *The perception of the ability of teaching staff to manage online courses will have a positive effect on the perception of the adaptation of the educational institution to online education.*

Figure 1 shows a conceptualization of the model regarding the effectiveness of online courses:

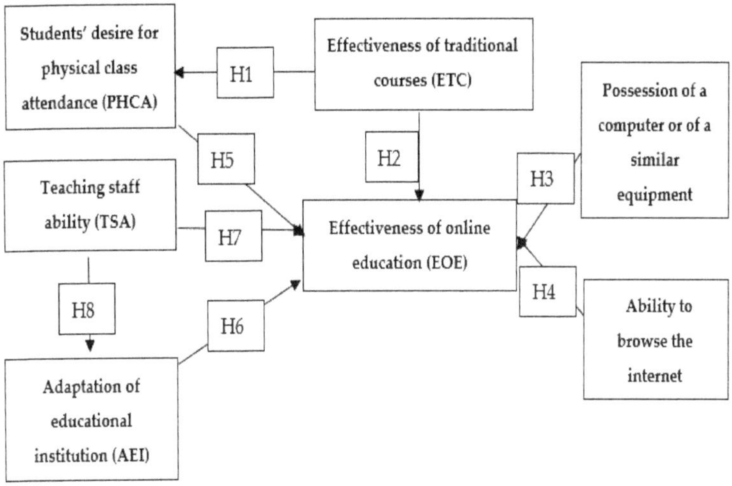

Figure 1. Conceptual model regarding the effectiveness of online courses.

As such, this study used the variables presented in Table 1.

Table 1. List of variables used in this study.

Variable	Description	Type of Variable
PHCA	Students' desire for physical class attendance.	Construct
ETC	Perceptions of university students and of high school students regarding the effectiveness of face-to-face courses in the classroom.	Construct
EOE	Perceptions of university students and of high school students regarding the effectiveness of online courses.	Construct
AEI	Perceptions of university students and of high school students regarding the adaptation of educational institutions to online education requirements.	Construct
TSA	Perceptions of university students and of high school students regarding the adaptation of teaching staff to the new requirements of online education.	Construct
Gender	A categorical variable, which takes the value 0 when the respondent's gender is male and the value 1 when the gender is female.	Categorical
Age	Represents the respondents' age.	Numerical
Ability to browse the internet	The respondents who "strongly agreed" or "agreed" with the statement "I am able to easily access the internet in accordance with the educational interests" were classified as respondents who had the capacity to use technologies that are able to access online courses and they were encoded with the value 1. Those who expressed "total disagreement" or "disagreement" with the statement from above were classified as university/high school students who had no technical knowledge required to access and participate in the online courses, and were encoded with the value 0.	Dummy/ binary
Possession of a computer or of similar equipment	The respondents who "strongly agreed" or "agreed" with the statement "*I have a computer which allows me to participate easily in online courses*" were encoded with the value 1, and those who expressed "total disagreement" or "disagreement" with the statement from above were encoded with the value 0.	Dummy/ binary

Source: authors' contribution.

3. Materials and Methods

3.1. Description of the Research Method

The current study included 784 participants: 575 women (74.34%), and 209 men (26.66%). The average age of the participants was 20.30 years. The participants in this study (n = 784) were undergraduate, postgraduate, and high school students. The recruitment of participants was performed through an e-mail sent to university students and high school students. The participants were instructed to take into account their general experience regarding education and online courses when answering the questions, instead of assessing the quality of the courses they attended online. Thus, the interested students followed a link to an online survey conducted through Google Forms, where they answered questions aimed at measuring their perception of the effectiveness of online education in general, and in the situation caused by the COVID-19 pandemic in particular.

3.2. The Sample

Although there is no general method for determining the optimal sample size, researchers have proposed recommendations and guidelines regarding appropriate sample sizes when performing a factor analysis. Field [58] recommended at least 10 subjects per variable to reduce sampling errors. Comrey and Lee [59] provided the following recommendations for appropriate sample sizes: 100 = poor, 200 = ok, 300 = good, 500 = very good, 1000 or more = excellent.

In the present research, the population analyzed included 784 students participating in online education from Alexandru Ioan Cuza University of Iași and Virgil Madgearu Technological Economic High School of Iași. The population was selected randomly. All participants had the opportunity to ask questions or communicate their concerns about the survey.

3.3. Work Stages

The survey was evaluated in two stages, as follows:

(1) Face validity: we made sure that the respondents understood the questions. In this respect, there were no misleading questions using multiple negations or unclear formulations;
(2) Construction validity testing: we used an analysis of exploratory factors and reliability, i.e., Cronbach's alpha tests.

In the first stage, we tested by descriptive statistics and regression analysis the differences between different groups of students (high school students, undergraduate students, postgraduate students, technically challenged and with technical skills, financially capable and financially incapable).

In the second stage, a model of structural equation was tested to analyze the impact of different beliefs on the perception of the effectiveness of online education.

A linear regression analysis was used to test the hypotheses H3 and H4, and an SEM model was used to test the hypotheses H1, H2, H5, H6, H7 and H8.

3.4. Measures

In this study we used a survey as a data collection tool. We used a five-point Likert scale. Thus, the reliability for each variable used in this study was analyzed by the calculation of Cronbach's alpha. Following the analysis of the exploratory factors, the following scales were constructed, presented in Table 2.

Table 2. Questions of the survey and outcomes of the factorial exploratory analysis.

Item/Latent Variable	Questions/Observed Variable	Cronbach's Alpha
		0.8738
PHCA	PHCA1. I wish to go back to school/university as soon as possible. PHCA2. I wish to see my teachers/professors again. PHCA3. I am happier when the courses take place in the school/academic environment. PHCA4. I wish to see my schoolmates again.	**0.8361**
ETC	ETC1. I feel that the physical attendance of the courses is necessary for an effective education. ETC2. I believe that online classes are less motivating than the classes taking place in a classroom. ETC3. I want to attend classes in a physical classroom.	**0.8458**
EOE	EOE1. I am willing to actively communicate electronically with my schoolmates and my teachers/professors. EOE. I could easily adapt to online courses without the direct assistance of the teaching staff. EOE3. I am capable of self-discipline and finding time to study at home, too. EOE4. I believe that a full course can be taught without difficulty through online programs/platforms. EOE5. Online education is efficient during COVID-19 pandemic.	**0.8277**
AEI	AEI1. The school/university was prepared to manage the situation of online teaching. AEI2. The school/university offered to high school/university students the required technique (user's guide) to access the online courses. AEI3. The school/university will be able to guarantee my safety in the future.	**0.7151**
TSA	TSA1. The teaching staff were more lenient with the high school/university students in the process of adaptation to online education. TSA2. The teaching staff manage to transmit effectively the main ideas and information to high school/university students through online programs. TSA3. The online communication with the teaching staff took place without difficulties.	**0.7621**

Source: our own calculations based on Stata statistical analysis software.

4. Results

4.1. Results of Regression Equation

The following multiple regression model was used to test the hypotheses H3 and H4 of this study, and to identify the factors influencing the perception of the effectiveness of online courses:

$$Y = \beta_0 + \beta_1 X_1 + \beta_2 X_2 + \ldots + \beta_n X_n + \varepsilon, \quad (1)$$

In this equation, Y is the dependent variable and X is the independent variable. The value of β_0 (which can be negative, positive or zero) is called the intercept, while the value of β_1 is called the "slope" or "regression coefficient".

As described in the introductory section, the statistical analysis was performed using the survey data. The statistical analysis was performed using the statistical program Stata.

The respondents who "strongly agreed" or "agreed" with the statement "I am able to easily access the internet in accordance with the educational interests" were classified as respondents who had the capacity to use the technologies at hand to access the online courses. Those who expressed "total disagreement" or "disagreement" with the statement from above were classified as university/high school students who did not have the required technical knowledge to access and participate in online courses.

The respondents who "strongly agreed" or "agreed" with the statement "I have a computer which allows me to easily participate in online courses" were classified as respondents used to online communication, and those who expressed "total disagreement" or "disagreement" with the statement from above were classified as university/high school students who did not have the required technical knowledge to access and participate in online courses.

Table 3 presents the descriptive statistics of the sample included in this study. The current study included 784 participants: 575 women (74.34%), and 209 men (26.66%). The average age of the participants was 20.30 years, with 70.16% of the respondents between 18 and 25 years old, 23.47% between 14 and 18 years old, and the remaining 50 respondents being over 25 years old. Regarding the categorization according to the education level, 248 (31.63%) were high school students, 405 (51.66%) were undergraduate students, while the remaining 131 (16.71%) were postgraduate students.

Table 3. Descriptive statistics.

			Number	Percentage (%)
Gender	Male		209	26.66
	Female		575	74.34
Age	14–18		184	23.47
	18–25		550	70.16
	over 25		50	6.37
Education level	High school	9th grade	40	5.10
		10th grade	89	11.35
		11th grade	63	8.04
		12th grade	56	7.14
	Undergraduate students	1st year	65	8.29
		2nd year	247	31.51
		3rd year	93	11.86
	Postgraduate students (master's studies)	1st year	56	7.14
		2nd year	75	9.57
Ability to browse the internet	Yes		55	7.00
	No		661	84.31
	Doesn't know/undecided		68	8.69
Possession of a computer or of similar equipment	Yes		629	80.22
	No		93	11.86
	Doesn't know/undecided		62	7.92

Source: our own calculations based on Stata statistical analysis software.

In this study, the level of analysis also involved obtaining descriptive statistics. As shown in Table 4, the perception of the scales differs among groups.

Table 4. Descriptive statistics according to the educational level.

Construct	High School		Undergraduate		Postgraduate	
	Mean	Std. Error	Mean	Std. Error	Mean	Std. Error
PHCA	3.571	0.068	4.034	0.043	3.818	0.077
ETC	3.527	0.063	4.051	0.040	3.768	0.077
EOE	3.734	0.064	3.192	0.054	3.495	0.084
AEI	2.579	0.075	2.790	0.047	3.123	0.095
TSA	2.908	0.063	2.975	0.048	3.438	0.065

This item is based on the five-point scale: 5 = strongly agree, 4 = agree, 3 = don't know/uncertain, 2 = disagree and 1 = strongly disagree; Source: our own calculations based on Stata statistical analysis software.

Table 5 presents descriptive statistics regarding the perceptions of university students and high school students considering the technology they have in order to access the online courses, as well as the ease of browsing the internet. According to the results, university students and high school students who declare that they are comfortable to browse the internet and have a computer or similar equipment allowing them to access the online courses will perceive online education as more effective. However, a regression equation is needed to show that this difference is also statistically significant in order to be able to test the hypotheses H3 and H4.

Table 5. Descriptive statistics according to groups of knowledge regarding the use of technology to access the courses.

Construct	Technologically Challenged (Cannot Browse the Internet Easily)		Has a Suitable Computer or Similar Equipment	
	Yes	No	Yes	No
	Mean	Mean	Mean	Mean
PHCA	3.852	3.636	3.845	3.881
ETC	3.748	3.600	3.753	3.795
EOE	3.425	2.036	3.426	2.344
AEI	2.804	3.236	2.790	3.333
TSA	2.937	2.072	2.939	2.322

This item is based on the five-point scale: 5 = strongly agree, 4 = agree, 3 = don't know/uncertain, 2 = disagree and 1 = strongly disagree. Source: our own calculations based on Stata statistical analysis software.

Table 6 shows the parallel correlations between the variables, which are generally weak, without a correlation above 0.411. Even if there are some variables significantly correlated with each other, we can state that there are no factors preventing consistent results. This is also confirmed by the verification of the variation inflation values (VIF), which are lower than 10.

Table 6. Variable testing and validation.

	VIF	EOE	Age	Gender	Ability to Browse the Internet	Possession of a Computer or of Similar Equipment
EOE	-	1.000				
Age	2.090	0.048	1.000			
Gender	1.083	0.120 *	0.145 *	1.000		
Ability to browse the internet	1.294	0.383 *	0.071	0.066	1.000	
Possession of a computer or of similar equipment	1.511	0.367 *	0.100 *	0.013	0.411 *	1.000

Note: * indicate the fact that $p < 0.05$. Source: our own calculations based on Stata statistical analysis software.

Table 7 shows the regression coefficients. The results of the equation show that university students and high school students who own a computer/laptop which allows them to access online courses perceive online courses as more effective than those who do not possess this technology which is vital in an online education environment. In addition, university students and high school students who are accustomed to the online communication, as well as those who declared themselves able to access online courses perceive online education as more effective than those who declared themselves new in online communication, or who have difficulties in accessing the internet. The significance level is in both cases 0.000, which is very good. The results of the regression also show that women perceive online education as more effective, compared with men who perceive it as less effective. In addition, undergraduate students perceive online education as slightly less effective.

Table 7. Variable testing and validation.

EOE	β	Err. Std.	t	F	Adj R^2
Age	0.009 (0.816)	0.011	0.84	49.09 ***	0.224
Gender	0.263 *** (0.001)	0.078	3.40		
Ability to browse the internet	1.032 *** (0.000)	0.142	7.24		
Possession of a computer or of similar equipment	0.859 *** (0.000)	0.119	7.20		
cons	1.414 *** (0.000)	0.209	6.77		

Note: *** indicate the fact that $p < 0.01$, respectively. Source: our own calculations based on Stata statistical analysis software.

4.2. Model of the Structural Equation of the Factors Influencing the Perception of University Students and High School Students of Online Education during the State of Emergency Caused by the COVID-19 Pandemic

After an analysis of the exploratory factors and a regression analysis, we proposed the research model (Figure 2), provided that the standardized loadings are at least 0.400. Moreover, there were no loadings on several factors (β > 0.35). Thus, the next step indicates the estimated measurement model using the confirmatory factor analysis (CFA).

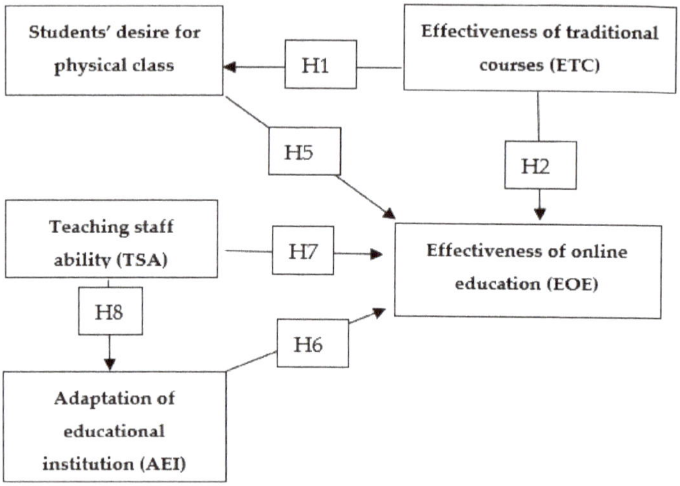

Figure 2. The structural model regarding effectiveness of online courses during the state of emergency caused by the COVID-19 pandemic.

We used several tests to evaluate the general model and to test hypotheses H1, H2, H5, H6, H7 and H8.

Goodness of fit measures were used to assess the overall structural model. For high school students chi^2/df is 412,250/126, RMSEA is 0.075, CFI is 0.890, TLI is 0.877, and SRMR is 0.083. For undergraduate students, the fit indices are as follows: chi^2/df is 432,772/126, RMSEA is 0.077, CFI is 0.907, TLI is 0.887 and SRMR is 0.067, and for postgraduate students chi^2/df is 287,591/126, RMSEA is 0.087, CFI is 0.908, TLI is 0.877 and SRMR is 0.083. The study results and the general classification indices for the proposed model are presented according to the education level in Table 8.

Table 8. Fit indices for the structural model.

Education Levels	Fit Indices				
	chi^2/df	RMSEA	CFI	TLI	SRMR
High school	412.250 /126	0.075	0.890	0.877	0.083
Undergraduate	432.772 /126	0.077	0.907	0.887	0.067
Postgraduate	287.591 /126	0.087	0.908	0.877	0.062

Source: our own calculations based on Stata statistical analysis software.

The AVE is used to determine convergent validity. An average variance extracted (AVE) that is higher than 0.5 is preferred, but a value of 0.4 can be accepted, as Fornell and Larcker argued that if AVE is less than 0.5, but composite reliability is higher than 0.6, the convergent validity of the construct is still adequate [60] (see Appendices A–C for the standardized loadings and AVE results).

Based on the results, the structural model by education levels is shown in Figures 3–5, together with the standardized statistical results of the structural equation.

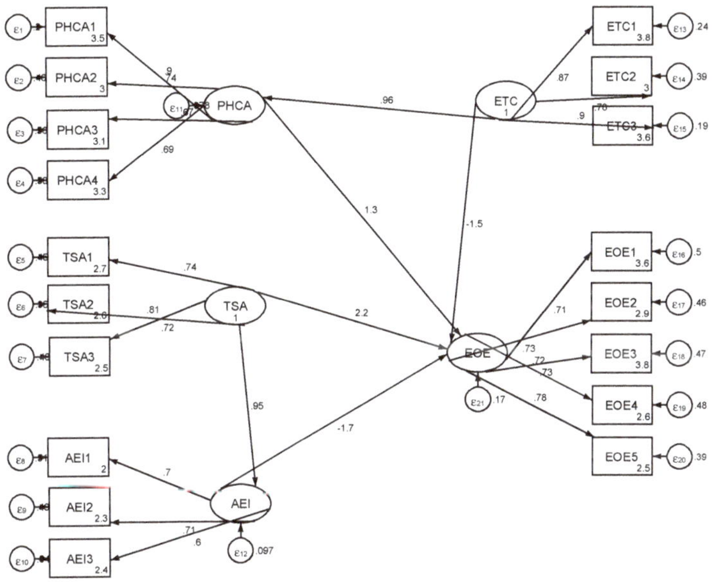

Figure 3. The structural model regarding effectiveness of online courses during the state of emergency caused by the COVID-19 pandemic for high school students.

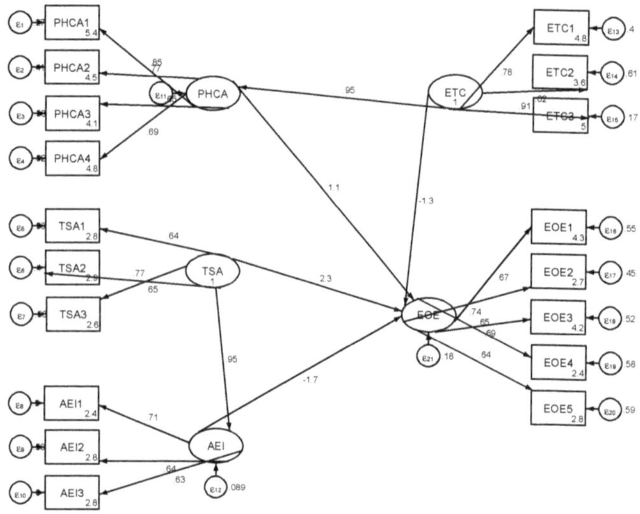

Figure 4. The structural model regarding effectiveness of online courses during the state of emergency caused by the COVID-19 pandemic for undergraduate students.

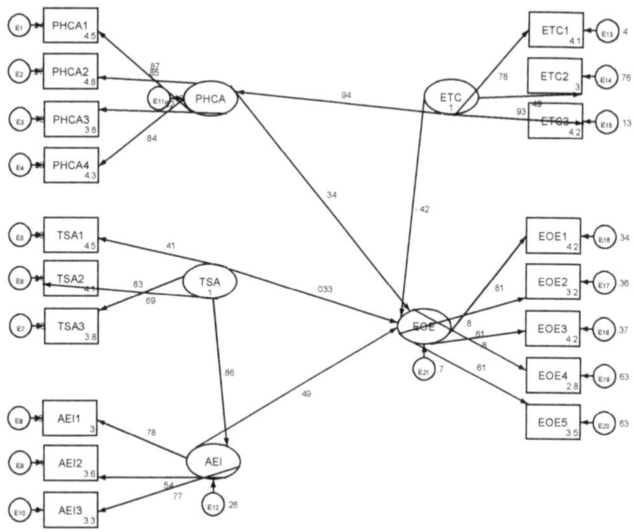

Figure 5. The structural model regarding effectiveness of online courses during the state of emergency caused by the COVID-19 pandemic for postgraduate students.

The structural model assessment helps to investigate the empirical data to confirm the hypotheses. Figure 1 presents the model studied. Figure 3 shows the relationships between variables for the sample containing the high school students. The findings confirm the significant effect ($\beta = -0.960$; $p = 0.000$) of the perception of effectiveness of traditional courses (ETC) on students' desire for physical class attendance (PHCA), of the perception of effectiveness of traditional courses (ETC) on the perception of effectiveness of online education (EOE), with β of -1.462, $p = 0.018$. Positive significant results also confirm the effect of PHCA on EOE ($\beta = 1.342$, $p = 0.032$), the effect of the perception of teaching staff ability (TSA) on the perception of effectiveness of online education (EOE) with a β of 2.238

($p = 0.090$) and the effect of TSA on the adaptation of the educational institution (AEI), $\beta = 0.950$ ($p = 0.000$).

Moreover, from the results, the effect of AEI on EOE is not statistically significant ($\beta = -1.709, p = 0.198$).

Figure 4 shows the relationships between variables for the sample containing the undergraduate students. The findings confirm the significant effect ($\beta = -0.948; p = 0.000$) of the perception of effectiveness of traditional courses (ETC) on students' desire for physical class attendance (PHCA), of the perception of effectiveness of traditional courses (ETC) on the perception of effectiveness of online education (EOE), with β of $-1.298, p = 0.002$. Positive significant results also confirm the effect of PHCA on EOE ($\beta = 1.088, p = 0.011$) and the effect of TSA on the adaptation of the educational institution (AEI), $\beta = 0.954$ ($p = 0.000$).

Moreover, from the results, the effect of AEI on EOE and on TSA on EOE are not statistically significant.

Figure 5 shows the relationships between variables for the sample containing the postgraduate students. The findings confirm the significant effect ($\beta = -0.938; p = 0.000$) of the perception of effectiveness of traditional courses (ETC) on students' desire for physical class attendance (PHCA). The only other statistically significant relationship is represented by the effect of TSA on AEI ($\beta = 0.858,\ p = 0.000$).

Moreover, from the results, the effect of ETC on EOE, PHCA on EOE, AEI on EOE and on TSA on EOE are not statistically significant.

Standardized regression coefficients show that hypotheses H1 and H8 are confirmed for all groups of students, hypotheses H2, H5 and H7 are partially confirmed, while hypothesis H6 cannot be confirmed. The coefficients are shown in Table 9.

Table 9. Coefficients (SEM output).

	Parameter Estimations					
	Coefficient			**Is the Hypothesis Supported?**		
	High School Students	**Undergraduate Students**	**Postgraduate Students**	**High School Students**	**Undergraduate Students**	**Postgraduate Students**
H1:ETC→PHCA	0.960 *** (0.000)	0.948 *** (0.000)	0.938 *** (0.000)	YES	YES	YES
H2:ETC→EOE	−1.462 ** (0.018)	−1.298 *** (0.002)	−0.422 (0.366)	YES	YES	NO
H5:PHCA→ EOE	1.342 ** (0.032)	1.088 ** (0.011)	0.337 (0.470)	YES	YES	NO
H6:AEI→EOE	−1.709 (0.198)	−1.737 (0.216)	0.493 (0.101)	NO	NO	NO
H7:TSA→EOE	2.238 * (0.090)	2.264 (0.105)	0.032 (0.915)	YES	NO	NO
H8:TSA→AEI	0.950 *** (0.000)	0.954 *** (0.000)	0.858 *** (0.000)	YES	YES	YES

Note: *, **, *** indicate the fact that $p < 0.10$, $p < 0.05$, $p < 0.01$, respectively. Source: our own calculations using Stata statistical analysis software.

5. Discussion

Theoretical and empirical approaches investigate online education and the perception of university students and/or high school students of online education. This study addresses an important and topical issue.

In this study we used multiple linear regression models and structural analysis (SEM), which tested the following hypotheses:

Hypothesis 1 (H1). *The perception of the effectiveness of face-to-face courses will influence students' desire for physical class attendance.*

This hypothesis was tested with the help of the structural model, and the results show that there is a statistically significant positive relationship between the perception of the effectiveness of face-to-face courses and students' desire for physical class attendance. This is available for all categories of students analyzed, with a standardized coefficient of 0.960 for high school students, 0.948 for undergraduate students, and 0.938 for postgraduate students. The hypothesis is thus confirmed.

Hypothesis 2 (H2). *The perception of the effectiveness of face-to-face courses will negatively influence the perception of the effectiveness of online courses.*

This hypothesis was tested with the help of the structural model, and the results show that there is a negative relationship between the perception of the effectiveness of face-to-face courses and the perception of effectiveness of online courses. However, this relationship is statistically significant ($p < 0.05$) only for high school students and for undergraduate students. However, the results show a statistically insignificant relationship for postgraduate students, with a p-Value of 0.366. The hypothesis is only partially confirmed.

Hypothesis 3 (H3). *The students who do not possess the technology needed for easy access to online courses will perceive them as ineffective compared with the students who possess the technology.*

This hypothesis was tested with the help of a regression equation. The regression coefficient is 0.859 and it is statistically significant. Therefore, we can see that respondents who own a computer, or similar good-quality equipment granting them access to online courses, perceive these courses as more effective. The hypothesis confirms research results suggesting that mobile devices and good internet access can lead to improvements by stimulating and involving students in the educational process [20,21].

Hypothesis 4 (H4). *The students who can browse the internet easily will perceive the effectiveness of online courses differently compared with the other students.*

This hypothesis was tested with the help of a regression equation. The results show that the respondents who said that they could easily browse the internet will perceive online courses as more effective. The regression coefficient is 1.032 and it is statistically significant. Consequently, hypothesis H4 is confirmed.

Hypothesis 5 (H5). *Students' desire for physical class attendance will generate the perception of ineffectiveness of online courses.*

This hypothesis was tested with the help of the structural model, and the results show that there is a statistically significant reverse relationship between the students' desire for physical class attendance and the perception of the effectiveness of online courses. Nevertheless, this is less true for the postgraduate students, where the relationship is not statistically significant. The hypothesis is thus only partially confirmed.

Hypothesis 6 (H6). *The capacity of educational institutions to manage online courses will have a positive effect on the students' perception of the effectiveness of online courses.*

This hypothesis was tested with the help of the structural model, and the results show that this relationship is statistically insignificant for all groups of students. This may suggest that the trust in educational institutions does not manifest itself within online courses, where the trust in teachers/professors is more important and where we can say that there is a disconnection between high school students and university students on the one hand, and the educational institution on the other hand. Consequently, we cannot state that Hypothesis H6 is confirmed, because the results are not statistically significant for any of the categories analyzed.

Hypothesis 7 (H7). *The capacity of the teaching staff to be more lenient will have a positive effect on the perception of the effectiveness of online courses.*

This hypothesis was tested with the help of the structural model, and the results show that there is a statistically significant positive relationship between the capacity of the teaching staff to be more lenient and the perception of the effectiveness of online courses for the high school students group. However, this is not true for some categories of students analyzed. For the groups of undergraduate and postgraduate students, this relationship is statistically insignificant. Hypothesis H7 is confirmed only for the group of high school students.

Hypothesis 8 (H8). *The perception of the ability of teaching staff to manage online courses will have a positive effect on the perception of the adaptation of the educational institution to online education.*

This hypothesis was tested with the help of the structural model, and the results show that there is a statistically significant positive relationship between the perception regarding teaching staff's ability to manage online courses and the perception regarding the adaptation of educational institution to online education. This is true for all categories of students analyzed. Hypothesis H8 is thus confirmed.

This study aimed to explore the perceptions of university students and high school students regarding online education during the state of emergency caused by the COVID-19 pandemic. The results confirmed previous research such as Bali and Liu [5] and Platt, Raile and Yu [6], suggesting that students have clear perceptions of online education and of the components they consider necessary for their success in this environment.

Our results show that the educational level does not necessarily have an influence on perceptions and attitudes regarding online education. However, the material level has a positive effect (expressed by the existence or not of a technology allowing them to participate in online courses), in terms of familiarity with online communication and skills in using a computer, a tablet or a laptop.

6. Conclusions

The coronavirus pandemic differs from one country to another, but for the educational environment it is extremely important to take measures aimed at protecting young students and, implicitly, their health, such as: providing students with equal and non-discriminatory education opportunities, equal training for all students, protection of students with various vulnerabilities, and the use of opportunities regarding the possibilities of reorganization of the instructional/educational process. A priority of the reorganization of the education system is to build teams made of students, teaching staff and people outside the education system in order to identify responses to current emerging requirements [10,57].

The distance learning model has attracted new experiences which require the transfer of activity planning from the traditional format to the online format. This change is expected to be temporary, based on the circumstances generated by the COVID-19 crisis. Once the crisis ends, the format of instructional/educational activities will return to its original state. The objective of the current circumstances is to allow temporary access to the contents of compulsory training programs for the current year of study and not to create a new education system [1].

Author Contributions: Conceptualization, G.I.B., V.N., A.A. and G.B.; methodology, G.I.B. and A.A.; software, A.A.; validation, G.I.B. and A.A.; formal analysis, G.I.B., V.N., A.A. and G.B.; investigation, A.A.; resources, G.I.B., V.N., A.A. and G.B.; data curation, V.N., A.A. and G.B.; writing—original draft preparation, G.I.B., V.N., A.A. and G.B.; writing—review and editing, G.I.B., V.N., A.A. and G.B.; visualization, G.I.B., V.N., A.A. and G.B.; supervision, G.I.B., V.N., A.A. and G.B.; project administration, G.I.B., V.N., A.A. and G.B.; funding acquisition, G.I.B., V.N., A.A. and G.B. All authors have read and agreed to the published version of the manuscript.

Funding: This research received no external funding.

Institutional Review Board Statement: Not applicable.

Informed Consent Statement: Not applicable.

Data Availability Statement: Not applicable.

Conflicts of Interest: The authors declare no conflict of interest.

Appendix A

Table A1. Variable validity testing (structural model for the group of high school students).

Items	Question	Loading (std.)	AVE
PHCA	PHCA1.	0.9	0.57
	PHCA2.	0.74	
	PHCA3.	0.67	
	PHCA4.	0.69	
ETC	ETC1.	0.87	0.72
	ETC2.	0.78	
	ETC3.	0.9	
EOE	EOE1.	0.71	0.54
	EOE2.	0.73	
	EOE3.	0.72	
	EOE4.	0.73	
	EOE5.	0.78	
AEI	AEI1.	0.7	0.45
	AEI2.	0.71	
	AEI3.	0.6	
TSA	TSA1.	0.74	0.57
	TSA2.	0.81	
	TSA3.	0.72	

Source: authors' elaboration using Stata statistical analysis software.

Appendix B

Table A2. Variable validity testing (structural model for the group of undergraduate students).

Items	Question	Loading (std.)	AVE
PHCA	PHCA1.	0.85	0.55
	PHCA2.	0.77	
	PHCA3.	0.65	
	PHCA4.	0.69	
ETC	ETC1.	0.78	0.6
	ETC2.	0.62	
	ETC3.	0.91	
EOE	EOE1.	0.67	0.46
	EOE2.	0.74	
	EOE3.	0.65	
	EOE4.	0.69	
	EOE5.	0.64	
AEI	AEI1.	0.71	0.44
	AEI2.	0.64	
	AEI3.	0.63	
TSA	TSA1.	0.64	0.47
	TSA2.	0.77	
	TSA3.	0.65	

Source: authors' elaboration using Stata statistical analysis software.

Appendix C

Table A3. Variable validity testing (structural model for the group of postgraduate students).

Items	Question	Loading (std.)	AVE
PHCA	PHCA1.	0.87	0.61
	PHCA2.	0.85	
	PHCA3.	0.52	
	PHCA4.	0.84	
ETC	ETC1.	0.78	0.57
	ETC2.	0.49	
	ETC3.	0.93	
EOE	EOE1.	0.81	0.54
	EOE2.	0.8	
	EOE3.	0.61	
	EOE4.	0.8	
	EOE5.	0.61	
AEI	AEI1.	0.78	0.5
	AEI2.	0.54	
	AEI3.	0.77	
TSA	TSA1.	0.83	0.45
	TSA2.	0.41	
	TSA3.	0.69	

Source: authors' elaboration using Stata statistical analysis software.

References

1. Hodges, C.; Moore, S.; Lockee, B.; Trust, T.; Bond, A. The Difference between Emergency Remote Teaching and Online Learning, Why IT Matters to Higher Education. Available online: https://er.educause.edu/articles/2020/3/the-difference-between-emergency-remote-teaching-and-online-learning (accessed on 12 July 2020).
2. Li, C.; Lalani, F. The COVID-19 Pandemic Has Changed Education Forever. This Is How. Available online: https://www.weforum.org/agenda/2020/04/coronavirus-education-global-covid19-online-digital-learning/ (accessed on 12 July 2020).
3. UNESCO. COVID-19 Impact on Education, Education: From Disruption to Recovery, COVID-19. Available online: https://en.unesco.org/covid19/educationresponse (accessed on 12 July 2020).
4. Bentley, K. Hybrid Learning Goes Mainstream amid Response to COVID-19. Available online: https://www.govtech.com/education/k-12/hybrid-learning-goes-mainstream-amid-response-to-covid-19.html (accessed on 12 July 2020).
5. Bali, S.; Liu, M. Students' perceptions toward online learning and face-to-face learning courses. *J. Phys. Conf. Ser.* **2018**, *1108*, 012094. [CrossRef]
6. Platt, C.A.; Raile, A.; Yu, N. Virtually the same? Student perceptions of the equivalence of online classes vs. face-to-face classes. *Merlot J. Online Learn. Teach.* **2014**, *10*, 489–494.
7. The World Bank. World Bank Education and COVID-19. Available online: https://www.worldbank.org/en/data/interactive/2020/03/24/world-bank-education-and-covid-19 (accessed on 1 July 2020).
8. UNICEF. Sondaj U-Report Privind Școala Online. O Treime din Elevii Români Nu Sunt Mulțumiți de Școala Online și Nu Urmează Cursurile Prin Intermediul Platformelor Virtuale. UNICEF România. Available online: https://www.unicef.org/romania/ro/pove%C8%99ti/sondaj-u-report-privind-%C8%99coala-online (accessed on 12 July 2020).
9. Ciobanu, C.L.; Ciobanu, N.M. E-learning Security Vulnerabilities. *Procedia Soc. Behav. Sci.* **2012**, *46*, 2297–2301. [CrossRef]
10. UNESCO. COVID-19 and Higher Education: Today and Tomorrow. Impact Analysis, Policy Responses and Recommendations. 2020. Available online: http://www.iesalc.unesco.org/en/wp-content/uploads/2020/04/COVID-19-EN-090420-2.pdf (accessed on 2 July 2020).
11. Koehler, M.J.; Mishra, P.; Hershey, K.; Peruski, L. With a little help from your students: A new model for faculty development and online course design. *J. Technol. Teach. Educ.* **2004**, *12*, 25–55.
12. Ocak, M What We Learned about Distance Education during Covid-19? EPALE–Electronic Platform for Adult Learning in Europe. 2020. Available online: https://epale.ec.europa.eu/en/blog/what-we-learned-about-distance-education-during-covid-19 (accessed on 12 July 2020).
13. Fortune, M.; Spielman, M.; Pangelinan, D. Students' perceptions of online or face-to-face learning and social media in hospitality, recreation and tourism. *Merlot J. Online Learn. Teach.* **2011**, *7*, 1–16.
14. Arkorful, V.; Abaidoo, N. The role of e-learning, advantages and disadvantages of its adoption in higher education. *Int. J. Educ. Res.* **2014**, *2*, 397–410.

15. Healy, S.; Block, M.; Judge, J. Adapted physical educators' perceptions of advantages and disadvantages of online teacher development. *Palaest. Sagamore J.* **2014**, *28*, 4.
16. UNICEF. UNICEF și Microsoft Lansează o Platformă Educațională Globală Pentru a Contribui la Gestionarea Crizei din Educație Provocată de COVID-19. Available online: https://www.unicef.org/romania/ro/topics/covid-19?items_per_page=10&page=8 (accessed on 12 July 2020).
17. Ghaderizefreh, S.; Hoover, M.L. Student satisfaction with online learning in a blended course. *Int. J. Digit. Soc.* **2018**, *9*, 3. [CrossRef]
18. Apuke, O.D.; Iyendo, T.O. University students usage of the internet resources for research and learning: Forms of access and perceptions of utility. *Heliyon* **2018**, *4*, e01052. [CrossRef]
19. Gan, C.; Li, H. Understanding the effects of gratifications on the continuance intention to use WeChat in China: A perspective on uses and gratifications. *Comput. Hum. Behav.* **2018**, *78*, 306–315. [CrossRef]
20. Ramirez, M.; Gordy, C. Tips for Students to Prepare for Online Learning, Online and Distance Education News. Available online: https://online-distance.ncsu.edu/tips-for-students-to-prepare-for-online-learning/ (accessed on 12 July 2020).
21. Santos, H.; Batista, J.; Marques, R.P. Digital transformation in higher education: The use of communication technologies by students. *Procedia Comput. Sci.* **2019**, *164*, 123–130. [CrossRef]
22. Dorn, E.; Hancock, B.; Sarakatsannis, J.; Viruleg, E. COVID-19 and Student Learning in the United States: The Hurt Could Last a Lifetime. Available online: https://www.mckinsey.com/industries/public-sector/our-insights/covid-19-and-student-learning-in-the-united-states-the-hurt-could-last-a-lifetime# (accessed on 12 July 2020).
23. Farrington, D. The Consequences of COVID-19 on the Education System: The Legal Perspective. Available online: https://www.coe.int/en/web/education/article-the-consequences-of-covid-19-on-the-education-system-the-legal-perspective (accessed on 12 July 2020).
24. UNESCO. Distance Learning Solutions, More on UNESCO's COVID-19 Education Response. Available online: https://en.unesco.org/covid19/educationresponse/solutions (accessed on 12 July 2020).
25. Peticilă, M. Părinții, Obligați să le Asigure Copiilor Calculator/Laptop/Tabletă/ Telefon Smart cu Internet Pentru A Participa la Cursuri Online, A Stabilit Ministerul Educației. Elevii, Obligați să Participe la Orele pe Net Stabilite de Școală. Available online: https://www.edupedu.ro/parintii-obligati-sa-le-asigure-copiilor-calculator-laptop-tableta-telefon-smart-cu-internet-pentru-a-participa-la-cursurile-online-a-stabilit-ministerul-educatiei-elevii-obligati-sa-participe-la/ (accessed on 12 July 2020).
26. Kebritchi, M.; Lipschuetz, A.; Santiague, L. Issues and challenges for teaching successful online courses in higher education: A Literature review. *J. Educ. Technol. Syst.* **2017**, *46*, 4–29. [CrossRef]
27. Hung, M.; Chou, C.; Chen, C.; Own, Z. Learner readiness for online learning: Scale development and student perceptions. *Comput. Educ.* **2010**, *55*, 1080–1090. [CrossRef]
28. Basilaia, G.; Kvavadze, D. Transition to online education in schools during a SARS-CoV-2 Coronavirus (COVID-19) pandemic in Georgia. *Pedagog. Res.* **2020**, *5*, 1–9. [CrossRef]
29. Neil, K. Using Social Distance to Strengthen University Communities. Available online: https://www.universityworldnews.com/post.php?story=20200504082923788 (accessed on 23 July 2020).
30. Reimers, F.; Schleicher, A.; Saavedra, J.; Tuominen, S. Supporting The Continuation of Teaching and Learning during the COVID-19 Pandemic Annotated Resources for Online Learning. Available online: http://www.oecd.org/education/Supporting-the-continuation-of-teaching-and-learning-during-the-COVID-19-pandemic.pdf (accessed on 2 July 2020).
31. Bender, L. Key Messages and Actions for COVID-19 Prevention and Control in Schools. Available online: https://www.who.int/docs/default-source/coronaviruse/key-messages-and-actions-for-covid-19-prevention-and-control-in-schools-march-2020.pdf?sfvrsn=baf81d52_4&gclid=Cj0KCQjw6uT4BRD5ARIsADwJQ197uVW8uB1-D26t0LB9t-P3ENYMn-m_Cbyk6fMh9mf2yZ1s-0Dxlt4aAikNEALw_wcB (accessed on 12 July 2020).
32. Castaneda, D.; Rentz, S. The Power of Discussion: Activating Learning Online (And in Person). Available online: https://elearningindustry.com/social-interaction-in-online-courses-discussion-activating-learning (accessed on 30 July 2020).
33. Watts, L. Synchronous and asynchronous communication in distance learning: A review of the literature. *Q. Rev. Distance Educ.* **2016**, *17*, 23–32.
34. Wazzan, M. 15 min of Fame: Online Learning in the Coronavirus Era. Available online: https://www.al-fanarmedia.org/2020/04/15-minutes-of-fame-online-learning-in-the-coronavirus-era/ (accessed on 2 July 2020).
35. Hobfoll, S.E.; Cameron, R.P.; Chapman, H.A.; Gallagher, R.W. Social Support and Social Coping in Couples. In *Handbook of Social Support and the Family*; The Springer Series on Stress and Coping; Pierce, G.R., Sarason, B.R., Sarason, I.G., Eds.; Springer: Boston, MA, USA, 1996.
36. Mishra, S. Social networks, social capital, social support and academic success in higher education: A systematic review with a special focus on 'underrepresented' students. *Educ. Res. Rev.* **2020**, *29*, 100307. [CrossRef]
37. Darling-Hammond, L.; Cook-Harvey, M. Educating the Whole Child: Improving School Climate to Support Student Success. Available online: https://learningpolicyinstitute.org/sites/default/files/product-files/Educating_Whole_Child_REPORT.pdf (accessed on 12 July 2020).
38. Mather, M.; Sarkans, A. Student perceptions of online and face-to-face learning. *Int. J. Curric. Instr.* **2018**, *10*, 61–76.
39. Alsaaty, F.; Carter, E.; Abrahams, D.; Alshameri, F. Traditional versus online learning in institutions of higher education: Minority business students' perceptions. *Bus. Manag. Res.* **2016**, *5*. [CrossRef]

40. Popescu, A.L. Misiune Imposibilă în Educație: Cum Muți Peste Noapte Școala în Online. Available online: https://romania.europalibera.org/a/misiune-imposibila-educatie-cum-muti-peste-noapte-scoala-online/30592445.html (accessed on 12 July 2020).
41. Gabor, A. Schools Aren't Ready for Online Learning. Available online: https://www.bloombergquint.com/gadfly/coronavirus-shutdowns-highlight-weakness-of-online-u-s-schooling (accessed on 30 July 2020).
42. Anderson, D.; Imdieke, S.; Standerford, N.S. Feedback please: Studying self in the online classroom. *Int. J. Instr.* **2011**, *4*, 3–15.
43. Manea, R. Ce ar putea fi învățarea online în lipsa școlii. Available online: https://www.edupedu.ro/ce-ar-putea-fi-invatarea-online-in-lipsa-scolii-ruxandra-manea-antreprenor-in-educatie/ (accessed on 12 July 2020).
44. Vu, C.T.; Hoang, A.D.; Than, V.Q.; Nguyen, M.T.; Dinh, V.H.; Le, Q.A.T.; Le, T.T.T.; Pham, H.H.; Nguyen, Y.C. Dataset of Vietnamese teachers' perspectives and perceived support during the COVID-19 pandemic. *Data. Brief.* **2020**, *31*, 105788. [CrossRef] [PubMed]
45. Hardy, L. 5 Skills That Online Teachers Are Constantly Developing, eLearning Best Practices. Available online: https://elearningindustry.com/5-skills-online-teachers-developing (accessed on 12 July 2020).
46. Bao, W. COVID -19 and online teaching in higher education: A case study of Peking University. *Hum. Behav. Emerg. Technol.* **2020**, 2. [CrossRef]
47. Cooper, S. 5 Strategies To Improve Your Online Teaching. eLearning Best Practices. Available online: https://elearningindustry.com/5-strategies-improve-your-online-teaching (accessed on 12 July 2020).
48. Niess, M.; Gillow-Wiles, H. Developing asynchronous online courses: Key instructional strategies in a social metacognitive constructivist learning trajectory. *Distance. Educ.* **2013**, *27*. Available online: http://www.ijede.ca/index.php/jde/article/view/831/1473 (accessed on 9 January 2021).
49. Duncan, L. The Transition from Face-to-Face to Online Learning: Maintaining an Engaging Experience. Available online: https://community.brightspace.com/s/article/The-Transition-from-Face-to-Face-to-Online-Learning-Maintaining-an-Engaging-Experience (accessed on 12 July 2020).
50. Cross, L.K. Graduate student perceptions of online advising. *NACADA J.* **2018**, *38*, 72–80. [CrossRef]
51. Song, H.; Kim, J.; Luo, W. Teacher–student relationship in online classes: A role of teacher self-disclosure. *Comput. Hum. Behav.* **2016**, *54*, 436–443. [CrossRef]
52. The Romanian Ministry of Education. Instrucțiunea Pentru Crearea și/sau întărirea Capacității Sistemului de învățământ Preuniversitar prin învățare on-line, din 21.04.2020, Text Publicat în Monitorul Oficial, Partea I nr. In 331 din 23 aprilie 2020. În vigoare de la 23 aprilie 2020. Available online: https://lege5.ro/Gratuit/gm3dqmrzhe2a/instructiunea-pentru-crearea-si-sau-intarirea-capacitatii-sistemului-de-invatamant-preuniversitar-prin-invatare-on-line-din-21042020 (accessed on 2 July 2020).
53. Educred Un Portal în Sprijinul Cadrelor Didactice. Available online: https://digital.educred.ro/home (accessed on 12 July 2020).
54. Treacy, B. Developing Strategies for Online Teaching and Learning, Professional Education. Available online: https://www.gse.harvard.edu/ppe/program/developing-strategies-online-teaching-and-learning (accessed on 12 July 2020).
55. Langa, C.; Tudor, L.S.; Stan, M.M.; Bulgaru, I. Educația Față cu Provocările Unei Situații Excepționale. Available online: https://www.upit.ro/ro/upit-pentru-comunitate/educatia-fata-cu-provocarile-unei-situatii-exceptionale (accessed on 12 July 2020).
56. Roșculeț, G. Contextul Învățării în Vreme De Izolare. Available online: https://oradenet.salvaticopiii.ro/blog/contextul-invatarii-in-vreme-de-izolare-2/ (accessed on 12 July 2020).
57. The Romanian Ministry of Education. Consultare Publică Lansată de Comisia Europeană Privind Revizuirea Planului De Acțiune Pentru Educația Digitală. Available online: https://www.edu.ro/consultare-public%7B/fontencoding%7BT5%7D/selectfont%7B/abreve%7D%7D-lansat%7B/fontencoding%7BT5%7D/selectfont%7B/abreve%7D%7D-de-comisia-european%7B (accessed on 12 July 2020).
58. Field, A. *Discovering Statistics Using SPSS*; Sage Publications Ltd.: London, UK, 2009; ISBN 978-1-84787-906-6.
59. Comrey, A.; Lee, H. *A First Course in Factor Analysis*; Erlbaum: Hillsdale, NJ, USA, 1992; ISBN 0-8058-1062-5.
60. Fornell, C.; Larcker, D. Evaluating Structural Equation Models with Unobservable Variables and Measurement Error. *J. Mark. Res.* **1981**, *18*, 39–50. [CrossRef]

sustainability

Review

Conducting Population Health Research during the COVID-19 Pandemic: Impacts and Recommendations

Amy R. Villarosa [1,2,3,*], Lucie M. Ramjan [1,3,4], Della Maneze [1,2,3] and Ajesh George [1,2,3,4,5]

1. Centre for Oral Health Outcomes and Research Translation (COHORT), Western Sydney University, Liverpool, NSW 2170, Australia; l.ramjan@westernsydney.edu.au (L.M.R.); d.maneze@westernsydney.edu.au (D.M.); a.george@westernsydney.edu.au (A.G.)
2. South Western Sydney Local Health District, Liverpool, NSW 2170, Australia
3. Ingham Institute for Applied Medical Research, Liverpool, NSW 2170, Australia
4. Translational Health Research Institute, Western Sydney University, Penrith, NSW 2751, Australia
5. Department of Dentistry and Oral Health, La Trobe University, Bendigo, VIC 3552, Australia

* Correspondence: amy.villarosa@westernsydney.edu.au

Abstract: The COVID-19 pandemic has resulted in many changes, including restrictions on indoor gatherings and visitation to residential aged care facilities, hospitals and certain communities. Coupled with potential restrictions imposed by health services and academic institutions, these changes may significantly impact the conduct of population health research. However, the continuance of population health research is beneficial for the provision of health services and sometimes imperative. This paper discusses the impact of COVID-19 restrictions on the conduct of population health research. This discussion unveils important ethical considerations, as well as potential impacts on recruitment methods, face-to-face data collection, data quality and validity. In addition, this paper explores potential recruitment and data collection methods that could replace face-to-face methods. The discussion is accompanied by reflections on the challenges experienced by the authors in their own research at an oral health service during the COVID-19 pandemic and alternative methods that were utilised in place of face-to-face methods. This paper concludes that, although COVID-19 presents challenges to the conduct of population health research, there is a range of alternative methods to face-to-face recruitment and data collection. These alternative methods should be considered in light of project aims to ensure data quality is not compromised.

Keywords: population health research; public health research; COVID-19; research methods

check for updates

Citation: Villarosa, A.R.; Ramjan, L.M.; Maneze, D.; George, A. Conducting Population Health Research during the COVID-19 Pandemic: Impacts and Recommendations. *Sustainability* **2021**, *13*, 3320. https://doi.org/10.3390/su13063320

Academic Editor: Haywantee Ramkissoon

Received: 15 February 2021
Accepted: 15 March 2021
Published: 17 March 2021

1. Introduction

Since the announcement of COVID-19 as a pandemic in early March, countries around the world have been instating measures to prevent its spread. Among developed countries, these measures have centered around restricting the movement and gathering of people [1,2]. In Australia, specific measures have included closures of state borders, staying at home where possible, restriction on the size of outdoor and indoor assemblies, minimum space requirements for each person in enclosed areas, and the requirement to keep a 1.5-metre distance from others [3–5]. Furthermore, measures have been taken to protect groups considered at higher risk of contracting COVID-19, such as older people, those with chronic diseases, hospitalised individuals and Aboriginal and Torres Strait Islander peoples. These measures have included restrictions of visitors to residential aged care facilities and hospitals [6,7] and restriction of movement into remote Aboriginal and Torres Strait Islander communities [8].

It is, therefore, understandable that such constraints have, and will continue to have, a significant impact on many aspects of human undertakings and social interactions, including conducting health research over this period of change. In Australia, national guidelines have been released regarding the safe conduct of clinical trials during COVID-19, which inform contingency planning and alternative models to mitigate risks, particularly if the trial involves participants who are symptomatic for COVID-19 [9]. However, there remains a lack of such guidance for other types of research studies, including those which utilise population health research methods. Population health research, also known as public health research, has been defined as research that investigates and analyses factors that affect the health of populations or population groups or that tests and evaluates interventions to improve population health [10]. This research can employ quantitative, qualitative and mixed methods methodologies. Population health research is essential to inform health policies and programs, to ensure minimisation of gaps and inequities in healthcare provision and reduce strain on health services through preventative measures, an important consideration during COVID-19 [10]. Thus, it is essential that, where ethical and feasible, measures are taken to ensure population health research continues, particularly during this period when a number of health issues related to the pandemic and its control measures are emerging.

Over recent decades, as technology advanced and new techniques emerged, the use of alternatives to face-to-face research, such as e-research methods and secondary data sources in population health research, have been heavily critiqued. These methods have been shown to provide both benefits, such as reduced time and costs for large samples, and challenges, such as low response rates [11,12]. However, given the global situation, there is increasing pressure to reduce the need for population health researchers to conduct face-to-face research with participants [13,14]. To ensure ethical conduct of research and adherence with guidelines from governing institutions, researchers may no longer have the option to recruit participants, deliver interventions or gather data in person, and may actually be required to outline contingency plans if this becomes the case [15,16]. Thus, it is essential for researchers to consider alternative methods to undertake population health research that do not rely on face-to-face contact. The aims of this article are, therefore, to:

(i) Explore the potential impact of COVID-19 on the conduct of population health research, and
(ii) Discuss potential alternative recruitment and data collection strategies

This discussion will be supported by existing literature and supplemented with reflections from an ongoing population health research project, the conduct of which has been impacted by the COVID-19 pandemic.

2. Review of the Literature

2.1. *Impact of COVID-19 on the Conduct of Population Health Research*

Due to ethical considerations, policy changes and new guidelines that have arisen from the COVID-19 pandemic, the conduct of population health research has been impacted in various ways [13–16]. This section will discuss the effects of these changes in light of the current literature and reflect upon the authors' own experience of COVID-19 during their research.

2.1.1. Research and Ethics

During times of public health emergency, it is imperative for health services to prioritise responses that address issues relating to the emergency [17]. However, concurrent research that explores the general health of the population during such times, or produces evidence that supports the response, should be given equal importance. Redirecting personnel and mobilizing equipment, facilities or other resources for outbreak response in pandemics, such as COVID-19, must be balanced with the ongoing need to facilitate research, both regarding the pandemic as well as research that is unrelated to the pandemic but could benefit the health of the general population [18]. In line with this, when assessing the ethical nature of their projects, researchers need to take measures to consider the value of their projects in view of the global context of COVID-19.

As a result, healthcare bodies and ethics committees are adopting different approaches when assessing new research projects, with some organisations choosing to postpone all face-to-face non-trial research activities and defer approval for commencement of new studies which are not related to COVID-19 [19]. Researchers will need to act in accordance with the policies and recommendations of their governing bodies during this time, and thus they may not be able to commence or continue their population health research projects.

2.1.2. Recruitment of Participants

Moreover, it has been identified that COVID-19 may be impacting vulnerable populations in a much more severe way than other population groups, and thus individuals from vulnerable population groups may experience more risk from participation in population health research [20]. However, the exclusion of such participants on the basis of being from a vulnerable population group is unethical, as any exclusion criteria must be based upon robust and current evidence [17]. Thus, the risks and benefits of involving individuals from vulnerable population groups in population health research must be carefully considered during this time, and it is vital to ensure that the benefits of participation outweigh the risks [21].

Although there is a range of recruitment methods that can be ethically employed in health research, there are several methods that could be affected by restrictions related to the COVID-19 pandemic. In the circumstance where the researcher is recruiting patients in a clinical setting that they are not a member of, changes in visitation policies may restrict the entry of the researcher to meet potential participants face-to-face [7,22]. Therefore, any recruitment methods dependent on face-to-face contact to inform potential participants and obtain consent may need to be reconsidered.

Furthermore, if the recruitment method involves clinical team members identifying eligible participants and explaining the study or being study participants themselves, there are additional factors to be considered [22]. Even prior to the pandemic, clinicians' workload and time availability already posed significant challenges to research participation [23]. This is now even more of a challenge, considering that the changes in context, practice and policy resulting from the COVID-19 pandemic have increased the workloads of clinical staff around the world. Along with the stress and burnout staff are currently experiencing, their capacity to participate in the research project may be further limited [24]. Thus, it is likely that researchers may be unable to rely upon recruitment methods or participant samples that heavily involve the input of clinicians.

In both clinical and public settings, convenience recruitment methods, such as placement of research advertisement materials in areas where people congregate, for example, clinic waiting rooms, universities or community centers, may also be affected by COVID-19 restrictions. This is due to the fact that physical distancing and closures of public spaces and community facilities could greatly reduce the number of people frequenting such areas [25]. In addition, many health services are providing temporary telehealth services for patients, circumventing the need to attend clinics in person at all [9]. In addition, many universities have taken measures to operate partially or completely online [13]. As a result, many convenience recruitment methods may not be feasible for the foreseeable future.

2.1.3. Data Collection

The safety of research participants and research staff is paramount, and due to the communicable nature of COVID-19, many institutions are mandating the suspension of face-to-face data collection methods altogether [14]. Unfortunately, many population health research methods traditionally rely on face to face contact. An example of one of these methods used in both qualitative and quantitative research is observation, which is often conducted face-to-face in the natural environment [26]. However, not only may a researcher have limited ability to access this natural environment, but furthermore, in light of physical distance measures, it may not be permissible or ethical for a researcher to observe a participant in close proximity. Data collection that requires researchers and participants to be in close contact, such as with biophysical measures, will also pose difficulties for similar reasons [27]. Such limitations to data collection methods may impact the types of outcomes researchers can explore.

Similar to guidance stipulated for clinical trials, in cases where face-to-face data collection may proceed, it may be within the researchers' duty of care to monitor participants for COVID symptoms and/or provide COVID testing [9]. This may be particularly crucial in instances where participants and research teams will be working with healthcare services or where group data collection methods, such as focus groups, will be used. This may require modification of study methods and allocation of additional resources to monitor symptoms and conduct COVID tests, and thus may have budget and staffing implications.

In addition, COVID-19 has had a significant socioeconomic impact resulting in job losses, financial instability, and family stressors [28] which may disproportionately increase respondent burden, particularly among vulnerable populations. Respondent burden is defined as how difficult, time-consuming or emotionally stressful an individual perceives participation in a research project to be, and may result in non-responses, or lower data quality [29]. This must also be considered in choosing data collection methods.

2.1.4. Data Quality

There are some data collection methods that, although easily conducted remotely, may obtain better data quality when conducted face-to-face with a researcher. For example, questionnaire data are often more complete and of higher quality when participants are assisted or supervised by a researcher [30].

There must also be considerations regarding how the current global context may impact data validity. Workforces around the world may be required to change their workplace practices [31]. These changes have been particularly extreme for those working in healthcare settings, including additional guidelines regarding hygiene and cleaning, use of personal protective equipment, procedures used during face-to-face consultations, postponement of non-emergency care and transition of some services to telehealth [32]. Given these changes to how healthcare is being delivered and received, it is important to consider comparability of data before and after such changes occurred and that such changes may impact the internal validity of longitudinal studies [33]. Furthermore, given that some practice changes, such as postponement of non-emergency consultations, are intended as temporary measures, researchers may need to consider the generalizability and/or relevance of any data collected while these measures are in place. This phenomenon, referred to as the interaction of history and treatment, is a major threat to the external validity of study findings [34].

Overall, there is the potential for COVID-19 to significantly impact the conduct of population health research projects. In some recent scoping reviews on population oral health strategies involving the broader workforce, only five out of 75 primary research studies followed qualitative or mixed methods, which all involved direct contact with participants [35–38]. A total of 33 out of the 34 quantitative intervention studies required direct contact with participants for delivery of the intervention and/or data collection [35–38]. A further 23 of the remaining 36 quantitative observational studies also required face-to-face contact for data collection, meaning a total of 81% of primary research studies in these scoping reviews would have been impacted by COVID-19 restrictions [35–38]. For many quantitative studies, face-to-face data collection methods may be the only appropriate way to collect data, particularly where participants are considered 'hard-to-reach', that is, from socially disadvantaged or underprivileged groups [39]. This could include those who are homeless and transient, those who have a chronic mental illness, those with low literacy levels, and indigenous peoples [39]. Furthermore, investigators may choose these methods to minimise nonresponse, or for sensitive topics [30]. This highlights that depending on the research area, there is the potential for the majority of studies to be impacted.

2.2. *Alternative Recruitment and Data Collection Strategies*

Despite the impacts identified above, particularly on recruitment of participants and data collection, there are many alternatives to face-to-face data collection that could be of merit during COVID-19. This section will discuss the benefits and limitations of these alternatives and draw upon reflection from the authors' research to describe how such alternatives could be implemented.

2.2.1. Recruitment Strategies

Although face-to-face and certain convenience recruitment methods may not be feasible during a pandemic, there is a range of other recruitment methods that may be used. These include distribution of advertisement material via emails, letters, phone calls and the use of social media platforms. Evidence suggests that second to face-to-face recruitment, telephone recruitment may result in the highest response rates from participants and thus should be considered as an alternative during COVID-19 [40]. Limitations are that such methods can be costly and time-consuming, with one study reporting that telephone recruitment resulted in 1680 h spent on the phone and a total cost of $79 USD per participant [40]. Mailing advertising materials was shown to have a similar response rate [40] while incurring lower costs, with one study citing costs of printing and postage to be around $52 USD per participant [41].

With the advancement of online technology, research has highlighted the merit of online recruitment methods in health research. Although emailing of advertising materials may result in lower response rates than telephone or postal mail methods, it enables the identification and contact of a higher number of eligible participants [42]. Furthermore, with the growing popularity of various online and social media platforms, which are now being accessed by millions of people, their value as a recruitment tool is increasingly recognised [42]. Research indicates recruitment via social media remains valid in the COVID-19 context and has been able to reach a large number of participants [43]. One of the most commonly used platforms for research recruitment is Facebook, with its advantages of being the largest social media platform and having the ability for targeted advertisements to be shown to people with specific demographic characteristics, thereby, increasing chances of identifying eligible individuals [42]. This may also be considerably more cost-effective than other methods, with costs per participant reported to be from $0.60 to $20 USD [42]. Similarly, Instagram, which is owned by Facebook, can create similar targeted advertisements, which can also be linked to Facebook. However, Instagram users are generally a younger demographic, thus desired participant demographics should be taken into consideration when using these social media platforms for recruitment [44]. Twitter is also a commonly used platform, where users can pay for posts to be promoted, that is, displayed in feeds of users who do not follow the posting account, however, it is not able to target advertisements based on specific demographic characteristics [44]. Studies have also reported advertisements on search engines, such as Google, to be an effective recruitment strategy, and may have similar costs when compared to advertisements on social media platforms [45]. Despite the known merits of online recruitment methods, it has also been identified that samples from social media strategies can be biased, reducing diversity in age, socioeconomic status, location and ethnic background, and often result in lower retention rates [40–42,45]. Thus, in designing an online recruitment strategy, researchers should consider the potential limitations of these methods and have strategies in place to ensure a representative sample [43].

2.2.2. Data Collection Strategies

As discussed previously, face-to-face quantitative data collection methods, such as observations, biophysical measures and face-to-face questionnaires, may no longer be feasible or ethical during this period. However, there are several alternative data collection options. An example of this is the retrospective review of medical records, particularly electronic medical records, preventing unnecessary visits to healthcare facilities, with the added advantage of increased accessibility to a larger volume of patient data [46]. Medical records can be an effective and valid source of data for outcomes, such as length of stay, discharge destination and diagnosis, and thus should be considered as alternative data sources during the COVID-19 pandemic [46,47]. However, there are some variations seen between medical records and other forms of data collection, and this should be considered, for example, when interpreting results [48,49]. Another potential alternative to observations or physical measures is self-report. Studies have shown this to be valid for measures, such as weight, height and physical activity, and therefore, could serve as valid proxy measures for physical measurements [50,51]. However, some inaccuracies are seen in self-report of less socially accepted behaviors, such as cigarette smoking status [52]. This phenomenon of self-report bias is well recognised in the literature, however, when these limitations are acknowledged and accounted for, self-report data can still provide significant contributions to the existing body of evidence [53]. Compared to using medical records and self-report data in place of observations and physical measures, adapting face-to-face questionnaires for mail, telephone or online modes seems much more straightforward. Despite the idea that face-to-face questionnaires provide higher data quality, evidence continues to emerge that telephone, mail and online questionnaires yield sufficiently similar data quality and thus may be viable alternatives during COVID-19 [54,55].

The use of alternative methods when face-to-face options are not possible is probably more controversial when it comes to qualitative methodologies. For example, the use of telephone interviews for qualitative research has been widely criticised, with suspected losses of contextual and nonverbal data due to the absence of visual cues, as well as a lack of rapport and probing, making interpretation more challenging [56]. However, there is a lack of evidence to support these claims, which currently suggests that quality of data from telephone interviews is similar to that from face-to-face interviews, thus telephone interviews could be a simple alternative to use during COVID-19 [56,57]. Additionally, as various online platforms become more accessible with advancements in technology, the popularity of conducting interviews via online means is increasing [58]. Email has been used as a platform for qualitative interviews for years, although it presents some challenges regarding trustworthiness of data, specifically credibility, and also requires asynchronous interview techniques to be considered [58,59]. However, it has also been recognised as a viable and convenient method that may be particularly useful during COVID-19 [58,59]. Moreover, synchronous online qualitative interviews have become more feasible in recent years due to widespread access to electronic devices and internet connections, improvements in internet speeds and improvements in instant messaging and video conferencing software, such as Skype [60–62]. These platforms overcome the challenges presented with asynchronous interviews, and although instant messaging does not allow for visual cues, video conferencing does [61,62]. These methods are particularly valuable during times of limited flexibility or access to research participants, such as COVID-19, and have the potential to produce data comparable to face-to-face interviews, as long as any technical difficulties are considered and addressed [60–62]. Online methods may also be viable alternatives for the conduct of focus groups during COVID-19. For example, online discussion boards have been effectively used as an alternative to face-to-face focus groups, despite having similar limitations to other asynchronous online qualitative methods [63]. Likewise, the use of video conferencing software, such as Skype, may have similar effectiveness for focus groups as when used with interviews [60,63,64].

3. Reflection

3.1. Impact of COVID-19 on the Conduct of the Authors' Project

The authors' own research project aims to develop and evaluate a guideline implementation strategy to facilitate the translation of children's healthy weight guidelines for public oral health staff into practice. All primary research associated with this project is centred around two phases: (i) Developing the guideline implementation strategy using co-design focus groups and input from an expert panel, and (ii) evaluation of the guideline implementation strategy using a pretest-posttest design and a validated questionnaire. Although ethical approval had already been obtained for the conduct of the project, this was done prior to the announcement of the pandemic, and the authors anticipated modification of study methods, and therefore, ethics amendments may be required. Data collection was scheduled to commence just as the pandemic was announced.

As the first phase was based around face-to-face focus groups with both public dental practitioners and parents of children in the community, the authors immediately had concerns regarding the safety and logistics of such gatherings in an enclosed space. Moreover, the implementation of children's healthy weight guidelines for public oral health staff was dependent on routine care being provided at the public oral health services. However, as a response to the COVID-19 pandemic, the participating public oral health services were only providing emergency care at the scheduled time of data collection, and many staff were being redeployed to COVID-19 testing clinics. As a result, the internal and external validity of any data collected during this time of altered practice would be greatly compromised if data collection were to proceed. Finally, as the authors intended to use convenience methods to recruit parents by distributing flyers in oral health service waiting rooms, it was anticipated that these recruitment methods would need modification. Therefore, it was determined that the development and evaluation of the guideline implementation strategy should be postponed until routine dental practice resumed. Once this time came, it was anticipated that the recruitment and data collection strategies would need to be modified to ensure the most effective methods were chosen, and risks to participants were minimised.

3.2. *Alternative Recruitment and Data Collection Strategies*

As stated in the previous section, primarily due to internal validity concerns, the development and evaluation of the guideline implementation strategy for the authors' research project were postponed until the dental services could resume practice as normal. Even with the inability to undertake the development and evaluation component as planned, there were still research activities that could be undertaken during this time. This included the development and evaluation of the questionnaire that would be used to assess the guideline implementation strategy. The authors initially planned to take a five-step validation process for the questionnaire: (i) Item generation using a review of the literature, (ii) face validation using a face-to-face reference group, (iii) construct validation using an online questionnaire with an expert panel, (iv) pilot testing and (v) factorial validation using a large online sample. Once the COVID-19 restrictions were instated, the authors scrutinised this process and realised that very few face-to-face steps were included in the process, and the face-to-face steps that were included could be easily modified to instead utilise alternative data collection strategies with minimal impact on data quality.

The main step requiring modification was step 2: Face validation using a face-to-face reference group. The first consideration was the recruitment method, which although it was already intended to be purposive recruitment via email, the original participation incentive was provision of refreshments during the time of the reference group. As this reference group would no longer be assembling face-to-face, the authors decided to reallocate the funds for the light refreshments to supermarket gift vouchers, and this change was approved by the governing ethics committee. This more direct method to reimburse time and effort of participation was also deemed necessary due to the potential of increased respondent burden that participants may be feeling during this time.

In addition, the data collection method also required modification. The reference group became a virtual, online reference group conducted via email. With the difficulties that the enforced lockdown presented for participants, including the need for child minding, homeschooling and redeployment of health staff, it was agreed that this method would increase flexibility, allowing a two-week period where participants could provide feedback at their convenience. As per the initial design, this reference group did not require synchronous discussion, rather, each participant was to provide written feedback on their own hard copy of the questionnaire. Thus, the asynchronous format of providing the same briefing and feedback via email was seen to have minimal impact on the quality of data obtained. Using these methods, the authors were able to successfully obtain complete feedback on the questionnaire from all four members of the reference group.

Shortly after the questionnaire validation was completed, practice at the dental health services resumed as normal, therefore, the development and evaluation of the guideline implementation strategy could proceed. However, the authors were increasingly aware that the risks of COVID-19 with face to face data collection may not change in the near future. As a result, it was concluded that alternative data collection strategies would be implemented for the remainder of the project to eliminate the need for face to face contact. For the focus groups, it was decided that internet videoconferencing would be the best way to go, as visual brainstorming was an essential component of the focus group. Despite concerns about accessibility of videoconferencing software for study participants, throughout recruitment, it was found that participants preferred participating via videoconferencing, with platforms, such as Zoom, becoming more familiar to them over lockdown periods. A total of four videoconference focus groups were conducted for the project, and the authors were able to still utilise conventional focus group techniques to collect rich data from all participants. While the evaluation phase of this study is yet to commence, it was concluded that administering the evaluation questionnaires via an online survey platform would be an appropriate alternative to hard copy questionnaires. To ensure response rates are as high as possible in this phase, study champions will be nominated at each study site to help promote and coordinate this step. It is hypothesised that the authors' careful consideration of the possible alternatives to data collection and recruitment contributed to the success of these substitute methods, which have traditionally been considered less desirable.

4. Conclusions

COVID-19 presents a range of challenges that can impact the conduct and quality of population health research. Regardless, there is a range of alternative strategies that can be employed to facilitate the safe and effective conduct of population health research during uncertain times. While all research methods have their own strengths and limitations, researchers should carefully consider the methods most appropriate for their study's specific needs without compromising the quality of their data.

Author Contributions: A.R.V. drafted the paper and provided reflective summaries. L.M.R., D.M. and A.G. also assisted in drafting and revising the entire paper. All authors have read and agreed to the published version of the manuscript.

Funding: The authors acknowledge funding from the National Health and Medical Research Council and Western Sydney University, grant number APP1168119, for the conduct of this project.

Institutional Review Board Statement: Not applicable.

Informed Consent Statement: Not applicable.

Data Availability Statement: Data sharing not applicable.

Conflicts of Interest: The authors declare no conflict of interest.

References

1. California Department of Public Health. COVID-19. Available online: https://www.cdph.ca.gov/Programs/CID/DCDC/Pages/Immunization/ncov2019.aspx (accessed on 19 May 2020).
2. Cabinet Office. Coronavirus Outbreak FAQs: What You Can and Can't Do. Available online: https://www.gov.uk/government/publications/coronavirus-outbreak-faqs-what-you-can-and-cant-do/coronavirus-outbreak-faqs-what-you-can-and-cant-do (accessed on 19 May 2020).
3. NSW Government. What You Can and Can't Do under the Rules. Available online: https://www.nsw.gov.au/covid-19/what-you-can-and-cant-do-under-rules (accessed on 19 May 2020).
4. Victoria State Government. Stay at Home Restrictions. Available online: https://www.dhhs.vic.gov.au/stay-home-restrictions-coronavirus (accessed on 19 May 2020).

5. Government of Western Australia. COVID-19 Coronavirus: Community Advice. Available online: https://www.wa.gov.au/organisation/department-of-the-premier-and-cabinet/covid-19-coronavirus-community-advice (accessed on 19 May 2020).
6. Australian Government. Coronavirus (COVID-19) Advice for People in Aged Care Facilities. Available online: https://www.health.gov.au/news/health-alerts/novel-coronavirus-2019-ncov-health-alert/advice-for-people-at-risk-of-coronavirus-covid-19/coronavirus-covid-19-advice-for-people-in-aged-care-facilities (accessed on 21 May 2020).
7. Government of Western Australia. COVID-19 Public Hospital Visitor Guidelines. Available online: https://healthywa.wa.gov.au/~{}/media/Files/Corporate/general%20documents/Infectious%20diseases/PDF/Coronavirus/COVID-19-Public-Hospital-Visitor-Guidelines.pdf (accessed on 21 May 2020).
8. Australian Government. Coronavirus (COVID-19) Advice for Aboriginal and Torres Strait Islander People and Remote Communities. Available online: https://www.health.gov.au/news/health-alerts/novel-coronavirus-2019-ncov-health-alert/advice-for-people-at-risk-of-coronavirus-covid-19/coronavirus-covid-19-advice-for-aboriginal-and-torres-strait-islander-peoples-and-remote-communities (accessed on 21 May 2020).
9. Australian Government. COVID-19 Temporary MBS Telehealth Services. Available online: http://www.mbsonline.gov.au/internet/mbsonline/publishing.nsf/Content/Factsheet-TempBB (accessed on 1 June 2020).
10. Centre for Epidemiology and Evidence. *Population Health Research Strategy 2018–2022*; NSW Ministry of Health: Sydney, NSW, Australia, 2019.
11. Benfield, J.; Szlemko, W. Internet-Based Data Collection: Promises and Realities. *J. Res. Pract.* **2006**, *2*, D1.
12. Knottnerus, J.A.; Tugwell, P. Requirements for utilizing health care—Based data sources for research. *J. Clin. Epidemiol.* **2011**, *64*, 1051–1053. [CrossRef] [PubMed]
13. Western Sydney University. Information on Coronavirus (COVID-19). Available online: https://www.westernsydney.edu.au/coronavirus-information.html (accessed on 1 June 2020).
14. American Psychological Association. Conducting Research during the COVID-19 Pandemic: Advice from Psychological Researchers on Protecting Participants, Animals and Research Plans. Available online: https://www.apa.org/news/apa/2020/03/conducting-research-covid-19 (accessed on 21 May 2020).
15. Charles Sturt University. Funding and Grants: COVID-19. Available online: https://research.csu.edu.au/funding-and-grants (accessed on 24 May 2020).
16. Australian Research Council. ARC Post Award Guidance Including Major Investments: Responding to the Impact of COVID-19. Available online: https://www.arc.gov.au/arc-post-award-guidance-including-major-investments-responding-impact-covid-19 (accessed on 24 May 2020).
17. WHO Working Group on Ethics & COVID-19. *Ethical Standards for Research during Public Health Emergencies: Distilling Existing Guidance to Support COVID-19 R&D*; World Health Organisation: Geneva, Switzerland, 2020.
18. Lurie, N.; Manolio, T.; Patterson, A.P.; Collins, F.; Frieden, T. Research as a Part of Public Health Emergency Response. *N. Engl. J. Med.* **2013**, *368*, 1251–1255. [CrossRef]
19. Austin Health. Contingency Plan for COVID19 Interruption to Non-Drug/Non-Device Studies. Available online: https://www.austin.org.au/Assets/Files/OfR_Austin%20Health%20COVID%20guidelines%20non-drug%20non-device%20studies_20200324.pdf (accessed on 1 June 2020).
20. United Nations. UN Working to Ensure Vulnerable Groups Not Left behind in COVID-19 Response. Available online: https://www.un.org/en/un-coronavirus-communications-team/un-working-ensure-vulnerable-groups-not-left-behind-covid-19 (accessed on 1 June 2020).
21. National Academies of Sciences, Engineering, and Medicine; Committee on Clinical Trials during the 2014–2015 Ebola Outbreak; Board on Global Health; Board on Health Scienves Policy; Health and Medicine Division. Conducting clinical research during an epidemic. In *Integrating Clinical Research into Epidemic Response: The Ebola Experience*; Keusch, G., McAdam, K.P., Busta, E.R., Mancher, M., Cuff, P.A., Eds.; National Academies Press: Washington, DC, USA, 2017.
22. Sydney Local Health Discrict. Recruitment and Privacy. Available online: https://www.slhd.nsw.gov.au/rpa/research/recruitment.html (accessed on 24 May 2020).
23. Asch, S.; Connor, S.E.; Hamilton, E.G.; Fox, S.A. Problems in recruiting community-based physicians for health services research. *J. Gen. Intern. Med.* **2000**, *15*, 591–599. [CrossRef]
24. Liu, Q.; Luo, D.; Haase, J.E.; Guo, Q.; Wang, X.Q.; Liu, S.; Xia, L.; Liu, Z.; Yang, J.; Yang, B.X. The experiences of health-care providers during the COVID-19 crisis in China: A qualitative study. *Lancet Glob. Health* **2020**, *8*, e790–e798. [CrossRef]
25. ACT Government. ACT Government Public Spaces Closed. Available online: https://www.covid19.act.gov.au/news-articles/act-government-public-spaces-closed (accessed on 1 June 2020).
26. Whitehead, D.; Whitehead, L. Sampling data and data collection in qualitative research. In *Nursing and Midwifery Research*, 5th ed.; Schneider, Z., Whitehead, D., LoBiondo-Wood, G., Haber, J., Eds.; Elsevier Australia: Chatswood, NSW, Australia, 2016; pp. 111–126.
27. Da Costa, C.; Schneider, Z. Quantitative data collection and study validity. In *Nursing and Midwifery Research*, 5th ed.; Schneider, Z., Whitehead, D., LoBiondo-Wood, G., Haber, J., Eds.; Elsevier Australia: Chatswood, NSW, Australia, 2016; pp. 181–196.
28. Nicola, M.; Alsafi, Z.; Sohrabi, C.; Kerwan, A.; Al-Jabir, A.; Iosifidis, C.; Agha, M.; Agha, R. The socio-economic implications of the coronavirus pandemic (COVID-19): A review. *Int. J. Surg.* **2020**, *78*, 185–193. [CrossRef]

29. Graf, I. Respondent Burden. In *Encyclopedia of Survey Research Methods*; Lavrakas, P.J., Ed.; SAGE Publications, Inc.: Thousand Oaks, CA, USA, 2008; pp. 739–740. [CrossRef]
30. Lavrakas, P.J. (Ed.) Face-to-face interviewing. In *Encyclopedia of Survey Research Methods*; SAGE Publications, Inc.: Thousand Oaks, CA, USA, 2008. [CrossRef]
31. Australian Government. Coronavirus Disease (COVID-19): Information for Employers. Available online: https://www.health.gov.au/sites/default/files/documents/2020/04/coronavirus-covid-19-information-for-employers_2.pdf (accessed on 9 June 2020).
32. Australian Government. Coronavirus (COVID-19) Advice for the Health and Aged Care Sector. Available online: https://www.health.gov.au/news/health-alerts/novel-coronavirus-2019-ncov-health-alert/coronavirus-covid-19-advice-for-the-health-and-aged-care-sector (accessed on 9 June 2020).
33. Shields, L.; Smyth, W. Common quantitative methods. In *Nursing and Midwifery Research*, 5th ed.; Schneider, Z., Whitehead, D., LoBiondo-Wood, G., Haber, J., Eds.; Elsevier Australia: Chatswood, NSW, Australia, 2016; pp. 143–164.
34. Ferguson, L. External Validity, Generalizability, and Knowledge Utilization. *J. Nurs. Scholarsh.* **2004**, *36*, 16–22. [CrossRef] [PubMed]
35. Kwok, C.; McIntyre, A.; Janzen, S.; Mays, R.; Teasell, R. Oral care post stroke: A scoping review. *J. Oral Rehabil.* **2015**, *42*, 65–74. [CrossRef] [PubMed]
36. Poudel, P.; Griffiths, R.; Wong, V.W.; Arora, A.; Flack, J.R.; Khoo, C.L.; George, A. Oral health knowledge, attitudes and care practices of people with diabetes: A systematic review. *BMC Public Health* **2018**, *18*, 577. [CrossRef]
37. Villarosa, A.C.; Villarosa, A.R.; Salamonson, Y.; Ramjan, L.M.; Sousa, M.S.; Srinivas, R.; Jones, N.; George, A. The role of indigenous health workers in promoting oral health during pregnancy: A scoping review. *BMC Public Health* **2018**, *18*, 381. [CrossRef] [PubMed]
38. Villarosa, A.R.; George, D.; Ramjan, L.M.; Srinivas, R.; George, A. The role of dental practitioners in addressing overweight and obesity among children: A scoping review of current interventions and strategies. *Obes. Res. Clin. Pract.* **2018**, *12*, 405–415. [CrossRef] [PubMed]
39. Bonevski, B.; Randell, M.; Paul, C.; Chapman, K.; Twyman, L.; Bryant, J.; Brozek, I.; Hughes, C. Reaching the hard-to-reach: A systematic review of strategies for improving health and medical research with socially disadvantaged groups. *BMC Med Res. Methodol.* **2014**, *14*, 42. [CrossRef] [PubMed]
40. Heerman, W.J.; Jackson, N.; Roumie, C.L.; Harris, P.A.; Rosenbloom, S.T.; Pulley, J.; Wilkins, C.H.; Williams, N.A.; Crenshaw, D.; Leak, C.; et al. Recruitment methods for survey research: Findings from the Mid-South Clinical Data Research Network. *Contemp. Clin. Trials* **2017**, *62*, 50–55. [CrossRef] [PubMed]
41. Peel, R.; Ren, S.; Hure, A.; Evans, T.J.; D'Este, C.A.; Abhayaratna, W.P.; Tonkin, A.M.; Hopper, I.; Thrift, A.G.; Levi, C.R.; et al. Evaluating recruitment strategies for AUSPICE, a large Australian community-based randomised controlled trial. *Med. J. Aust.* **2019**, *210*, 409–415. [CrossRef]
42. Amon, K.L.; Campbell, A.J.; Hawke, C.; Steinbeck, K. Facebook as a Recruitment Tool for Adolescent Health Research: A Systematic Review. *Acad. Pediatrics* **2014**, *14*, 439–447.e4. [CrossRef]
43. Ali, S.H.; Foreman, J.; Capasso, A.; Jones, A.M.; Tozan, Y.; DiClemente, R.J. Social media as a recruitment platform for a nationwide online survey of COVID-19 knowledge, beliefs, and practices in the United States: Methodology and feasibility analysis. *BMC Med. Res. Methodol.* **2020**, *20*, 116. [CrossRef]
44. Arigo, D.; Pagoto, S.; Carter-Harris, L.; Lillic, S.E.; Nebeker, C. Using social media for health research: Methodological and ethical considerations for recruitment and intervention delivery. *Digit. Health* **2018**, *4*, 2055207618771757. [CrossRef]
45. Lane, T.S.; Armin, J.; Gordon, J.S. Online Recruitment Methods for Web-Based and Mobile Health Studies: A Review of the Literature. *J. Med. Internet Res.* **2015**, *17*, e183. [CrossRef] [PubMed]
46. Sarkies, M.N.; Bowles, K.A.; Skinner, E.H.; Mitchell, D.; Haas, R.; Ho, M.; Salter, K.; May, K.; Markham, D.; O'Brien, L.; et al. Data Collection Methods in Health Services Research. *Appl. Clin. Inform.* **2015**, *6*, 96–109. [CrossRef]
47. Tirschwell, D.L.; Longstreth, W.T., Jr. Validating administrative data in stroke research. *Stroke* **2002**, *33*, 2465–2470. [CrossRef]
48. Greiver, M.; Barnsley, J.; Glazier, R.H.; Harvey, B.J.; Moineddin, R. Measuring data reliability for preventive services in electronic medical records. *BMC Health Serv. Res.* **2012**, *12*, 116. [CrossRef]
49. Kennedy, A.D.M.; Leigh-Brown, A.P.; Torgerson, D.J.; Campbell, J.; Grant, A. Resource use data by patient report or hospital records: Do they agree? *BMC Health Serv. Res.* **2002**, *2*, 2. [CrossRef]
50. Davies, A.; Wellard-Cole, L.; Rangan, A.; Allman-Farinelli, M. Validity of self-reported weight and height for BMI classification: A cross-sectional study among young adults. *Nutrition* **2020**, *71*, 110622. [CrossRef] [PubMed]
51. Gosney, J.L.; Scott, J.A.; Snook, E.M.; Motl, R.W. Physical Activity and Multiple Sclerosis: Validity of Self-Report and Objective Measures. *Fam. Community Health* **2007**, *30*, 144–150. [CrossRef]
52. Wong, S.L.; Shields, M.; Leatherdale, S.; Malaison, E.; Hammond, D. Assessment of validity of self-reported smoking status. *Health Rep.* **2012**, *23*, D1.
53. Althubaiti, A. Information bias in health research: Definition, pitfalls, and adjustment methods. *J. Multidiscip. Healthc.* **2016**, *9*, 211–217. [CrossRef] [PubMed]
54. De Leeuw, E. (Ed.) *Data Quality in Mail, Telephone and Face to Face Surveys*; T. T. Publikaties: Amsterdam, The Netherlands, 1993; Volume 38, pp. 80–81.

55. Teo, T. Online and paper-based survey data: Are they equivalent? *Br. J. Educ. Technol.* **2013**, *44*, E196–E198. [CrossRef]
56. Novick, G. Is there a bias against telephone interviews in qualitative research? *Res. Nurs. Health* **2008**, *31*, 391–398. [CrossRef]
57. Vogl, S. Telephone Versus Face-to-Face Interviews: Mode Effect on Semistructured Interviews with Children. *Sociol. Methodol.* **2013**, *43*, 133–177. [CrossRef]
58. James, N.; Busher, H. Credibility, authenticity and voice: Dilemmas in online interviewing. *Qual. Res.* **2006**, *6*, 403–420. [CrossRef]
59. Meho, L.I. E-mail interviewing in qualitative research: A methodological discussion. *J. Am. Soc. Inf. Sci. Technol.* **2006**, *57*, 1284–1295. [CrossRef]
60. Janghorban, R.; Roudsari, R.L.; Taghipour, A. Skype interviewing: The new generation of online synchronous interview in qualitative research. *Int. J. Qual. Stud. Health Well-Being* **2014**, *9*, 24152. [CrossRef] [PubMed]
61. Mirick, R.; Wladkowski, S. Skype in Qualitative Interviews: Participant and Researcher Perspectives. *Qual. Rep.* **2019**, *24*, 3061–3072.
62. Pearce, G.; Thøgersen-Ntoumani, C.; Duda, J.L. The development of synchronous text-based instant messaging as an online interviewing tool. *Int. J. Soc. Res. Methodol.* **2014**, *17*, 677–692. [CrossRef]
63. Lally, R.; Eisenhauer, C.; Buckland, S.; Kupzyk, K. Feasibility of Synchronous Online Focus Groups of Rural Breast Cancer Survivors on Web-Based Distress Self-Management. *Oncol. Nurs. Forum* **2018**, *45*, E111–E124. [CrossRef] [PubMed]
64. Rosenthal, M. Qualitative research methods: Why, when, and how to conduct interviews and focus groups in pharmacy research. *Curr. Pharm. Teach. Learn.* **2016**, *8*, 509–516. [CrossRef]

sustainability

Article

The Critical Factors of Student Performance in MOOCs for Sustainable Education: A Case of Chinese Universities

Qin Yang [1] and Young-Chan Lee [2,*]

1 College of Economics, Jiaxing University, Jiaxing 314001, China; yangqin05@yeah.net
2 Department of Business Administration, College of Management and Economics, Dongguk University, Gyeongju 38066, Korea
* Correspondence: chanlee@dongguk.ac.kr

Abstract: COVID-19 has drastically changed human life, both in the current era and, no doubt, in the years to come. One of the hot topics during this period is whether and how COVID-19 will change education patterns in the next few years. This study aimed to investigate the relationship between massive open online courses (MOOCs) and students' learning performance. To that end, this study combined the gamification concept with the modified information systems (IS) success model and the expectation–confirmation model (ECM) for building a research model. We surveyed 586 students through an online survey and tested the hypotheses. The research results indicated that MOOCs' information quality (IQ), system quality (SQ), and service quality (SEQ) had positive influences on the confirmation, which then affected MOOCs' usefulness, satisfaction, and gamification. Students' continued usage intention and course performance were positively influenced by usefulness and gamification. Students' continued usage intention of MOOCs also positively affected course performance. The relationship between gamification and satisfaction, however, was not proved. The research results also showed that gamification, which was composed of entertainment, challenge, and social interaction, played a critical role in improving students' continued usage intention of MOOCs and their course performance.

Keywords: MOOCs; MOOC; higher education; sustainable education; IS success model; expectation–confirmation model; gamification; continued usage intention; course performance; student performance; COVID-19; Chinese universities

Citation: Yang, Q.; Lee, Y.-C. The Critical Factors of Student Performance in MOOCs for Sustainable Education: A Case of Chinese Universities. *Sustainability* **2021**, *13*, 8089. https://doi.org/10.3390/su13148089

Academic Editors: Miltiadis D. Lytras, Abdulrahman Housawi and Basim Alsaywid

Received: 30 May 2021
Accepted: 13 July 2021
Published: 20 July 2021

1. Introduction

The outbreak of COVID-19 spread worldwide at an unprecedented rate, becoming the first global pandemic over the past 100 years, causing severe harm and threatening human lives. However, the impacts and damages of the pandemic on the economy, politics, culture, education, and ideology remain unassessed. At the beginning of 2020, for the first time in history, worldwide universities closed offline classes as an emergency measure to prevent the increase in the virus infection rate. Thanks to the massive open online course (MOOC), hundreds of millions of students worldwide can continue their studies. Universities could work in a usual fashion during the lockdown period by providing temporary access to learning on MOOC platforms. During the pandemic, Chinese universities had to organize teaching and learning online. In December of 2020, Tsinghua University and the UNESCO Institute for Information Technologies in Education co-hosted the Global MOOC Conference. The minister of the Ministry of Education of the People's Republic of China (MOE) pointed out that MOOCs in China started in 2013, after 7 years of rapid development, with more than 32,000 courses online. The number of courses and the application scale currently rank first among countries in the world. Leading Chinese universities created and offered MOOCs, and students from other universities could enroll in these courses. Teachers evaluated available courses and chose the most appropriate course for students first. Typically, each course lasts 11–14 weeks. Students must pass quizzes and a final exam to receive a certificate from the MOOC provider (e.g., https://www.icourse163.org/ accessed on 21 June 2021) and eventually obtain credit from their university. Teachers usually provide extra instruction and consultation as a supplement to ensure students' learning outcomes.

Education institutions have made tremendous efforts, and teachers have created online courses and delivered them to students through the Internet. The "virtual semester" has started in many universities, and various classes are launched and provided online in a well-organized manner. In the field of education, many countries and regions, including China, have accumulated experience in the past efforts and exploration to combat COVID-19 outbreaks, but many unprecedented problems, meanwhile, have been revealed. Humans are bound to coexist with COVID-19 for a long time to come. In order to restore and rebuild the ecosystem of education, maintain the vitality of the education system, and promote the healthy and sustainable development of education in the post-pandemic era, it is not only necessary to reflect on and examine the experience accumulated in the fight against COVID-19 but also to conduct systematic and comprehensive plans from the perspective of education sustainability in the future. As an important mode of online education, MOOC gradually occupies an important position in traditional education, has its own unique advantages, and tends to take over the sole dominance of traditional education. MOOC has emerged from the rapid development of technology and the urgent need for education [1]. The MOOC's distinguishing feature is its convenience and extensive sharing, making it a tool for improving the quality of education with high efficiency and proper education at a low cost. In addition to recorded lectures and assignments, many MOOCs provide tools that enable students to interact with the teachers and communicate with online classmates through social media and online communities. MOOCs can also provide immediate feedback, online examinations, and assignments. Interaction between students and professors can occur within an online community through synchronous and asynchronous online learning activities, and students' social, interactive, and collaborative skills will improve [2]. More than this, students can select their schedule for studying with the assistance of their teacher. After COVID-19, MOOC has been widely discussed in academic and practical areas as an innovative tool for sustainable education.

There is a debate because people wonder whether online classes offered by schools will be attractive to students. With the outbreak of COVID-19, most of our daily activities have been transferred from offline to online. Many educational institutions worldwide provide MOOCs, including edX, Moodle, CourseSites, Udemy, Versal, etc. According to Class Central, a search engine and reviews site for free online courses popularly known as MOOCs, by 2019, MOOCs had reached 110 million learners. Most importantly, the top three MOOC platforms (Coursera, edX, and FutureLearn) registered as many new learners in April 2020 as in the whole of 2019. Worldwide platforms provide more than 2500 courses, 11 online degrees, and 170 micro-credentials. The use of MOOCs during COVID-19 has allowed educational institutions to ensure teaching work runs smoothly on the rails. It enables students to have greater autonomy in preparing and conducting their learning. MOOCs differ significantly in terms of their purposes and the mechanisms that power them. Main categories of MOOCs usually include: First, connectivist MOOCs (cMOOCs) are a relatively loose online community for learners, mainly focusing on a shared learning interest. cMOOCs, on the other hand, use various social media tools and personal interaction tools for learning and sharing knowledge among the participants. Second, eXtended MOOCs (xMOOCs) are additional supplementary courses at universities. Third, cooperative MOOCs often combine characteristics of both xMOOCs and cMOOCs. Lastly, gamification cooperative MOOCs (gcMOOCs) [3,4] add gamification designs to the cooperative MOOCs.

The focus of previous research on MOOCs has been varied. Some key topics include barriers and intentions for using MOOCs [5], MOOC scholarship [6], challenges and motivation for teachers and students [7], and engagement and retention [8]. Lee et al. [9] discussed MOOCs from the perspective of self-regulated learning. Furthermore, many studies have discussed the ECM in explaining and exploring learners' continued usage intentions. For instance, Dai et al. [1] extended the ECM and proposed a theoretical model including cognitive and affective variables to discuss and explain users' intentions to stay in MOOC learning. Zhou [10] added "social influence" as a new construct to the original ECM in the MOOCs' learning context.

Gamification is a prevalent concept in mobile and technological systems that adopts a game mechanism to facilitate desired behaviors and bring positive performance outcomes. It has been proven that gamification in MOOCs positively influences students' motivation [11]. Among the diverse research theories that were used for discussing MOOCs, the gamification theory was a notable one. Antonaci et al. [12] provided a gamification design process with six phases. In designing gamified MOOCs, learner, social, openness, course design, and technology were considered. Antonaci et al. [13] summarized the effects of game elements on learners' behavior in an online learning environment, such as performance, motivation, engagement, collaboration, and social awareness. Dominguez et al. [14] and Dicheva et al. [15] indicated that gamification of online learning could increase users' motivation for studying if adequately designed and implemented. Thus, gamification is a favorable mechanism that can motivate users of information systems. Malone [16] emphasized the concept of gamification. This study focused on stimulating more people to consider game-based elements in education because game mechanisms influence users' learning motivation and engagement. As gamification gains increasing attention in education research, many scholars have emphasized the benefits that gamified designs bring to learners. Their findings provide breakthroughs for future studies in online learning and gamification research [17–19]. In order to maintain and increase students' interest and engagement, the gamified design of MOOCs should include emotional incentives by giving students challenging cases, entertaining functions, and social interaction opportunities. Although those studies noted that gamification affected students' positive experience with the MOOC, to the best knowledge of the authors, no study has been arranged to examine the relationship between gamification and overall effects of MOOCs. To address this research gap, the current study adopted an integrated framework based on the IS success model, expectation–confirmation model (ECM), and gamification concept to investigate the relationship between MOOCs and students' course performance empirically. For the analysis and research hypothesis tests, we used covariance-based structural equation modeling (CB-SEM). This paper is organized as follows. The description of MOOC and the theoretical basis of this study are systematically reviewed and arranged in Sections 2.1–2.5. Then we propose a research model and hypotheses in Section 2.6. Next, we describe the data collection and research methods in Sections 2.7 and 2.8, followed by data analysis results in Section 3. Finally, we discuss the major findings, implications, and limitations in the final part of the paper.

2. Materials and Methods

2.1. MOOCs (Massive Open Online Courses)

The media used in MOOCs is open to the public. Learners are encouraged to share and contribute materials. Course modules are often 5–10 min, and course content is edited when needed. As to traditional online courses, most institutions have their own online course platform and restrict the media. Students search for courses within an institution's library database. Course modules are typically 45–60 min long. In MOOCs, lectures are usually pre-recorded, and courses are open-ended with no due dates. As for traditional online courses, in addition to recorded lectures, the instructors sometimes conduct live lectures. Courses frequently end on the due date. The learners of MOOC include not only school students but also on-job students. As with traditional online courses, learners are often students of the institution. In addition, unlike online courses, which focus more on content, MOOCs focus more on context.

Wu and Chen [20] suggested that MOOCs are attractive to students who have strong motivation and who hold the perception that MOOCs are useful for individual users. To further examine and discuss the factors influencing students' perceptions and intentions to use MOOCs, many studies have paid attention to the antecedents of learners' behavioral intentions from various perspectives [5,10,21,22]. Although past research has successfully revealed various valuable and interesting implications and provided views that support the importance of an actively engaged MOOC, it is necessary to investigate MOOCs' effect on students' continued usage intentions and course performance.

2.2. Overview of a Chinese MOOC Platform

In MOOCs, each course runs periodically, and the entire learning process consists of multiple sessions: attending lectures, participating in discussions, submitting assignments, pop quizzes, and a final exam. Courses are managed and operated by the Academic Affairs Office of each university. Universities create courses and designate the teachers to manage the courses.

Usually, there are 9 sections involved in one course, from course selection to certificate issuance, as shown in Figure 1.

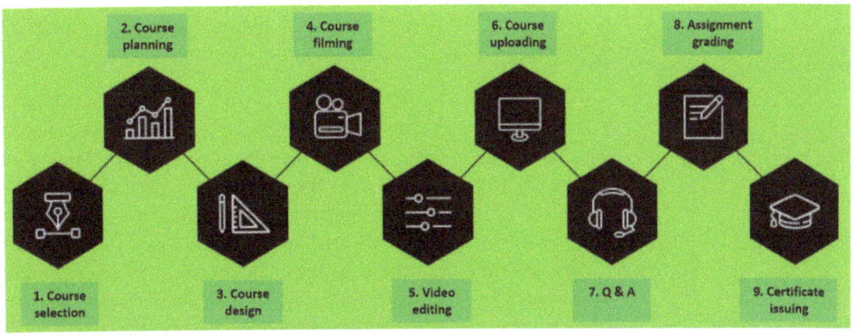

Figure 1. Processes for managing one course. *Source:* https://www.icourse163.org/, accessed on 21 June 2021.

In addition, students should finish authentication first so that teachers can manage the students in an organized manner, as shown in Figure 2:

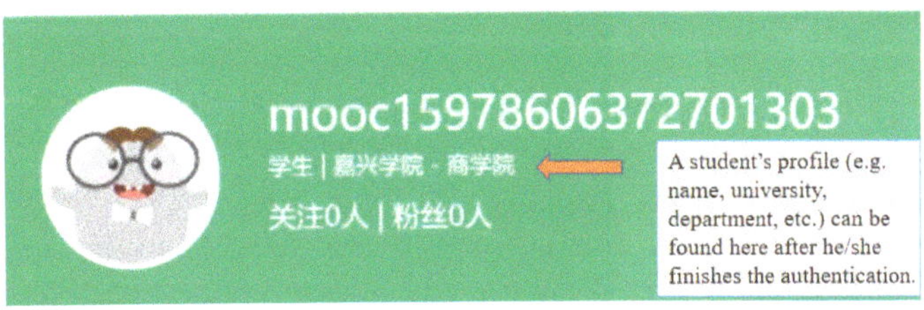

Figure 2. Student's profile sample. *Source:* https://www.icourse163.org/, accessed on 21 June 2021.

The design and structure of a course include 4 parts: a bulletin board, courseware (video and PPT files), assignments and quizzes, and a discussion board. Consider the course Financial Risk Management, as shown in Figure 3.

Figure 3. Design and structure of a sample course. *Source:* https://www.icourse163.org/, accessed on 21 June 2021.

The first figure on the left shows the bulletin board of Financial Risk Management, which includes the course outline, description of each class, and final exam description. The second figure presents the courseware of the course, where students can find lecture videos from chapter 1 to the last chapter and obtain PPT and/or PDF files. The third figure indicates the assignments and quizzes of the course. Students should finish and pass these assignments and quizzes before the deadline. The last figure demonstrates the discussion board of the course. The discussion board is divided into three sections: general discussion, Q&A, and the Learning Exchange Zone. On the discussion board, students can communicate with the teachers and their classmates.

2.3. IS Success Model

The implementation success of an information system can be influenced by a lot of factors, such as environmental factors, system factors, end-users' characteristics, etc. [23] DeLone and McLean [24,25] proposed the IS success model, an integrated model that comprehensively explains IS success. The model presents, identifies, describes, examines, and explores the relationships between the essential dimensions. For decades, research of information systems and technology has adopted the IS success model. From 1995 to 2019, various studies have applied the IS success model and provided beneficial results [16,26–30].

In online education, well-established research models and theories such as the technology acceptance model (TAM), the theory of reasoned action (TRA), the theory of planned behavior (TPB), and the unified theory of acceptance and use of technology (UTAUT) have been discussed so far. Rogers' [31] innovation diffusion theory (IDT) was also a remarkable one. In addition, some researchers examined students' behaviors in online education by employing the IS success model. For example, Freeze et al. [32] and Aparicio et al. [33] investigated e-learning system success using the IS success model. Cidral et al. [34] proposed a research model integrating the IS success model and success factors in e-learning systems. Albelbisi [35] verified the role of quality factors in self-regulated learning in MOOCs, and Albelbisi [36] also developed and validated MOOCs' success factors using Malaysian samples.

Information quality (IQ), system quality (SQ), and service quality (SEQ) are three independent variables in the IS success model. IQ deals with the supplied information from the information system with accuracy, correlation, and conformance. SQ usually refers to the usefulness, stability, function, reliability, complexity, compatibility, and demonstrability of the IS. SEQ discusses whether information systems provide assured, responsive, and empathetic service. Most previous studies usually included IQ, SQ, SEQ, intention to use, user satisfaction, and net benefits when researching information technology and systems [37–41]. MOOC platform provides content courses, a learning environment, and services for potential learners through the Internet. IQ of the MOOC platform refers to information provided on the MOOC platform having good quality when contents are perceived as useful, reliable, complete, and understandable. SQ of the MOOC platform refers to good system technical performance, accessibility, availability, and usability. SEQ of the MOOC platform refers to the responsiveness, competence, and empathy of the MOOC platform service representatives. From previous studies using these IS success variables (IQ, SQ, SEQ) [39,41,42], we presume that this model is often adopted in a variety of fields since it provides a comprehensive understanding of the context in which the users use the system. Hence, it is also important to explore and examine how these factors influence success on MOOC platforms. MOOC platforms are attracting attention as a new online education tool, and it is important to keep these platforms available to students. This study focused on improving students' course performance by continuously using the MOOCs platform. We proposed that the information system success model can be adopted since it is useful to measure how students perceive the usefulness and gamified designs of a MOOC platform, students' satisfaction with a MOOC platform, their continued use intention toward a MOOC platform, and their learning performance resulting from the involvement. MOOCs presuppose the importance of their content, scheme, and instant service as a representative online education tool. These attributes have been neglected in the evaluation of MOOCs. As a result, it is critical to explore and explain MOOCs' qualities, which may significantly influence learners' perceptions and behaviors.

2.4. *Expectation–Confirmation Model*

When considering post-purchase or post-adoption satisfaction, previous studies have used various constructs. Some representative antecedents of continued usage intentions include perceived usefulness, satisfaction, perceived ease of use, trust, etc. Many studies have discussed users' continued usage intentions of information technologies through the expectation–confirmation model (ECM). Hsu and Lin [43], Leung and Chen [44], and Gupta et al. [45] investigated the relationship between pre-adoption experiences and post-adoption continued usage intentions in the field of paid mobile apps and e-health technology. Wu et al. [40] explored online impulse buying through the expectation–confirmation model, an epistemic research model that explores post-adoption satisfaction and continued usage of the information system. Bhattacherjee [46] demonstrated that expectations formed by accepting a specific information system would lead to confirmation by comparing the characteristics and performance. When users confirm the performance, their satisfaction will be followed. Thus, in turn, continued usage intentions will be induced.

2.5. *Gamification*

In IS research, gamification is an emerging concept that uses gamified designs and non-gamified technologies to make tasks more engaging. Gamification has a profound impact on many different types of information systems, both in theory and action. British-born computer programmer and inventor Nick Pelling initially proposed the concept of gamification in 2002. It started to gain attention and hit mainstream academic circles around 2010. Research in the health system, crowdsourcing, education, etc., has been applied to gamification. One of the predominant problems of MOOCs is students' low retention rates and completion rates [47,48]. This phenomenon emphasizes the importance of introducing a gamification mechanism to increase the retention and completion rate. Levels, badges, leaderboards, points, feedback, challenges, likes, and social features, communication channels, narratives, etc., are the most widely used gamified elements in MOOCs [13,33]. Figure 4 shows a "Lucky Card" game provided by a MOOC provider during exam week, and Figure 5 shows a badge.

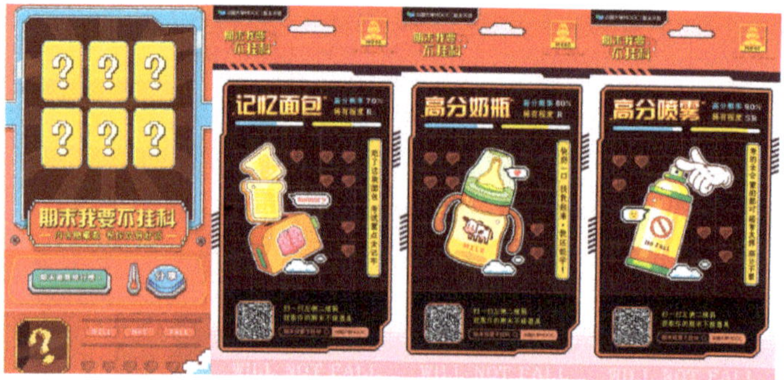

Figure 4. Lucky Card game of a MOOC.

Figure 5. The badge provided by a MOOC. *Source: https://www.icourse163.org/*, accessed on 21 June 2021.

The Lucky Card game is a "Will Not Fall" game with a lucky meaning. By engaging this game, students can draw different cards randomly. Each card represents a "weapon" for the final examination, such as the "memory bread" card, which means you will be able to memorize all the contents of the exam. "Energy milk" card means you will be full of energy. "Not fall spray" card means you will outdo yourself in the exam. These cards are supposed to be lucky in Chinese culture. Furthermore, students can share this game with their friends on social networks and have various discussions together. These entertaining games can help students reduce the pressure of studying psychologically. In addition to that, students will be rewarded (points gained) when they complete predefined activities of a weekly mission, such as watching video lectures and finishing quizzes. The digital badges are also incentives to motivate students to continue learning.

Dominguez et al. [14] indicated the benefits and consequences of gamified elements and designs in an e-learning platform, and Dicheva et al. [15] prepared systematic mapping research to discuss gamified elements in education. An et al. [49] found that learners had various attitudes concerning the gamification of MOOCs. Academics and industry are becoming more interested in the gamification of MOOCs. What influence does gamified design have on MOOCs? Among the effects, the most remarkable one must be an increase in participation. When students participate in the discussion forum, they have an opportunity to gain "likes" from classmates and teachers. This is a way to increase students' participation. Furthermore, group interactions are available for students as a tool for getting feedback and making contributions. In addition, time pressure is another gamified element for encouraging students' participation. Time pressure means MOOC platforms give students a time limit to finish a certain activity to push them to complete a required task. For students, creating time pressure can encourage them to participate in learning positively because the time pressure is connected to their learning goals (e.g., finishing the current learning stage and moving to a higher level). By adding the gamification concept to MOOCs, students' completion rate will increase, and students will be willing to spend more time on the MOOCs. Ortega-Arranz al. [19] and Khalil et al. [50] showed that students' retention increased by employing gamification, and their motivation and engagement were improved. Saputro et al. [51] found that students' intrinsic motivation in MOOC increased under the framework of fulfilling rules and suitability of the instruments and game elements. The current study tried to add the gamification concept to the IS success model and expectation–confirmation model framework based on the previous studies and findings.

2.6. Research Model and Hypotheses

2.6.1. MOOC Qualities and User Confirmation

DeLone and McLean [24] proposed a research model explaining information system success. The original research model includes six constructs, system quality, information quality, use, user satisfaction, and individual and organizational impact. The updated IS success model was proposed by them later in 2003. The updated model [25] added a new construct, service quality, to the original model. In addition, the "use" in the basic model was divided into two separate variables: "intention to use" and "use" for measuring and explaining IS success. In online education, Hassanzadeh et al. [52], and Marjanovic et al. [53] applied the IS success model for exploring the e-learning platform. Lin [54] indicated that the platform's SQ, IQ, and SEQ significantly influenced actual online learning platform use through users' satisfaction and intention to use. The ECM is built on consumer behavior studies and integrates theoretical and empirical findings from previous IS literature. Bhattacherjee [55] suggested that users' continued usage intention (CI) is influenced by their satisfaction with the IS and users' perception of its usefulness. The users' dissatisfaction will result in the discontinuance of the IS if performance disconfirms the expectations.

By using MOOCs and evaluating their performance, learners constitute perceptions in their minds and compare MOOCs with other learning tools, corresponding to their expectations. The qualities of MOOCs are connected with potential users' expectations of MOOCs. The higher the users' perceived qualities of the MOOCs and confirmed expectations, the more positive perceptions and satisfaction will be induced. The study of Roca et al. [56] proved that the IQ, SQ, and SEQ of an e-learning system positively affected expectation–confirmation of the e-learning system.

Similarly, Chung et al. [57] also suggested that an information system's confirmation was positively influenced by its IQ, SQ, and design quality (DQ). In addition, in smartphone classes, Shin et al. [58] showed that the system's perceived content quality and perceived service quality positively influenced users' confirmation. Based on previous studies, this paper hypothesizes that MOOCs' information quality (IQ), system quality (SQ), and service quality (SEQ) have positive influence on MOOCs' quality confirmation:

Hypothesis 1 (H1). *Information quality of the MOOC positively affects students' confirmation of MOOC qualities.*

Hypothesis 2 (H2). *System quality of the MOOC positively affects students' confirmation of MOOC qualities.*

Hypothesis 3 (H3). *Service quality of the MOOC positively affects students' confirmation of MOOC qualities.*

2.6.2. User Confirmation, MOOC Platform Usefulness, User Satisfaction, and Gamification

Bhattacherjee [46] showed that users' confirmation of their original expectations is a critical predictor of users' usefulness perception formation. Information system confirmation deals with the degree to which the information systems satisfy the user's expectations. The perceived usefulness of an information system implies the user's expectation of that information system, influencing users' post-adoption behavior. To be more specific, if any user confirms his/her expectations, their perceived usefulness will be formed. In addition, ECM proposed that users' satisfaction with an information system is induced by their confirmation of experience and the perceived usefulness of the system. Previous studies have proved the relationships between users' confirmation, perceived usefulness, and users' satisfaction [1,45,59,60]. Pozón-López et al. [61] showed that satisfaction is affected by MOOC quality and other qualities such as usefulness. In MOOCs, if students find that the actual performance is better than their original expectations, their confirmation of the platform will be positive. In turn, the positive confirmation will result in a higher level of user satisfaction. On the contrary, if students believe that the system's performance falls short of their expectations, their confirmation will be negative. Consequently, the negative confirmation will lead to a lower level of user satisfaction in the subsequent stages. According to Bhattacherjee and Lin [55], user satisfaction is an emotional evaluation of an information system based on previous experience.

In addition to the above constructs, gamification is a new critical component of the information system. Although gamified elements' influences have been explored and discussed in several fields, studies on gamification in MOOCs are still limited. Some researchers have recently addressed the relationship between gamification and expected benefits [2,51,62,63]. Extrinsic motivation could be considered and explained in terms of gamified elements in MOOCs. While learning, learners engage in gamified MOOCs to get points and obtain badges and other prizes. Recently, studies discussing the effectiveness of the gamified design of the online educational system have been growing. However, it is necessary and valuable to conduct empirical studies to explore the relationship between gamification (e.g., badges, points, forum participation, "likes", time pressure, social networking interaction, etc.) and learners' behaviors and learning performance. Students' perceptions of gamified elements of the MOOCs are their expectations of the system, and the expectations will influence their post-adoption behaviors. If students' expectations can be confirmed, the perception of the gamified design will be formed as well. Based on previous gamification literature, we hypothesized that:

Hypothesis 4 (H4). *Students' confirmation of MOOC qualities positively affects MOOC usefulness.*

Hypothesis 5 (H5). *Students' confirmation of MOOC qualities positively affects MOOC gamification perceptions.*

Hypothesis 6 (H6). *Students' confirmation of MOOC qualities positively affects their satisfaction with the MOOC.*

Hypothesis 7 (H7). *MOOC usefulness positively affects students' satisfaction with the MOOC.*

Hypothesis 8 (H8). *MOOC gamification positively affects students' satisfaction with the MOOC.*

2.6.3. MOOC Usefulness, Satisfaction, Gamification, and Continuance Intention

In MOOCs, the platform's perceived usefulness deals with the degree to which MOOCs can be useful and improve students' learning. Ajzen [64] believed that users' subjective perceptions influence their behaviors, even though the perceptions they hold are tendentious or incorrect. As a result, instead of an objective assessment, the perceived feeling is supposed to be relevant. According to Bhattacherjee [46], users' satisfaction with an information system is an active, neutral, or passive feeling. Expectation–confirmation studies state that users' satisfaction with an information system is a prerequisite for building users' continued usage intentions. Bhattacherjee [46], Joo et al. [65], Li and Fang [66], and Bölen and Özen [67] proved the positive relationship between perceived usefulness and IS continued usage intention, as well as satisfaction and IS continued usage intention. Hence, in MOOCs, students are expected to use MOOCs continuously if satisfied with the platform. On the other hand, the direct relationship between gamification and users' continued usage intentions has not been empirically verified yet. However, several studies have proved that users' benefits and continued usage will increase in a gamified environment [68,69]. Hence, we hypothesized that students' perceptions of MOOCs' gamified design will positively influence their continued usage intentions in the MOOC context.

Hypothesis 9 (H9). *MOOC usefulness positively affects students' continued usage intention of MOOC.*

Hypothesis 10 (H10). *MOOC gamification positively affects students' continued usage intention of MOOC.*

Hypothesis 11 (H11). *Satisfaction with the MOOC positively affects students' continued usage intention of MOOC.*

2.6.4. Usefulness, Gamification, Continued Usage of MOOCs, and Course Performance

Course performance in MOOCs refers to the "perceived outcome" of learning through MOOCs. Specifically, it deals with the ability and possibility to learn well with higher efficiency and less effort and time. Tam and Oliveira [70] investigated individual performance in the context of m-banking. However, insufficient empirical research has studied the salient relationship between the IS' perceived usefulness and individual performance and the IS' continued usage intention and individual performance. Usefulness keeps ahead of user satisfaction, and users' positive experience with usefulness will induce greater user satisfaction. Increased IS user satisfaction leads to increased IS usage intentions, which improves individual performance [71]. Increased use of MOOCs may raise student awareness of the benefits of MOOCs. In a MOOC learning context, students who adopt digital systems and have digital skills can improve their learning performance. This study investigated whether students could gain a clear understanding of the classes through using MOOCs, measured whether students could easily achieve the learning goals asserted by the course via MOOCs, and discussed if it is easier to accomplish the assignments by using MOOCs.

Gamified design is another critical element of MOOCs. Gerdenitsch et al. [72] examined how gamification affects work enjoyment, productivity, and the moderating role of leadership responsibility. Feng et al. [73] discussed whether gamification could increase consumers' engagement with fitness apps. Groening and Binnewies [74] examined the relationship between digital achievements, motivation, and performance. The research results show that digital achievements could improve motivation and enhance performance. The findings provide significant inspiration for future IS research. MOOCs employ gamified design to induce and increase students' engagement and participation. A gamified MOOC may change students' study patterns and communication patterns. Thus, if students feel satisfied with the novel studying tools and the interaction patterns, their motivation and engagement will increase. In turn, students' learning performance will be improved compared with those online education tools without gamified design. Based on previous studies, we hypothesized that:

Hypothesis 12 (H12). *MOOC usefulness positively affects students' course performance.*

Hypothesis 13 (H13). *MOOC gamification positively affects students' course performance.*

Hypothesis 14 (H14). *Students' continued usage intention of the MOOC positively affects students' course performance.*

2.6.5. Gamification: Sociality, Entertainment, and Challenges

As discussed above, gamification is a way to increase students' motivation and engagement and, finally, affect students' course performance. The gamification concept has also gained attention from researchers and practitioners. Legaki et al. [75] proved that the challenge-based gamified elements in an online learning platform positively affect students' learning compared to conventional teaching tools. The concept of gamification has been employed in many research areas. For example, Tobon et al. [76] examined the relationship between gamified features and online customer behavior, and Rodrigues et al. [77] explored gamification in the field of e-banking. Specifically, Rodrigues et al. [77] investigated how bank users face gamified smart applications, mainly focused on entertainment and ease of use elements. Their findings reveal that the system's ease of use and entertainment had an impact on e-banking usage. Social interaction is another critical component of gamification. Fang et al. [21] stated that online learning platforms' social interaction design positively influences users' study focus. Students could feel a sense of entirely participating or breathtakingly engaging in the MOOCs' learning [78]. If applied to MOOCs, social interaction may stimulate frequent use of the platform and possibly improve students' learning performance. In addition, the social interaction design may improve the students' MOOC user experience. Without face-to-face communication, students still maintain social connections with professors and classmates. Thus, we propose that gamification may include vibrant designs by providing students with challenges, entertaining atmosphere, and social interaction in MOOCs.

Hypothesis 15 (H15). *Gamification is a second-order construct that is comprised of socialness, challenge, and entertainment.*

Based on the hypotheses described above and previous studies, this study presents a research model as shown in Figure 6.

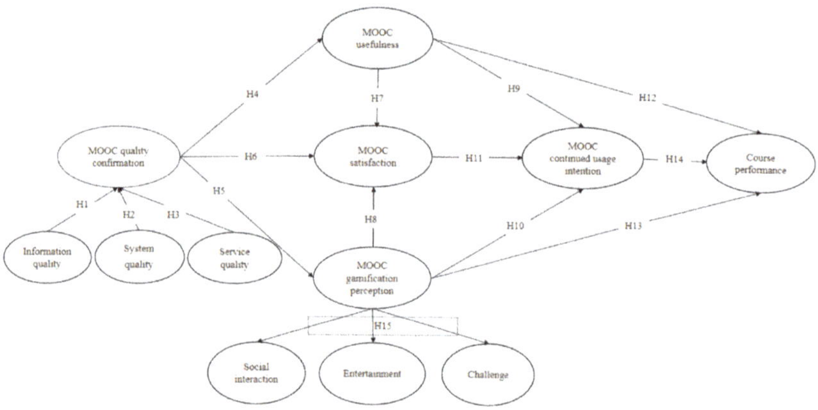

Figure 6. Proposed research model.

2.7. Data Collection

This study conducted a questionnaire survey to collect data and investigate the relationship between the constructs. After completing an extensive review of IS literature, we created the survey items and translated the original ones into Chinese. Then, we translated them into English again to ensure the questionnaire's accuracy and consistency. We distributed the survey questionnaires to universities in China and collected 629 questionnaires. All of the students attended MOOCs in the spring semester of 2020. After excluding 43 questionnaires with incomplete responses, 586 valid questionnaires were used for analysis.

The respondents' demographic characteristics are summarized in Table 1. Table 1 shows 209 (35.7%) males and 377 (64.3%) females. In terms of educational backgrounds, most of those who responded were college students (482, 82.3%). In terms of past MOOC user experience, 136 (23.2%) have used them for less than a month, 127 (21.7%) have learned in them for 1 to 3 months, and 107 (18.3%) have used them for more than 12 months. When it comes to the average amount of time spent on MOOCs each week, most participants

(260, 44.4%) spend 1 to 4 h per week on them. Finally, 160 (27.3%) respondents came from the country's western region, 215 (36.3%) from the country's central region, and 211 (36%) from the country's eastern region.

Table 1. Demographic distribution.

Variables	Classification	Number	Percentage
Gender	Male	209	35.7
	Female	377	64.3
Education background	Undergraduate	482	82.3
	Graduate school (M.A.)	20	3.4
	Graduate school (Ph.D. and above)	84	14.3
Experience	Less than 1 month	136	23.2
	1–3 months	127	21.7
	3–6 months	117	20
	6–9 months	64	10.9
	9–12 months	35	6
	12 months and above	107	18.3
Average using time/week	1–4 h	260	44.4
	5–9 h	151	25.8
	10–14 h	87	14.8
	15 h and above	88	15
Residency	Western	160	27.3
	Central	215	36.7
	Eastern	211	36

2.8. Measurement Items

The measurement items were developed through an extensive literature review (see Appendix A). The survey aimed at measuring students' perceptions of MOOCs positively or negatively, which includes the following variables: information quality (IQ), system quality (SQ), service quality (SEQ) of the MOOC platform, students' confirmation, perceptions of the MOOC platform's gamified characteristics, perceived usefulness of MOOCs, students' satisfaction with MOOCs, students' MOOC platform continued usage intentions, and students' course performance through learning on MOOC platforms.

This study employed a 5-point Likert scale, ranging from 1 to 5 (1 represents an inclination to strongly disagree, 5 represents strongly agree). We analyzed the collected data in 3 steps. First, outliers, normality, and missing values were checked. Second, we tested the validity and reliability of the measurement items. Finally, we examined correlations between the constructs, tested the model fit, checked common method bias, and tested research hypotheses.

3. Analysis and Results

This study analyzed the data using the covariance-based structural equation model (CB-SEM). The measurement model and structural model were evaluated, respectively.

3.1. Measurement Model

We evaluated the measurement model by checking reliability and convergent and discriminant validity. In Table 2, all values of standardized item loadings for the indicator were $\geq$0.7. All of the composite reliability (CR) values were greater than 0.8. All values of Cronbach's α for each construct were $\geq$0.8. All average variance extracted (AVE) values employed for assessing the convergent validity were $\geq$0.5. Consequently, Table 2 indicates that all values met the minimum requirement, and the measurement model has good reliability and convergent validity.

Table 2. Factor loading, CR, AVE, and Cronbach's α values.

Constructs	First-Order Constructs	Item	Item Loading	CR	AVE	Cronbach's α
MOOC platform quality confirmation	N/A	MCON1 MCON2 MCON3	0.814 0.777 0.794	0.838	0.632	0.837
Information quality	N/A	IQ1 IQ2 IQ3 IQ4	0.771 0.731 0.725 0.779	0.839	0.565	0.838
Service quality	N/A	SEQ1 SEQ2 SEQ3 SEQ4	0.784 0.763 0.75 0.811	0.859	0.605	0.858
System quality	N/A	SQ1 SQ2 SQ3 SQ4	0.818 0.767 0.752 0.784	0.862	0.61	0.861
MOOC usefulness	N/A	MUSE1 MUSE2 MUSE3	0.845 0.768 0.825	0.854	0.662	0.853
MOOC satisfaction	N/A	MSAT1 MSAT2 MSAT3	0.799 0.7 0.805	0.813	0.592	0.809
Gamification perceptions	Social interaction	SOC1 SOC2 SOC3 SOC4	0.834 0.754 0.794 0.839	0.881	0.65	0.88
	Challenge	CHA1 CHA2 CHA3	0.826 0.709 0.827	0.831	0.623	0.827
	Entertainment	ENT1 ENT2 ENT3 ENT4	0.814 0.722 0.733 0.792	0.85	0.587	0.849
MOOC continued usage intention	N/A	MCI1 MCI2 MCI3	0.795 0.756 0.818	0.833	0.624	0.831
Course performance	N/A	CP1 CP2 CP3 CP4	0.781 0.739 0.739 0.761	0.841	0.57	0.841

Pearson's correlation coefficients are shown in Table 3. Next, we compared the square root of AVE with the variables' correlation coefficients to assess the discriminant validity. As shown in Table 3, the AVE's square root is significantly larger than each variable's correlation coefficient. Consequently, the scale has good discriminant validity.

Table 3. Discriminant validity test.

	IQ	SQ	SEQ	MCON	MUSE	GAM	MSAT	MCI	CP
IQ	0.752								
SQ	0.433 **	0.781							
SEQ	0.457 **	0.418 **	0.778						
MCON	0.392 **	0.338 **	0.366 **	0.795					
MUSE	0.318 **	0.288 **	0.263 **	0.394 **	0.814				
GAM	0.366 **	0.312 **	0.323 **	0.328 **	0.178 **	0.767			
MSAT	0.353 **	0.307 **	0.325 **	0.532 **	0.514 **	0.284 **	0.769		
MCI	0.356 **	0.352 **	0.364 **	0.405 **	0.435 **	0.425 **	0.501 **	0.790	
CP	0.403 **	0.361 **	0.414 **	0.393 **	0.454 **	0.506 **	0.414 **	0.575 **	0.755

NOTE: ** $p < 0.01$. MCON = MOOC platform quality confirmation; IQ = Information quality; SQ = System quality; SEQ = Service quality; MUSE = MOOC usefulness; MSAT = MOOC satisfaction; GAM = Gamification; SOC = Social interaction; CHA = Challenge; ENT = Entertainment; MCI = MOOC continued usage intention; CP = Course performance.

Podsakoff and Organ [79] and Podsakoff et al. [80] suggested that there were possibilities for common method variance (CMV) in the self-reported data. We applied the Harman one-factor statistical analysis to examine the common method bias in the collected data. First, we checked if the measurements were affected by common method variance through Harman's single factor test and examined whether relationships between constructs were observed to inflate or deflate. This test includes a factor analysis of all the involved items. In addition, we performed exploratory factor analysis without rotation and checked whether all factors were returned. Previous research suggested that if one factor's total variance exceeds 50%, it shows common method bias in the study. However, as shown in Table 4, one factor explained only 27.899% of the total variance. Thus, we can conclude that there is no common method bias in the collected data.

Table 4. Common method bias test.

Component	Initial Eigen Values		
	Total	% of Variance	Cumulative %
	11.591	27.719	27.719
1	Extraction Sums of Squared Loadings		
	Total	% of Variance	Cumulative %
	10.881	27.899	27.899

Finally, we compared the recommended values and actual values of model fitness. Table 5 shows the values of χ^2, degrees of freedom (df), goodness-of-fit index (GFI), adjusted goodness-of-fit index (AGFI), normed fit index (NFI), comparative fit index (CFI), parsimony goodness fit index (PGFI), root mean square residual (RMR), and root mean square of approximation (RMSEA). All actual values met the standard of recommended values, which indicated a good model fit.

Table 5. Measurement model fit.

Fit Indices	χ^2/df	GFI	AGFI	NFI	CFI	RMR	RMSEA
Recommended Value	<3.0	>0.9	>0.8	>0.9	>0.9	<0.08	<0.08
Value indices	1.506	0.920	0.909	0.918	0.971	0.069	0.029

3.2. Structural Model

This section examined the R^2 of MOOC platform confirmation, usefulness, satisfaction, continued usage intention, course performance, gamification (social interaction, entertainment, and challenge), which suggested the explanatory power of the independent variables. The path coefficients and R^2 showed how well the data supported the proposed research model. It is recommended that the R^2 value of 75%, 50%, and 25% may indicate the substantial, moderate, and weak level of the construct's explanation power (s).

Information quality ($\beta = 0.312$ and p-value < 0.001), system quality ($\beta = 0.172$ and p-value $= 0.001$), and service quality ($\beta = 0.217$ and p-value < 0.001) have significant positive effects on MOOC quality confirmation. Three qualities explained MOOC quality confirmation with $R^2 = 33.8\%$. Thus, Hypotheses 1, 2, and 3 are supported. MOOC quality confirmation significantly positively affected MOOC usefulness ($\beta = 0.489$ and p-value < 0.001) with $R^2 = 23.9\%$. Thus, Hypothesis 4 is supported. MOOC quality confirmation significantly positively affected MOOC gamification perceptions ($\beta = 0.463$ and p-value < 0.001) with $R^2 = 21.5\%$. Therefore, Hypothesis 5 is supported. In addition, MOOC quality confirmation ($\beta = 0.453$ and p-value < 0.001) and MOOC usefulness ($\beta = 0.375$ and p-value < 0.001) have significant positive effects on students' satisfaction with MOOCs ($R^2 = 55.3\%$). However, gamification did not have a statistically significant effect on satisfaction. Thus, Hypothesis 6 and Hypothesis 7 are supported, but Hypothesis 8 is not. Regarding students' continued usage intention of the MOOCs, MOOC usefulness ($\beta = 0.23$ and p-value < 0.001), MOOC gamification perception ($\beta = 0.365$ and p-value < 0.001), and MOOC satisfaction ($\beta = 0.335$ and p-value < 0.001) have significant positive effects on students' continued usage intention ($R^2 = 51.8\%$). Therefore, Hypotheses 9, 10, and 11 are supported. In addition, MOOC usefulness ($\beta = 0.275$ and p-value < 0.001), MOOC gamification perception ($\beta = 0.392$ and p-value < 0.001), and MOOC continued usage intention ($\beta = 0.331$ and p-value < 0.001) significantly affected course performance with $R^2 = 62.1\%$. As a result, Hypotheses 12, 13, and 14 are supported. Lastly, as a second-order construct, gamification explained 67.5% of the variation of social interaction, 53.3% of the challenge, and 55.3% of entertainment. To summarize, based on these findings, gamification was proven to be a reflective second-order construct of social interaction, challenge, and entertainment. Thus, Hypothesis 15 is supported. Details can be found in Table 6 and Figure 7.

Table 6. Hypothesis test results.

Hypothesis	Path			β	p-Value	R^2	Remarks
H1	IQ	→	MCON	0.312	***		Supported
H2	SQ	→	MCON	0.172	0.001	0.338	Supported
H3	SEQ	→	MCON	0.217	***		Supported
H4	MCON	→	MUSE	0.489	***	0.239	Supported
H5	MCON	→	GAM	0.463	***	0.215	Supported
H6	MCON	→	MSAT	0.453	***		Supported
H7	MUSE	→	MSAT	0.375	***	0.553	Supported
H8	GAM	→	MSAT	0.062	0.189		Not supported
H9	MUSE	→	MCI	0.23	***		Supported
H10	GAM	→	MCI	0.365	***	0.518	Supported
H11	MSAT	→	MCI	0.335	***		Supported
H12	MUSE	→	CP	0.275	***		Supported
H13	GAM	→	CP	0.392	***	0.621	Supported
H14	MCI	→	CP	0.331	***		Supported
		→	SOC	0.675	***	0.675	
H15	GAM	→	CHA	0.553	***	0.533	Supported
		→	ENT	0.533	***	0.553	

NOTE: *** $p < 0.001$.

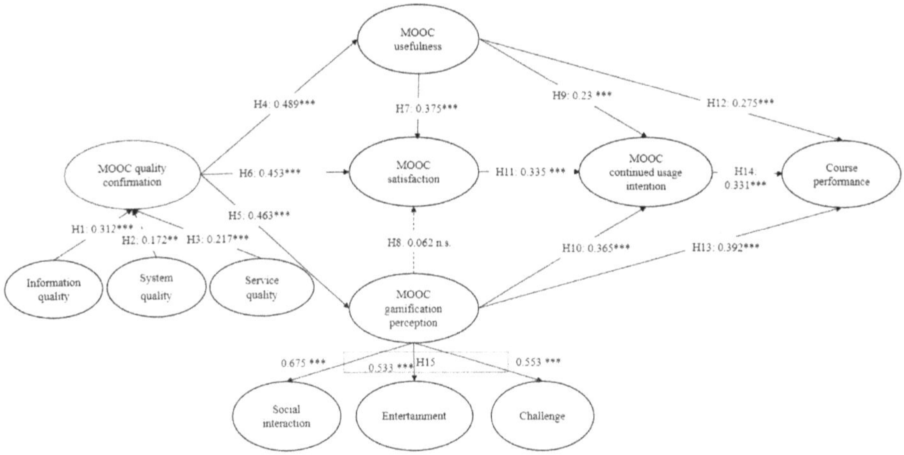

Figure 7. Path coefficients of the research model. NOTE: ** $p < 0.01$; *** $p < 0.001$.

4. Discussion

Campus closure is an important intervention that helps safeguard public health and respond to the spread of epidemic infections. According to the Policy Brief: Education during COVID-19 and beyond provided by the United Nations, the COVID-19 pandemic has created the largest disruption of education systems in history, affecting nearly 1.6 billion learners in more than 190 countries all continents. Closures of campuses and other learning spaces have impacted 94% of the world's student population, up to 99% in under-developed countries. The sudden outbreak of COVID-19 triggered a global education crisis and, at the same time, forced governments and schools to take the initiative to change education. In response to the ongoing situation of COVID-19, many countries have adopted a variety of flexible teaching and learning approaches in their education systems, among which MOOC is one of the main approaches. The MOOC, as a subset of distance education, has been focused on providing a more flexible and innovative educational experience in terms of time and space than school education by utilizing various types of technology. During the epidemic prevention and lockdown period, many countries closed schools and changed to MOOC mode and thereby accelerated the process of informatization and modernization of education.

The COVID-19 epidemic will not be eliminated for a long period in the future. Human beings are forced to move into and survive in the post-pandemic era with the coronavirus. The way of production, living, and survival also needs to be changed accordingly. Education is crucial to the sustainable development of human society. We have a mission and a task to adapt to coexist with the epidemic, work toward its improvement, and build a better world for future generations. Education in the digital information era is critical to the future development of the quality of human life and the world's sustainability. New digital technologies in education have far-reaching effects on formal education and informal learning regarding educational goals and objectives, educational ecology and the environment, learning processes, teaching and learning processes, and educational governance and policies.

We surveyed university students who had taken massive open online courses (MOOCs) during the COVID-19 pandemic to investigate the relationship between MOOCs and students' course performance. All hypotheses were supported except Hypothesis 8 (gamification satisfaction). It indicates that, among the three MOOC platform qualities, the MOOC platform's information quality was the most critical factor in affecting the quality confirmation of the MOOC platform. In addition, the MOOC platform's quality confirmation significantly influenced students' perceived usefulness of the MOOC platform, satisfaction with MOOCs and students' perceived gamified features. Moreover, only when students perceive MOOCs' usefulness and gamification, and when they are satisfied with the MOOC platform will they consider using MOOCs for future studies. Furthermore, with students' perceived usefulness and gamified MOOCs' features increasing, their learning performance will improve significantly. Finally, if students intend to use MOOCs for their future learning, their learning performance will also considerably improve. The results indicate that students' continued use intention and learning improved a positive attitude toward MOOCs. It is worth mentioning that we combined gamification with the IS success model and ECM in this study. The findings support the notion that the gamification of MOOCs consists of social interaction, entertainment, and challenges. Furthermore, gamification has a significant impact on students' intentions to continue using MOOCs and improve course performance.

4.1. Theoretical Implications

The results of this study could provide future studies with some theoretical implications. Firstly, our study utilized an integrated framework combined with DeLone and McLean's [24,25] IS success model and Bhattacherjee's ECM [17] and introduced MOOCs' gamified designs to the research model. Next, we introduced information quality, system quality, and service quality to define MOOC platform qualities. Then, we investigated the overall MOOC platform quality confirmation through the ECM framework. The integrated framework ensures that the MOOC platform quality confirmation links students' perceived characteristics of the MOOC platform, students' post usage intentions, and learning performance.

Secondly, most previous research on MOOCs has primarily focused on the users' intention to adopt or the users' intention to re-adopt. This study expanded the research model by syncing the MOOC platform's continued usage intention and course performance and empirically testing the influence of MOOCs' cognitive perceptions on course performance. The results show that MOOCs significantly impacted students' course performance through various factors. This study measured the characteristics of MOOCs and users' behaviors. Another difference between this study and previous studies is that it attempted to consider MOOCs as a learning tool and a combination of technology and a learning option during COVID-19. To the best of the authors' knowledge, this study may be novel research in measuring how MOOCs influence students' course performance under the context of a public health emergency.

4.2. Practical Implications

This study also makes several contributions to MOOC platform managers about how to improve their platforms and services. First, the results indicate that MOOC platform qualities had significant influences on the MOOC platform quality confirmation. Information quality had the most significant effect, followed by service quality and system quality. From this perspective, we concluded that the information quality of the MOOC platforms exceeded students' expectations. For this reason, MOOC platform managers should improve information quality continuously. MOOC platform managers should pay attention to online service or call center service construction as MOOCs' service quality is lower than students' expectations. One of the reasons for this phenomenon might be found in the increasing number of users during COVID-19, which meant that users could not get consulting services in a timely fashion. The results also indicate that system quality significantly influences the confirmation of a slight estimate. This emphasizes the importance of system quality in the management of the MOOC platform. MOOC platforms are frequently paralyzed during COVID-19 because tens of millions of students attend online classes at the same time. Improvements to the MOOC platform's quality will increase students' pre-expectations.

Second, the results show that students' perceived usefulness and confirmation of the MOOC platform significantly influenced students' satisfaction with the MOOC platform. If students believe MOOCs are simple to grasp, useful, well-managed, and beneficial to their studies, it will influence the level of students' satisfaction with MOOC platforms. MOOC platform managers should improve and upgrade the system and update course materials to provide a better user experience. By increasing the usefulness of MOOC platforms, students will be more likely to use MOOCs for continuous learning. Furthermore, useful course materials, well-constructed content, and an easily navigated system will significantly improve students' course performance.

Third, the results prove that MOOCs' continued usage intentions positively influenced students' course performance. Thus, MOOC platform managers should develop mechanisms to stimulate students' course completion rates. Lastly, the results indicate that gamification was composed of social interaction, entertainment, and challenges. Gamified mechanisms significantly influence students' continued usage intentions and their course performance. During COVID-19, students have to take all of their classes online and spend 6~8 h per day on online learning. Therefore, it is necessary to make MOOCs more exciting and attractive to students. As previous research has shown, one of the primary issues with MOOC learning is the low completion rate of students [81]. Gamification mechanisms such as redeemable points, team leaderboards, badges, peer grading, emoticon feedback, peer interaction, etc., will increase the course completion rate. In addition to the current gamified features, MOOC managers should develop other attractive and useful gamified designs to improve students' MOOCs' continued usage intentions and course performance.

5. Conclusions

Despite the theoretical and practical implications, this study has some limitations, as follows. First, this study's results and implications should not be generalized because the survey was conducted in China only. We consider that comparative research in another cultural context could contribute more effectively in the future. Second, we developed a questionnaire based on previous studies and obtained empirical data from online learners. Personal interviews or focus group interviews may be chosen for future studies to create new constructs and items. Third, we did not consider control variables in this study. Future research may examine the influences of control variables such as demographic variables and personality traits, which might influence the research results. Lastly, this study showed that gamification did not significantly influence students' satisfaction with MOOCs. However, previous studies have reported that gamification in IS is a critical attribute to users' satisfaction. Thus, we need to conduct further investigations into the relationship between gamification and satisfaction in future studies.

Author Contributions: Conceptualization, Q.Y. and Y.-C.L.; methodology, Q.Y. and Y.-C.L.; software, Q.Y.; validation, Q.Y. and Y.-C.L.; formal analysis, Q.Y. and Y.-C.L.; investigation, Q.Y.; data curation, Q.Y. and Y.-C.L.; writing—original draft preparation, Q.Y.; writing—review and editing, Y.-C.L.; visualization, Q.Y. and Y.-C.L.; supervision, Y.-C.L. Both authors have read and agreed to the published version of the manuscript.

Funding: This research received no external funding.

Data Availability Statement: The data presented in this study are available upon request from the corresponding author. Data are not publicly available due to a confidentiality agreement with participants.

Conflicts of Interest: The authors declare no conflict of interest.

Appendix A

Questionnaire items.

Factors	Items	Sources
Information quality (IQ)	The MOOC platform provides useful information. The information provided by MOOC platform is understandable. The information provided by MOOC platform is interesting. The information provided by MOOC platform is reliable.	[17]
System quality (SQ)	The MOOC platform is easy to navigate. The MOOC platform allows me to find easily the information I am looking for. The MOOC platform is well structured. The MOOC platform is easy to use.	
Service quality (SEQ)	The support staff is always highly willing to help whenever I need support with the MOOC platform. The support staff provides personal attention when I experience problems with the MOOC platform. The support staff provides services related to the MOOC platform at the promised time. The support staff has sufficient knowledge to answer my questions in respect of the MOOC platform.	
MOOC quality confirmation (MCON)	My experience with using MOOC platform is better than I expected. The service level provided by MOOC platform is better than I expected. Content on the MOOC platform is better than I expected.	[57]
MOOC usefulness (MUSE)	Using the MOOC platform can improve my study performance. Using the MOOC platform can increase my study effectiveness. I think the MOOC platform is useful to me	[56]
MOOC satisfaction (MSAT)	I am satisfied with the performance of the MOOC platform. I am pleased with the experience of using the MOOC platform. My decision to use the MOOC platform is a wise one.	
Social interaction (SOC)	I open up more to others via the MOOC than in other communication modes. I have a network of friends I made via studying through MOOC. Studying through MOOC enables me to connect with friends in my real life. Studying through MOOC enables me to keep in touch with friends in my real life.	[82]
Challenge (CHA)	The MOOC platform provides "hints" in text that helps me overcome the challenges. The MOOC platform provides "online support" that helps me overcome the challenges. The MOOC platform provides video or audio auxiliaries that help me overcome the challenges.	[17]
Entertainment (ENT)	For an online education website, MOOC features and applications are funny. For an online education website, MOOC features and applications are thrilling. For an online education website, MOOC features and applications are exciting. For an online education website, MOOC features and applications are delightful.	[83]
MOOC continued usage intention (MCI)	I will use the MOOC platform on a regular basis in the future. I will frequently use the MOOC platform in the future. I will strongly recommend my friends to use MOOC platform.	[56]
Course performance (CP)	I have gained a clear understanding about the classes through using MOOC platform. I can easily achieve the learning goals asserted by this course via MOOC platform. By using MOOC platform, it is easier to accomplish the assignments. I am capable in learning how to make good use of MOOC platform.	[84]

References

1. Dai, H.M.; Teo, T.; Rappa, N.A.; Huang, F. Explaining Chinese university students' continuance learning intention in the MOOC setting: A modified expectation confirmation model perspective. *Comput. Educ.* **2020**, *150*, 1–16. [CrossRef]
2. Zainuddin, Z.; Chu, S.K.W.; Shujahat, M.; Perera, C.J. The impact of gamification on learning and instruction: A systematic review of empirical evidence. *Educ. Res. Rev.* **2020**, *30*, 100326. [CrossRef]
3. Jarnac de Freitas, M.; Mira da Silva, M. Systematic literature review about gamification in MOOCs. *Open Learn. J. Open. Dis. eLearn.* **2020**, 1–23. [CrossRef]

4. Martínez-Núñez, M.; Fidalgo-Blanco, Á.; Borrás-Gené, O. New challenges for the motivation and learning in engineering education using gamification in MOOC. *Int. J. Eng. Educ.* **2016**, *32*, 501–512.
5. Mohan, M.M.; Upadhyaya, P.; Pillai, K.R. Intention and barriers to use MOOCs: An investigation among the post graduate students in India. *Educ. Inf. Technol.* **2020**, *25*, 5017–5031. [CrossRef]
6. Ebben, M.; Murphy, J.S. Unpacking MOOC scholarly discourse: A review of nascent MOOC scholarship. *Learn. Media Technol.* **2014**, *39*, 328–345. [CrossRef]
7. Hew, K.F.; Cheung, W.S. Students' and instructors' use of massive open online courses (MOOCs): Motivations and challenges. *Educ. Res. Rev.* **2014**, *12*, 45–58. [CrossRef]
8. Paton, R.M.; Fluck, A.E.; Scanlan, J.D. Engagement and retention in VET MOOCs and online courses: A systematic review of literature from 2013 to 2017. *Comput. Educ.* **2018**, *125*, 191–201. [CrossRef]
9. Lee, D.; Watson, S.L.; Watson, W.R. Systematic literature review on self-regulated learning in massive open online courses. *Australas. J. Educ. Technol.* **2019**, *35*, 28–41. [CrossRef]
10. Zhou, J. Exploring the factors affecting learners' continuance intention of MOOCs for online collaborative learning: An extended ECM perspective. *Australas. J. Educ. Technol.* **2017**, *33*, 123–135. [CrossRef]
11. Barata, G.; Gama, S.; Jorge, J.; Gonçalves, D. Studying student differentiation in gamified education: A long-term study. *Comput Hum. Behav.* **2017**, *71*, 550–585. [CrossRef]
12. Antonaci, A.; Klemke, R.; Kreijns, K.; Specht, M. Get Gamification of MOOC right! *Int. J. Serious Games* **2018**, *5*, 61–78. [CrossRef]
13. Antonaci, A.; Klemke, R.; Specht, M. The effects of gamification in online learning environments: A systematic literature review. *Informatics* **2019**, *6*, 32. [CrossRef]
14. Domínguez, A.; Saenz-De-Navarrete, J.; De-Marcos, L.; Fernández-Sanz, L.; Pagés, C.; Martínez-Herráiz, J.J. Gamifying learning experiences: Practical implications and outcomes. *Comput. Educ.* **2013**, *63*, 380–392. [CrossRef]
15. Dicheva, D.; Dichev, C.; Agre, G.; Angelova, G. Gamification in education: A systematic mapping study. *J. Educ. Technol. Soc.* **2015**, *18*, 75–88.
16. Malone, T.W. September. What makes things fun to learn? Heuristics for designing instructional computer games. In Proceedings of the 3rd ACM SIGSMALL symposium and the first SIGPC symposium on Small systems, New York, NY, USA, 18–19 September 1980; pp. 162–169.
17. Aparicio, M.; Oliveira, T.; Bacao, F.; Painho, M. Gamification: A key determinant of massive open online course (MOOC) success. *Inf. Manag.* **2019**, *56*, 39–54. [CrossRef]
18. Chang, J.W.; Wei, H.Y. Exploring engaging gamification mechanics in massive online open courses. *Educ. Technol. Soc.* **2016**, *19*, 177–203.
19. Ortega-Arranz, A.; Bote-Lorenzo, M.L.; Asensio-Pérez, J.I.; Martínez-Monés, A.; Gómez-Sánchez, E.; Dimitriadis, Y. To reward and beyond: Analyzing the effect of reward-based strategies in a MOOC. *Comput. Educ.* **2019**, *142*, 103639. [CrossRef]
20. Wu, B.; Chen, X. Continuance intention to use MOOCs: Integrating the technology acceptance model (TAM) and task technology fit (TTF) model. *Comput. Hum. Behav.* **2017**, *67*, 221–232. [CrossRef]
21. Fang, J.; Tang, L.; Yang, J.; Peng, M. Social interaction in MOOCs: The mediating effects of immersive experience and psychological needs satisfaction. *Telemat. Inform.* **2019**, *39*, 75–91. [CrossRef]
22. Sanchez-Gordon, S.; Luján-Mora, S. Research challenges in accessible MOOCs: A systematic literature review 2008–2016. *Univers. Access Inf. Soc.* **2018**, *17*, 775–789. [CrossRef]
23. Yoon, Y.; Guimaraes, T.; O'Neal, Q. Exploring the factors associated with expert systems success. *MIS Q.* **1995**, *19*, 83–106. [CrossRef]
24. DeLone, W.H.; McLean, E.R. Information systems success: The quest for the dependent variable. *Inf. Syst. Res.* **1992**, *3*, 60–95. [CrossRef]
25. Delone, W.H.; McLean, E.R. The DeLone and McLean model of information systems success: A ten-year update. *J. Manag. Inf. Syst.* **2003**, *19*, 9–30.
26. Jang, J.H.; Kim, J.K.; Hwang, Y.H. Influence of hotel information system quality on system use and user satisfaction. *J. Qual. Assur. Hosp. Tour.* **2006**, *7*, 41–58. [CrossRef]
27. Lin, H.F.; Lee, G.G. Determinants of success for online communities: An empirical study. *Behav. Inf. Technol.* **2006**, *25*, 479–488. [CrossRef]
28. Seddon, P.; Kiew, M.Y. A partial test and development of DeLone and McLean's model of IS success. *Australas. J. Inf. Syst.* **1996**, *4*. [CrossRef]
29. Sharma, S.K.; Sharma, M. Examining the role of trust and quality dimensions in the actual usage of mobile banking services: An empirical investigation. *Int. J. Inf. Manag.* **2019**, *44*, 65–75. [CrossRef]
30. Wang, W.T.; Lai, Y.J. Examining the adoption of KMS in organizations from an integrated perspective of technology, individual, and organization. *Comput. Hum. Behav.* **2014**, *38*, 55–67. [CrossRef]
31. Rogers, E.M. *Diffusion of Innovations*, 5th ed.; Free Press: New York, NY, USA, 2003.
32. Freeze, R.D.; Alshare, K.A.; Lane, P.L.; Wen, H.J. IS success model in e-learning context based on students' perceptions. *J. Inf. Syst. Educ.* **2010**, *21*, 173–184.
33. Aparicio, M.; Bacao, F.; Oliveira, T. Grit in the path to e-learning success. *Comput. Hum. Behav.* **2017**, *66*, 388–399. [CrossRef]

34. Cidral, W.A.; Oliveira, T.; Di Felice, M.; Aparicio, M. E-learning success determinants: Brazilian empirical study. *Comput. Educ.* **2018**, *122*, 273–290. [CrossRef]
35. Albelbisi, N.A. The role of quality factors in supporting self-regulated learning (SRL) skills in MOOC environment. *Educ. Inf. Technol.* **2019**, *24*, 1681–1698. [CrossRef]
36. Albelbisi, N.A. Development and validation of the MOOC success scale (MOOC-SS). *Educ. Inf. Technol.* **2020**, *25*, 4535–4555. [CrossRef]
37. Isaac, O.; Abdullah, Z.; Ramayah, T.; Mutahar, A.M. Factors determining user satisfaction of internet usage among public sector employees in Yemen. *Int. J. Technol. Learn. Innov. Dev.* **2018**, *10*, 37–68. [CrossRef]
38. Urbach, N.; Smolnik, S.; Riempp, G. An empirical investigation of employee portal success. *J. Strateg. Inf. Syst.* **2010**, *19*, 184–206. [CrossRef]
39. Wang, Y.S.; Wang, H.Y.; Shee, D.Y. Measuring e-learning systems success in an organizational context: Scale development and validation. *Comput. Hum. Behav.* **2007**, *23*, 1792–1808. [CrossRef]
40. Wu, L.; Chiu, M.L.; Chen, K.W. Defining the determinants of online impulse buying through a shopping process of integrating perceived risk, expectation-confirmation model, and flow theory issues. *Int. J. Inf. Manag.* **2020**, *52*, 102099. [CrossRef]
41. Zheng, Y.; Zhao, K.; Stylianou, A. The impacts of information quality and system quality on users' continuance intention in information-exchange virtual communities: An empirical investigation. *Decis. Support Syst.* **2013**, *56*, 513–524. [CrossRef]
42. Wu, J.H.; Wang, Y.M. Measuring KMS success: A respecification of the DeLone and McLean's model. *Inf. Manag.* **2006**, *43*, 728–739. [CrossRef]
43. Hsu, C.L.; Lin, J.C.C. What drives purchase intention for paid mobile apps?–An expectation confirmation model with perceived value. *Electron. Commer. Res. Appl.* **2015**, *14*, 46–57. [CrossRef]
44. Leung, L.; Chen, C. E-health/m-health adoption and lifestyle improvements: Exploring the roles of technology readiness, the expectation-confirmation model, and health-related information activities. *Telecomm. Policy* **2019**, *43*, 563–575. [CrossRef]
45. Gupta, A.; Yousaf, A.; Mishra, A. How pre-adoption expectancies shape post-adoption continuance intentions: An extended expectation-confirmation model. *Int. J. Inf. Manag.* **2020**, *52*, 102094. [CrossRef]
46. Bhattacherjee, A. Understanding information systems continuance: An expectation-confirmation model. *MIS Q.* **2001**, *25*, 351–370. [CrossRef]
47. Herodotou, C.; Mystakidis, S. Addressing the retention gap in MOOCs: Towards a motivational framework for MOOCs instructional design. In Proceedings of the 6th Biennial EARLI Conference for Research on Learning and Instruction, Limassol, Cyprus, 25–29 August 2015.
48. Hone, K.S.; El Said, G.R. Exploring the factors affecting MOOC retention: A survey study. *Comput. Educ.* **2016**, *98*, 157–168. [CrossRef]
49. An, Y.; Zhu, M.; Bonk, C.J.; Lin, L. Exploring instructors' perspectives, practices, and perceived support needs and barriers related to the gamification of MOOCs. *J. Comput. High Educ.* **2020**, *33*, 1–21. [CrossRef]
50. Khalil, M.; Ebner, M. Clustering patterns of engagement in Massive Open Online Courses (MOOCs): The use of learning analytics to reveal student categories. *J. Comput. High Educ.* **2017**, *29*, 114–132. [CrossRef]
51. Saputro, R.E.; Salam, S.; Zakaria, M.H.; Anwar, T. A gamification framework to enhance students' intrinsic motivation on MOOC. *Telkomnika* **2019**, *17*, 170–178. [CrossRef]
52. Hassanzadeh, A.; Kanaani, F.; Elahi, S. A model for measuring e-learning systems success in universities. *Expert Syst. Appl.* **2012**, *39*, 10959–10966. [CrossRef]
53. Marjanovic, U.; Delić, M.; Lalic, B. Developing a model to assess the success of e-learning systems: Evidence from a manufacturing company in transitional economy. *Inf. Syst. eBus. Manag.* **2016**, *14*, 253–272. [CrossRef]
54. Lin, H.F. Measuring online learning systems success: Applying the updated DeLone and McLean model. *Cyberpsychol. Behav.* **2007**, *10*, 817–820. [CrossRef]
55. Bhattacherjee, A.; Lin, C.P. A unified model of IT continuance: Three complementary perspectives and crossover effects. *Eur. J. Inf. Syst.* **2015**, *24*, 364–373. [CrossRef]
56. Roca, J.C.; Chiu, C.M.; Martínez, F.J. Understanding e-learning continuance intention: An extension of the Technology Acceptance Model. *Int. J. Hum. Comput. Stud.* **2006**, *64*, 683–696. [CrossRef]
57. Chung, N.; Lee, H.; Lee, S.J.; Koo, C. The influence of tourism website on tourists' behavior to determine destination selection: A case study of creative economy in Korea. *Technol. Forecast. Soc. Chang.* **2015**, *96*, 130–143. [CrossRef]
58. Shin, D.H.; Shin, Y.J.; Choo, H.; Beom, K. Smartphones as smart pedagogical tools: Implications for smartphones as u-learning devices. *Comput. Hum. Behav.* **2011**, *27*, 2207–2214. [CrossRef]
59. Jumaan, I.A.M.; Hashim, N.H.; Al-Ghazali, B.M. The role of cognitive absorption in predicting mobile internet users' continuance intention: An extension of the expectation-confirmation model. *Technol. Soc.* **2020**, *63*, 101355. [CrossRef]
60. Wang, C.; Teo, T.S.; Liu, L. Perceived value and continuance intention in mobile government service in China. *Telemat. Inform.* **2020**, *48*, 101348. [CrossRef]
61. Pozón-López, I.; Higueras-Castillo, E.; Muñoz-Leiva, F.; Liébana-Cabanillas, F.J. Perceived user satisfaction and intention to use massive open online courses (MOOCs). *J. Comput. High Educ.* **2020**, *33*, 1–36. [CrossRef]
62. Nasirzadeh, E.; Fathian, M. Investigating the Effect of Gamification Elements on Bank Customers to Personalize Gamified Systems. *Int. J. Hum. Comput. Stud.* **2020**, *144*, 102469. [CrossRef]

63. Putz, L.M.; Hofbauer, F.; Treiblmaier, H. Can gamification help to improve education? Findings from a longitudinal study. *Comput. Hum. Behav.* **2020**, *110*, 106392. [CrossRef]
64. Ajzen, I. The theory of planned behavior. *Organ. Behav. Hum. Decis. Process.* **1991**, *50*, 179–211. [CrossRef]
65. Joo, Y.J.; Park, S.; Shin, E.K. Students' expectation, satisfaction, and continuance intention to use digital textbooks. *Comput. Hum. Behav.* **2017**, *69*, 83–90. [CrossRef]
66. Li, C.Y.; Fang, Y.H. Predicting continuance intention toward mobile branded apps through satisfaction and attachment. *Telemat. Inform.* **2019**, *43*, 101248. [CrossRef]
67. Bölen, M.C.; Özen, Ü. Understanding the factors affecting consumers' continuance intention in mobile shopping: The case of private shopping clubs. *Int. J. Mob. Commun.* **2020**, *18*, 101–129. [CrossRef]
68. Hassan, L.; Dias, A.; Hamari, J. How motivational feedback increases user's benefits and continued use: A study on gamification, quantified-self and social networking. *Int. J. Inf. Manag.* **2019**, *46*, 151–162. [CrossRef]
69. Jun, F.; Jiao, J.; Lin, P. Influence of virtual CSR gamification design elements on customers' continuance intention of participating in social value co-creation: The mediation effect of psychological benefit. *Asia Pac. J. Mark. Logist.* **2020**, *32*, 1305–1326. [CrossRef]
70. Tam, C.; Oliveira, T. Understanding the impact of m-banking on individual performance: DeLone & McLean and TTF perspective. *Comput. Hum. Behav.* **2016**, *61*, 233–244.
71. Mohammadyari, S.; Singh, H. Understanding the effect of e-learning on individual performance: The role of digital literacy. *Comput. Educ.* **2015**, *82*, 11–25. [CrossRef]
72. Gerdenitsch, C.; Sellitsch, D.; Besser, M.; Burger, S.; Stegmann, C.; Tscheligi, M.; Kriglstein, S. Work gamification: Effects on enjoyment, productivity and the role of leadership. *Electron. Commer. Res. Appl.* **2020**, *43*, 100994. [CrossRef]
73. Feng, W.; Tu, R.; Hsieh, P. Can gamification increase consumers' engagement in fitness apps? The moderating role of commensurability of the game elements. *J. Retail. Consum. Serv.* **2020**, *57*, 1–9. [CrossRef]
74. Groening, C.; Binnewies, C. "Achievement unlocked!"—The impact of digital achievements as a gamification element on motivation and performance. *Comput. Hum. Behav.* **2019**, *97*, 151–166. [CrossRef]
75. Legaki, N.Z.; Xi, N.; Hamari, J.; Karpouzis, K.; Assimakopoulos, V. The effect of challenge-based gamification on learning: An experiment in the context of statistics education. *Int. J. Hum. Comput. Stud.* **2020**, *144*, 102496. [CrossRef]
76. Tobon, S.; Ruiz-Alba, J.L.; García-Madariaga, J. Gamification and online consumer decisions: Is the game over? *Decis. Support Syst.* **2020**, *128*, 113167. [CrossRef]
77. Rodrigues, L.F.; Oliveira, A.; Costa, C.J. Does ease-of-use contributes to the perception of enjoyment? A case of gamification in e-banking. *Comput. Hum. Behav.* **2016**, *61*, 114–126. [CrossRef]
78. Zhang, Q.; Peck, K.L.; Hristova, A.; Jablokow, K.W.; Hoffman, V.; Park, E.; Bayeck, R.Y. Exploring the communication preferences of MOOC learners and the value of preference-based groups: Is grouping enough? *Educ. Technol. Res. Dev.* **2016**, *64*, 809–837. [CrossRef]
79. Podsakoff, P.M.; Organ, D.W. Self-Reports in Organizational Research: Problems and Prospects. *J. Manag.* **1986**, *12*, 531–544. [CrossRef]
80. Podsakoff, P.M.; MacKenzie, S.B.; Lee, J.-Y.; Podsakoff, N.P. Common method biases in behavioral research: A critical review of the literature and recommended remedies. *J. Appl. Psychol.* **2003**, *88*, 879–903. [CrossRef] [PubMed]
81. Pursel, B.K.; Zhang, L.; Jablokow, K.W.; Choi, G.W.; Velegol, D. Understanding MOOC students: Motivations and behaviours indicative of MOOC completion. *J. Comput. Assist. Learn.* **2016**, *32*, 202–217. [CrossRef]
82. Li, H.; Liu, Y.; Xu, X.; Heikkilä, J.; van der Heijden, H. Modeling hedonic is continuance through the uses and gratifications theory: An empirical study in online games. *Comput. Hum. Behav.* **2015**, *48*, 261–272. [CrossRef]
83. Rauniar, R.; Rawski, G.; Yang, J.; Johnson, B. Technology acceptance model (TAM) and social media usage: An empirical study on Facebook. *J. Enterp. Inf. Manag.* **2014**, *27*, 6–30. [CrossRef]
84. Lin, W.S. Perceived fit and satisfaction on web learning performance: IS continuance intention and task-technology fit perspectives. *Int. J. Hum. Comput. Stud.* **2012**, *70*, 498–507. [CrossRef]

sustainability

MDPI

Article

Cyber-Physical System of Psychophysiological Support of Professional Self-Realization in Professions of the 'Man-Nature' Type in the Formation of Specialists for Sustainable Development

Mariia Ruda [1,*], Oleksandr Moroz [2], Olha Kuz [1] and Taras Boyko [3]

[1] Department of Ecological Safety and Nature Protection Activity, Viacheslav Chornovil Institute of Sustainable Development, Lviv Polytechnic National University, 79000 Lviv, Ukraine; olha.n.kuz@lpnu.ua
[2] Department of Tourism, Viacheslav Chornovil Institute of Sustainable Development, Lviv Polytechnic National University, 79000 Lviv, Ukraine; olyakuzon@gmail.com
[3] Department of Intelligent Mechatronics and Robotics, Institute of Computer Technologies, Automation and Metrology, Lviv Polytechnic National University, 79000 Lviv, Ukraine; tgbo@ukr.net
* Correspondence: mariia.v.ruda@lpnu.ua; Tel.: +380-98-79-81-281

Abstract: The article deals with the issue of developing a cyber-physical system of a psychological profile of a specialist in professions of the 'man-nature' type in the formation of specialists for sustainable development. The proposed cyber-physical system of psychophysiological support of professional self-realization in professions of the 'man-nature' type is presented in the form of a hierarchical structure, the upper level of which reflects the vision of the customer (client) and contains the necessary properties and their significance coefficients, and the lower level is a set of characteristics that affect the corresponding properties as well as the corresponding weight values. The cyber-physical system makes it possible to clearly represent the psychophysiological profile of a specialist when working on building a standard of a specialist using the method of expert assessments. This system also allows you to easily automate the process of creating a reference model of a specialist for professions of the 'man-nature' type and will provide estimates of the significance of each of the obtained characteristics. The novelty of the cyber-physical system lies in the integration of points of view on the professional self-realization of a specialist on the part of the customer and on the part of psychologists. The significance of the upper-level elements comes from experts (customers) for each profile, while the weights of the relationship between characteristics and a certain quality are constant regardless of the specialty and can be obtained from expert psychologists in advance. Based on the analysis of a formalized description of available psychodiagnostic techniques, a generalized model of psychodiagnostic techniques and a structural model of a set of psychodiagnostic techniques have been developed. This approach differs from the known ones in a formalized representation, as well as taking into account the vector of resources necessary for the implementation of the technique. The proposed approach made it possible for the first time to formulate the problem of choosing a set of psychodiagnostic techniques as an optimization problem. The developed complex of algorithms, containing all the characteristics of cybernetic systems, is controllable and consists of a set of subsystems that act as mechanisms for the restoration and implementation of self-preserving reactions. That is, this complex of algorithms is effective, which is characterized by a certain set of parameters. Hence, the key role of the system of transmission and processing of information in training specialists for sustainable development becomes clear, and the importance of intellectualization of management processes is explained.

Keywords: psychophysiological standard; professional-defining qualities; specialist professiogram; environmental engineer; employee psychophysiological profile; psychophysiological status; education for sustainable development

Citation: Ruda, M.; Moroz, O.; Kuz, O.; Boyko, T. Cyber-Physical System of Psychophysiological Support of Professional Self-Realization in Professions of the 'Man-Nature' Type in the Formation of Specialists for Sustainable Development. *Sustainability* **2021**, *13*, 7858. https://doi.org/10.3390/su13147858

Academic Editors: Miltiadis D. Lytras, Abdulrahman Housawi and Basim Alsaywid

Received: 7 May 2021
Accepted: 16 June 2021
Published: 14 July 2021

1. Introduction

Socio-economic transformations that take place in society are constantly changing the requirements for the personality of a modern specialist. The ability to be the subject of his professional development and independently find solutions to socially and professionally significant problems in a changing reality is becoming increasingly important for him. In addition to professional knowledge and skills, specialists today must also have special abilities, skills, and personality qualities that provide flexibility and dynamism of professional behavior, creativity in professional activities, independence in the search for and processing new information, and gaining professional experience. Of particular importance is the acquisition of the ability to make adequate decisions in 'non-standard» situations, in conditions of time pressure, as well as the development of optimal interaction skills with participants in the production process in joint professional activities. That is why specialists in professions such as the 'man-nature' type need the special attention of scientists. This is due, in addition to the enormous social significance of such professions, which is inherent in them, also to the fact that they are naturally closely integrated into professions of all other types. For example, a filmmaker, an actor (creative professions), a foreman, a workshop manager ('man-engineering'), an environmental worker ('man-nature'), a chief accountant, a director of a publishing house ('man-sign'), etc., in addition to their main type of professions, they also refer to professions of the 'man-man' type.

The work features and requirements for specialists in these professions have undergone significant changes over the past 5 years due to the introduction of the concept of sustainable development in Ukraine. A thorough morphological analysis of the concept of 'sustainable development' was carried out in the works of V. Bogolyubov ([1], pp. 28–37), L. Melnyk [2], Yu. Skyba ([3], pp. 95–104) and other researchers [4,5]. At the same time, as can be concluded from the studies of foreign experts D. Wortman [6], D. Tilbury [7], R. McKeown [8], R. Anderson [9], and Ukrainian researchers A.I. Pometun [10], O.Ye. Vysotska [11], V.R. Ilchenko [12], V.Ya. Shvets [13], I.O. Soloshyn [14], S.D. Rudyshyn [15], there is a significant contradiction in the training of specialists of different professions related to the 'man-nature' type: on the one hand, socio-economic changes occurring in society cause increased requirements for such specialists, their activity and responsibility in relation to their own professional and personal development; on the other hand, the training of these specialists is mostly formal, to a large extent. The analysis of literary sources [16–20] shows that most modern definitions of training specialists for sustainable development and education for sustainable development are contextual, formulated in the form of too-broad judgments, and they often lack the generic affiliation of the concept. Psychological-empirical study of the problem by a number of authors [21–25] made it possible to consider professional self-realization as an ability for self-change and as 'self-movement' of a specialist in the space of a profession; as a psychological readiness for constant work on oneself, an active search for opportunities for development; as a human ability, according to L.O. Kalmykova [26], 'to rethink one's own essence', to create oneself and, finally, to achieve the desired effect; as the ability of a person to transform himself-through self-education, self-development, the interiorization of values and norms, etc.–into a full-fledged member of the community of professionals to obtain the outstanding achievements in the profession. In addition, the movement of the individual in the professional space and his achievement of the peak of professional self-realization is determined, according to the study by E.F. Zeyer [27,28] and E.E. Symanyuk [29], by three meaning-making factors: age-related psychological changes, continuing professional education and the development and transformation of leading activities. These three factors concentrate in themselves both socio-economic conditions and the individual's own mental (socio-professional) activity, which, according to K. Whitaker [30], is the 'tool' of the profession. However, an analysis of the psychological, pedagogical, and specialized literature indicates that the problem of the formation of the psychophysiological standard of the profession of the 'man-nature' type is not sufficiently reflected in educational research. In this regard, the need for integration of numerous but quite heterogeneous, mixed studies related to various aspects of the formation of modern professionals of 'man-nature' type becomes apparent. There is a need to combine purely psychological and psychophysiological aspects of this problem. It is also necessary to develop a holistic system of psychophysiological support for all stages of becoming professionals of this type in training specialists for sustainable development. According to the problem posed, the aim of the study was to develop a cyber-physical system of psychophysiological support for becoming a specialist in professions of the 'man-nature' type in the formation of specialists for sustainable development.

2. Materials and Methods

The results presented in the article are based on experimental studies conducted at the V. Chornovil Institute for Sustainable Development (V.Chornovil ISTD) Lviv Polytechnic National University and the laboratory of psychophysiology and occupational health of the State Institution Sanitary and Epidemiological Station on the Lviv Railway of the Ministry of Health of Ukraine. A total of 55 people took part in these studies (40 environmental engineers, former graduates from the V. Chornovil ISTD, and 15 students of the same higher education institution). A total of 546 psychophysiological studies and 789 empirical studies were conducted using the methods described below.

A total of 40 environmental engineers, former graduates from the V. Chernovil ISTD, and 15 students of the same higher education institution were involved in psychophysiological research according to three groups of methods: methods for assessing professionally important psychophysiological qualities-they determine the readiness of the body for emergency action in conditions of monotonously acting factors, reliability of the body in a state of fatigue, attention switching speed, emotional stability, visual and auditory memory. These methods make it possible to determine sensorimotor reactions; methods for assessing the functional state (by means of which you can study the characteristics and psycho-emotional state of the body; methods for studying personality and psycho-emotional state) they assess the functional state of the organism: hemodynamic parameters, condition of the cardiovascular system, fatigue of the visual analyzer, state of long-term and short-term memory, reaction to a moving object, state of the nervous system. The study results are presented below.

Also, the participants of the experiment (55 people) were also involved in empirical research using the psychophysiological complex *Tolerance*, which allows evaluating 10 psychophysiological qualities (this computer system allows you to significantly speed up the testing procedure and preparation of conclusions on professional suitability; it enables to carry out a systematic multi-level examination of operators, to standardize the examination procedure, to objectify test results, to minimize errors and bias in personnel). In addition, the empirical studies were conducted with the help of the *Filter* technical complex with a set of techniques for assessing professionally important psychophysiological qualities and a functional state; by means of personal computers with a set of necessary techniques for studying personality characteristics and the psychoemotional state of the experimental subjects (MMPI, Raven, Shmishek-Leonhard, Spilberger-Khanin, Lusher). The study results are presented below.

The second stage of the work was the selection of groups of environmental engineers, master's students with relatively high (group A) and relatively low (group B) levels of professional success. Representatives of these groups are typical for the majority of workers/students in this profession by age, professional experience, educational background. Testing was carried out by 21 indicators. To determine the cognitive potential CP (psychophysiological professional suitability), we used methods for assessing 12 professionally important psychophysiological qualities: readiness for emergency actions (REA), reaction to a moving object (RMO), sensorimotor reactions, strength of the nervous system (SNS) mobility of the nervous processes (MNP) attention stability, switching attention, emotional stability, logical thinking, visual memory, long-term memory, and operative memory (short-term memory). To determine the biological age (BA), methods of assessing the functional state were used: the Genchi coefficient, systolic pressure, diastolic pressure, pulse, tilt, the Romberg stance, height, weight.

Professional success (*Ekho*) was determined by a 10-point system according to other non-psychophysiological criteria, such as learning success (training rating), the absence of accidents, shortcomings in work, violations of labor discipline, the number of vouchers issued, and tourists sent on vacation, etc., the presence of awards or prizes, innovation proposals, initiatives, competence, responsibility, exemplary attitude to their duties, etc.

The practical significance of the study results lies in the wide possibility of their practical use in order to increase the efficiency of professional development of specialists of the 'man-nature' type professions in training specialists for sustainable development in the Baltic Sea region.

The scientific novelty of the work lies in the fact that for the first time, theoretical and methodological principles have been developed for constructing a standard of a profession of the 'man-nature' type.

3. Results

The cyber-physical system consists of indicators and four interrelated stages: (1) standardization and weighing; (2) identification of problems and inconsistencies; (3) analysis of effects; and (4) feedback and improvement.

Stage I Standardization and weighing involve comparing the most significant indicators of the four-component structure of the personality model to assess the professional suitability of a specialist.

An integrated approach to assessing the professional suitability of a specialist involves, on the one hand, the study of the profession for which the selection is made, and on the other hand, a psychological study of the personality of the applicants. Therefore, it is advisable to use the method of two profiles as the basis for algorithms and procedures for constructing a standard of a professional worker [31].

The integral profile of a professional worker may consist of various components, according to which professional selection is carried out (intellectual, psychological, psychophysiological, technological, social).

For each component, its own professional defining qualities (PDQs) are determined, their assessment is carried out, and applicants are diagnosed. The professiogram of a specialist of the 'man-nature' type profession can be described using a set of tuples:

$$\begin{gathered}\left\langle d^1_{q_1}, d^1_{q_2}, \ldots, d^1_{q_l}\right\rangle, \left\langle d^2_{q_1}, d^2_{q_2}, \ldots, d^2_{q_u}\right\rangle,\\ \left\langle d^3_{q_1}, d^3_{q_2}, \ldots, d^3_{q_a}\right\rangle, \left\langle b_{q_1}, b_{q_2}, \ldots, b_{q_k}\right\rangle,\\ \left\langle t_{q_1}, t_{q_2}, \ldots, t_{q_n}\right\rangle, \left\langle s_{q_1}, s_{q_2}, \ldots, s_{q_m}\right\rangle,\\ \left\langle v_{q_1}, v_{q_2}, \ldots, v_{q_r}\right\rangle, \left\langle z_{q_1}, z_{q_2}, \ldots, z_{q_s}\right\rangle,\end{gathered} \tag{1}$$

or by one aggregate tuple:

$$\left\langle \begin{array}{cc} d^1_{q_1}, d^1_{q_2}, \ldots, d^1_{q_l}, d^2_{q_1}, d^2_{q_2}, \ldots, d^2_{q_u}, d^3_{q_1}, d^3_{q_2}, \ldots, & d^3_{q_a}, \\ b_{q_1}, b_{q_2}, \ldots, b_{q_k}, t_{q_1}, t_{q_2}, \ldots, t_{q_n}, s_{q_1}, s_{q_2}, \ldots, & s_{q_m}, \\ v_{q_1}, v_{q_2}, \ldots, v_{q_r}, z_{q_1}, z_{q_2}, \ldots, z_{q_s} & \end{array} \right\rangle, \tag{2}$$

where $D_1 = \left\{d^1_{q_1}, d^1_{q_2}, \ldots, d^1_{q_l}\right\}$ is a set of requirements imposed on a specialist in a profession of the 'man-nature' type in the field of knowledge; $D_2 = \left\{d^2_{q_1}, d^2_{q_2}, \ldots, d^2_{q_u}\right\}$ is a set of requirements imposed on a specialist in the field of specialized knowledge; $D_3 = \left\{d^3_{q_1}, d^3_{q_2}, \ldots, d^3_{q_a}\right\}$ is a set of requirements for intellectual qualities of a specialist; $B = \left\{b_{q_1}, b_{q_2}, \ldots, b_{q_k}\right\}$ is a set of requirements to a specialist to effectively perform the specified functions (psychophysiological requirements); $T = \left\{t_{q_1}, t_{q_2}, \ldots, t_{q_n}\right\}$ is a set of requirements to a specialist in the field of technical training; $S = \left\{s_{q_1}, s_{q_2}, \ldots, s_{q_m}\right\}$ is a set of requirements to a specialist in the field of social relations; $V = \left\{v_{q_1}, v_{q_2}, \ldots, v_{q_r}\right\}$ is a set of psychological qualities that a specialist must possess in order to effectively perform the specified functions; $Z = \{z_{q_1}, z_{q_2}, \ldots, z_{q_s}\}$ is a set of undesirable factors that impede one or another type of professional activities.

Stage II Identification of problems and inconsistencies.

Under real-life conditions, most often, only some of the listed components are subject to consideration. Most commonly, during professional selection, only medical and psychological selections are carried out in stages. Psychophysiological selection and control of professional knowledge are either not carried out at all or are carried out formally.

This division covers the largest set of characteristics that need to be diagnosed. It is the determination of this set of parameters that causes the most significant difficulties. While medical control and the assessment of professional knowledge have legal status, that is, on the basis of the data obtained, an official decision is possible, conducting psychological and psychophysiological selection de jure is not fixed. However, making the correct decision is possible only on the basis of the results of all four components of this process [32].

Currently, the four-component structure of the personality model is widely known [33]:

- Social and professional orientation (inclinations, interests, relations, expectations, mindset, motives);
- professional competence (professional knowledge, abilities, skills, qualifications);
- professionally important qualities (psychological qualities of a personality that determine the productivity of activities-scope, quality, effectiveness, etc.);
- professionally significant physiological qualities (reactivity, energetics, neurotism, extroversion, adaptive tracking, etc.).

The system-forming factor of the personality is their socio-professional orientation, which is characterized by a system of dominant needs and motives. The components of a professional direction are motives, value system, professional approach, and socio-professional status [34].

Professional competence is often understood as a set of professional knowledge and abilities, as well as ways to perform professional activities. The following main components are distinguished: social and legal competence, specific competence, personal competence, autocompetence, extreme professional competence [35].

In view of this, the specialist's model can be represented in the form of a hierarchical structure (Figure 1), which makes it possible to decompose the qualities of a professional worker using individual characteristics [32].

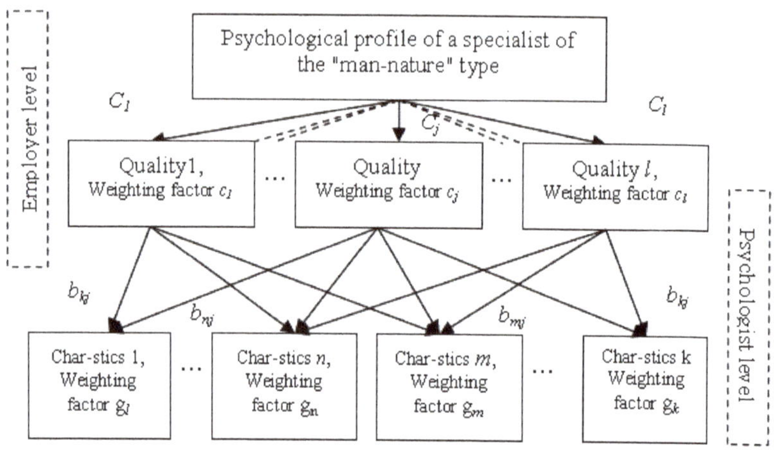

Figure 1. A multi-level model of the psychological profile of a specialist.

The intermediate stage—theoretically contains the basic theoretical principles of the cyber-physical system:

- Meaningfully, psychophysiological support of the professional development of a specialist is a system of successive measures aimed at the formation and development of a person's professional orientation and competence, professionally important qualities, and psychophysiological properties in order to achieve a high individually possible professional level and to realize his personal potential in a certain professional activity;
- The course of professional development of a specialist is determined by two groups of factors: external (socio-economic conditions, the content of educational, professional and professional activities, stimulation of career growth, vital events, etc.); internal (biopsychic characteristics, professional activity and motivation, the need for professional self-realization, professional crisis phenomena, etc.);
- The main difficulties and negative phenomena that can hinder the process of professional development are the phenomena of crises of professional development, professional deformations, and 'burnout';
- The subject matter of work in professions of the 'man-nature' type consists in the ability to infer cause-and-effect relationships, analytical thinking, the ability to predict and evaluate the changing factors of the abiotic environment and take into account anthropogenic factors, analysis of the ecosystem as a system (systematic thinking), rich imagination, thinking on feet (operative thinking), observation, endurance, persistence, responsibility, stability, distribution and voluntary switching of attention, visual-figurative, and visual memory;
- Psychophysiological support of the professional development of a specialist should be based on the principles of scientific character, consistency and continuity, optimization, consistency, individualization, adequacy, focus on the maximum possible professional achievements; it can be carried out in four main directions: according to the stages of professional development, according to its components, according to a certain profession and according to the area of activities.

Stage III Analysis of effects.

A psychophysiological profile of a specialist for a profession of the 'man-nature' type can be described as follows: $\langle b_1, b_2, \ldots, b_l \rangle$ and $\langle c_1, c_2, \ldots, c_l \rangle$—for the employer's level; $\langle x_1, x_2, \ldots, x_k \rangle$ and $\langle g_1, g_2, \ldots, g_k \rangle$—for the psychologist's level.

In general, the specialist model will take the form:

$$M = \langle B, C, X, G \rangle. \quad (3)$$

where $B = \{b_1, b_2, \ldots, b_l\}$ is a set of qualities that describe the psychophysiological profile of the specialist; $C = \{c_1, c_2, \ldots, c_l\}$ is a set of weighting factors of qualities; $X = \{x_1, x_2, \ldots, x_k\}$ are characteristics that describe the whole set B; $G = \{g_1, g_2, \ldots, g_k\}$ are corresponding characteristic weighting factors.

The given model makes it possible to demonstrate the psychophysiological profile of a professional worker when constructing a standard psychophysiological profile of a specialist of the 'man-nature' type using the method of expert assessments. It can be used to automate the process of creating a standard of specialty.

One of the main components of professionalization is the determination of the professional suitability of a person for future activities on the basis of his individual psychophysiological qualities. A person's professional activity requires a combination of qualities specific to each individual specialty (intellectual, psychological, psychophysiological characteristics). Their identification, as stated earlier, is one of the greatest difficulties. If the standard is not built correctly, then all further efforts are in vain. Therefore, almost all known methods for solving problems of professional selection are associated with providing this preparatory stage.

In our study, all models, algorithms, and procedures are considered in relation to the psychological and psychophysiological components.

The input data for the appropriate creation of the nomenclature of qualities and the construction of a generalized psychophysiological profile of a specialist is: $\{F\} = \{f_1, f_2, \ldots, f_l\}$ is a set of functions (specific professional tasks for a specialist of the 'man-nature' type), the execution of which is necessary for his successful work. For each function (task), a set of methods is formed for its performance, $\{R_1\} = \{r_{11}, r_{12}, \ldots, r_{1p}\}$, $\{R_2\} = \{r_{21}, r_{22}, \ldots, r_{2p}\}$, ...,$\{R_l\} = \left\{r_{l1}, r_{l2}, \ldots, r_{lp}\right\}$.

Using these data as the base, a set of requirements is formed (preliminary), which must be met by a professional worker for the most efficient performance of duties in this specialty. Based on these sets of qualities, a special questionnaire is formed that covers all groups of qualities necessary to build a standard of the 'man-nature' type specialist of a particular specialty.

After this, a survey of experts is conducted $\{E\} = \{E_1, E_2, \ldots, E_k\}$, who are the most experienced specialists in a specific professional field of activities. Based on these data, the psychophysiological logical standard of a specialist is built (Figure 2).

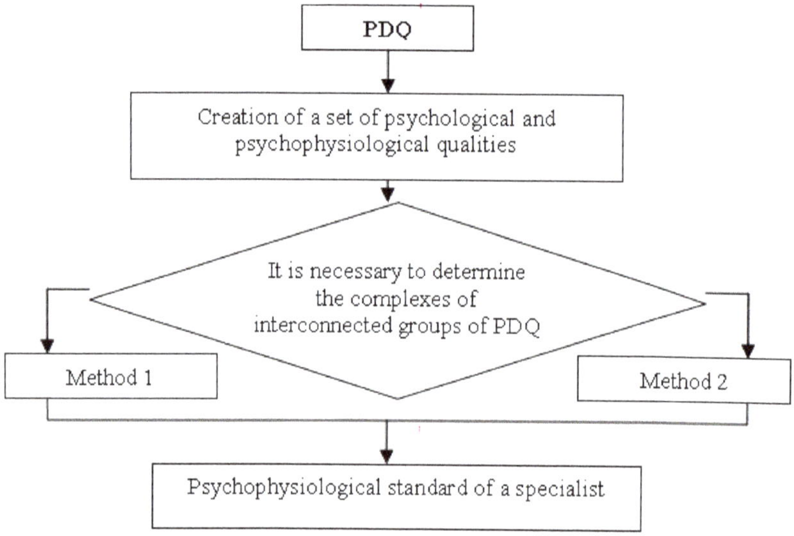

Figure 2. The structure of the method for constructing a generalized psychophysiological profile of a specialist of the 'man-nature' type.

The process of constructing a psychophysiological standard of a professional worker of the 'man-nature' type profession can be implemented using one of two methods [32], the choice of which depends on the required degree of research depth of complexes of interconnected groups of PDQ.

Due to the fact that the activities of a specialist in the 'man-nature' type profession are associated with increased risk and responsibility, it is, therefore, necessary to use a method that involves determining the complexes of interconnected PDQ groups.

Developing a model of the psychophysiological standard of the 'man-nature' type profession is based on the use of the method of expert assessments and factor analysis [36]. The purpose of factor analysis is to determine the relationships between variables (the 'objective R-classification') and the reduction in the number of variables needed to describe the data. Using factor analysis, it is possible to identify hidden variables that are responsible for the presence of linear statistical correlations between the observed variables.

The method for constructing a profile model of the 'man-nature' type profession is described as follows [37]:

1. Forming a list of psychological and psychophysiological qualities. Each j-th expert records a number of properties (characteristics) $\Omega_i = (\omega_{i1}, \omega_{i2}, \ldots, \omega_{il_i})$, which, in his opinion, should cover all the necessary characteristics of a specialist of the 'man-nature' type profession. The sets $\Omega_0 = \bigcap_{i=1}^{k} \Omega_i$, $\Omega_0^1 = \bigcup_{i=1}^{k} \Omega_i$, $\Omega_0^{11} = \Omega_0^1 \backslash \Omega_0$ are formed and analyzed.

It should be borne in mind that Ω_0 is the set of necessary qualities (characteristics) and Ω_0^{11} is the set of the desired qualities.

2. Each of the experts *Ej, (j = 1,2,... k)* puts for each element of the set their coefficients of significance in the range from −3 to +3, forming the vector $\vec{a}_j = (a_{j1}, a_{j2}, \ldots a_{jN})$ from those considerations that will reveal along with the positive qualities that increase work efficiency, also negative ones that reduce the efficiency of the specialist.

3. Having been received from the experts, the estimates are normalized by the Formula (4):

$$a'_{ji} = \frac{a_{ji}}{\sum\limits_{i=1}^{N} a_{ji}}, \tag{4}$$

where $j = 1, 2, \ldots k$, $i = 1, 2, \ldots N$ in order for the relationship $\sum\limits_{i=1}^{N} a'_{ji} = 1, j = 1, 2, \ldots, k$ to be fulfilled, as a result of which we obtain a matrix:

$$E = \begin{pmatrix} a'_{11}, a'_{12}, \ldots, a'_{1N} \\ a'_{21}, a'_{22}, \ldots, a'_{2N} \\ \cdots\cdots\cdots\cdots\cdots\cdots \\ a'_{j1}, a'_{j2}, \ldots, a'_{jN} \\ \cdots\cdots\cdots\cdots\cdots\cdots \\ a'_{k1}, a'_{k2}, \ldots, a'_{kN} \end{pmatrix} \tag{5}$$

where N is the number of psychological and psychophysiological qualities, and k is the number of experts.

4. After constructing the matrix E, the consistency of experts' judgments is checked.

To assess the consistency, it is proposed to use simple and user-friendly criteria that take into account the opinions of experts:

- The maximum ratio of half the scatter of indicators to the arithmetic mean:

$$\Delta_i = \max_j \frac{\beta_j \left| a'_{ji} - \overline{a'_i} \right|}{2\overline{a'_i}}, \; il = 1, 2, \ldots, k, \tag{6}$$

where β_j is the coefficient of competence of the j-th expert;

- Relative deviation of the arithmetic mean from the geometric mean:

$$\Delta_{2i} = 1 - \frac{c_{gi}}{c_{hi}}, i = 1, 2, \ldots, k, \tag{7}$$

where c_{hi}-arithmetic mean; c_{gi} the geometric mean value is calculated by the following Formula (8):

$$c_i = \sqrt[\sum\limits_{j=1}^{k} \beta_j]{\prod_{j=1}^{k} a'^{\beta_j}_{ji}} = \exp\left(\frac{\sum\limits_{j=1}^{k} \beta_j \ln a'_{ji}}{\sum\limits_{j=1}^{k} \beta_j} \right). \tag{8}$$

If the opinions of the experts are not agreed, there is a transition to paragraph 2. If they are not agreed a second time, it is recommended that the expert group be disbanded and re-created. The consistency of experts' judgments is determined by the Formulas (5)–(8).

5. Using the Hadamard product of the matrix E by the matrix of expert competence coefficients for a certain group of qualities, we obtain the matrix:

$$E' = E \cdot D. \tag{9}$$

where matrix D is the matrix of expert competence coefficients formed from the following considerations. For each j-th expert who participates in the professional selection, a set of competence coefficients is formed:

$$d_j = \left\{d_{j1}, d_{j2}, \ldots, d_{jk}, \ldots, d_{jh_j}\right\} \tag{10}$$

where d_{jh_j} is the competence coefficient of the j-th expert in the h_j-th group of psychological and psychophysiological qualities (characteristics).

$$\underbrace{b_1 \ldots b_g}_{d_{i1}} \underbrace{b_j \ldots b_l}_{d_{ik}} \ldots \underbrace{b_h \ldots b_N}_{d_{ih}} \tag{11}$$

where b_j... b_l is a group of psychological and psychophysiological qualities of the 'man-nature' type professions.

This division of competences is due to the assumption that the experts may not have the same competence in all professions of the 'man-nature' type.

6. Using factor analysis, namely, the principal-factor method [38], to highlight the key factors and cut off the least significant, we can leave only the qualities that are really significant for a given specialty (position).

Intermediate stage quantitative-empirical suggests that psychophysiological support of professional development should be based on the results of empirical research aimed at quantitative-empirical concretization of psychological and psychophysiological characteristics and factors of the professional development of a specialist in professions of this type; in particular, the developed methodological and practical aspects of such support are based on empirical results obtained during the study on the professional orientation, competence and motivation of specialists, their characterological, neurodynamic, temperamental and emotional properties, myokinetic abilities, peculiarities of thinking, imagination, attention, emotional sphere and self-regulation.

In this study, we aimed to select the best methods for diagnosing PDQ and build a psychophysiological standard of the 'man-nature' type profession, that is, to determine the necessary level of PDQ development and the boundaries of permissible deviations of the level of development of psychophysiological qualities in real profiles of an employee (RPE) or an applicant.

The first step is the selection of methods for studying professionally important qualities of the 'man-nature' type profession, which would ensure adequate assessment with sufficient reliability [39]. To build the correct model, it was necessary to make a number of refinements and explanations. Thus, the general characteristics of the 'man-nature' type professions are as follows:

- Means and objects of production of the 'man-nature' type profession are objects of living and abiotic nature, as well as anthropogenic and biological processes;
- Main activities–study, research, analysis, design, extraction, cultivation, care, prevention;
- The main requirements of the profession to the employee are: the ability to establish cause-and-effect relationships, analytical thinking, the ability to predict and assess the changing factors of the abiotic environment and take into account anthropogenic factors, analysis of the ecosystem as a system (systemic thinking), rich imagination, operational thinking, observation, endurance, persistency, responsibility, steadiness of mind, distribution and random switching of attention, visual-imagery, and visual memory;
- Professional activities of various workers of the 'man-nature' type professions require the use of a wide range of knowledge and intellectual skills from automation of processes and knowledge of the technological cycle of production to knowledge of the laws of growth and development of living organisms and biotopes, at the same time, significant changes in the nature and content of the professions of the 'man-nature' group requires the development of rational methods for conducting career-oriented

and vocational counseling work with school students, taking into account the specifics of these changes;

- Objects of labor are: climatope, edaphotope, hydrotope, as well as biocoenoses of different taxonomic rank;
- The tools of specialists of the 'man-nature' type professions are very diverse and heterogeneous. They can be classified as follows: hand tools and fixtures; mechanical tools and devices; machines with manual and automatic control; devices designed to detect signs that are not directly perceived by the senses.
- The working conditions for specialists of this type of profession may be the most diverse: daytime work, shift working; night work, shift working; irregular (long) working hours; work with the predominance of static components in actions (prolonged stay in one working position: sitting, standing); work that requires movement;
- For the professions of the 'man-nature' group, the preferred activities that have informational features are as follows: observation, control as gnostic actions (observation, measuring, inspection, examination by touch, taste test), solving operational and production tasks and other mental actions (planning, experimenting), processing of materials (processing, sorting, stacking, monitoring, measuring), control of mechanisms, machines, maintenance and care (for equipment, animals, plants), organization of a team and education of people;
- Based on working conditions, for professions of the 'man-nature' type, the following occupational hazards can be identified: overheating/hypothermia of the body, exposure to chemicals or toxic substances, stress, and work in conditions of uncertainty.

Group A included 20 professionally successful workers, environmental engineers-graduates from the V. Chernovil ISTD, with indicators of professional success being from 8 to 10 points. In addition, five master's students were in this group, with indicators of education quality being from 9 to 10 points.

Group B included 20 employees who have little work experience or just learning the profession of environmental engineer. They are graduates from the V. Chernovil ISTD whose activities success is beyond doubt by the same criteria, with indicators of professional success being from 0 to 4 points. This group also included 10 students, whose indicators of education quality are from 5 to 7 points.

We found the optimal level of PDQ development by constructing the averaged profiles of environmental engineers of both groups. The next step was to find common arithmetic means that meet the criteria of optimal psychophysiological readiness (OPR) for the profession of environmental engineer (Table 1).

Table 1. Dependence of professional success on neurodynamic qualities of environmental engineers (fragment of the table).

Belonging to the Study Group	Item No. of the Study Subject	Year of Birth	Age from Birth Certificate	Physiological Age	Rnp-1	Cnc	Cognitive Potential	Professional Success	Cognitive Potential
Group A	1	1984	35	32	8.6	121	200.00	10	**8**
	2	1985	34	30	10.6	108	199.00	8	**7**
	3	1985	34	31	12	71	239.00	7	**6**
	4	1986	33	32	9.2	96	244.00	8	**7**
	5	1984	35	32	2	92	198.00	10	**8**
	21	1999	20	18	7.6	92	199.00	10	**8**
	22	1999	20	19	14.6	83	196.00	8	**7**
	23	1998	21	18	9.3	99	309.00	6	**6**
	24	1999	20	18	14.2	89	201.00	9	**8**
	25	1998	21	18	15	72	344.00	5	**4**

Table 1. *Cont.*

Belonging to the Study Group	Item No. of the Study Subject	Year of Birth	Age from Birth Certificate	Physiological Age	Rnp-1	Cnc	Cognitive Potential	Professional Success	Cognitive Potential
Group B	26	1989	30	32	18.1	74	305.00	5	**4**
	27	1987	32	35	16.8	82	303.00	6	**5**
	28	1990	29	32	13.1	102	212.00	8	**7**
	29	1987	32	32	8.9	128	216.00	8	**8**
	46	1999	20	22	9.9	89	211.00	8	**7**
	47	1999	20	23	22.6	114	201.00	9	**5**
	48	1998	21	23	7.6	122	198.00	10	**9**
	49	1998	21	22	8.2	96	213.00	9	**8**
	50	1999	20	23	18.1	86	324.00	6	**4**

Thus, optimal psychophysiological readiness OPR acquires a quantitative characteristic and allows us to compare profiles, real and standard, which makes it possible to determine the degree of applicants' suitability (DAS), using the level of profile adequacy as criteria.

After statistical processing of test results, the following was found:

- The coefficient of correlation of the readiness for emergency actions (REA) with professional success is 0.9.

Let us estimate the correlation density by calculating the Pearson correlation coefficient $\bar{r}$ = 0.9, according to the Chaddock table. The coefficient of correlation of the readiness for emergency actions (REA) with professional success is very high. Considering $\bar{r}^2$ = 0.81, then REA is 81% dependent on professional success and 19% on other factors. For the significance level α = 0.05, we check the null hypothesis H_0: r = 0 according to the competing hypothesis H_1: $r \neq 0$. We calculate the empirical value of the criterion: $T_{emp} = \bar{r}\frac{\sqrt{n-2}}{1-\bar{r}^2} = 0.9\frac{\sqrt{55-2}}{\sqrt{1-0.9}} = 16.365$.

For a given level of significance and the number of degrees of freedom k = 55 − 2 = 53 by using the table of critical points of Student's distribution, we find that t_{kr} = 1.96. Since $T_{emp} > t_{kr}$, we reject the null hypothesis H_0 and conclude that the sample correlation coefficient is significant and random variables are correlated. We calculate the confidence interval for the general correlation coefficient: $\bar{r} - t_{kr}\frac{1-\bar{r}^2}{\sqrt{n}} \leq r \leq \bar{r} + t_{kr}\frac{1-\bar{r}^2}{\sqrt{n}}$. $0.9 - 0.05 \leq r \leq 0.9 + 0.05$

- The coefficient of correlation of the reaction to a moving object (RMO) with professional success is 0.7. The confidence interval for the general correlation coefficient: $0.7 - 0.13 \leq r \leq 0.7 + 0.13$;
- The coefficient of sensorimotor reactions correlation with professional success is 0.9. The confidence interval for the general correlation coefficient: $0.9 - 0.05 \leq r \leq 0.9 + 0.05$;
- The coefficient of the strength of nervous system (SNS) correlation with professional success is 0.6. The confidence interval for the general correlation coefficient: $0.6 - 0.17 \leq r \leq 0.6 + 0.17$;
- The coefficient of the mobility of nervous processes (MNP) correlation with professional success is 0.6. The confidence interval for the general correlation coefficient: $0.6 - 0.17 \leq r \leq 0.6 + 0.17$;
- The coefficient of the emotional stability (robustness) and switching attention (refocusing) correlation with professional success is 0.9. The confidence interval for the general correlation coefficient: $0.9 - 0.05 \leq r \leq 0.9 + 0.05$;
- The coefficient of logical thinking correlation with professional success is 0.77. The confidence interval for the general correlation coefficient: $0.77 - 0.11 \leq r \leq 0.77 + 0.11$;

- The coefficient of visual memory correlation with professional success is 0.6. The confidence interval for the general correlation coefficient: $0.6 - 0.17 \leq r \leq 0.6 + 0.17$;
- The coefficient of long-term memory correlation with professional success is 0.3. The confidence interval for the general correlation coefficient: $0.3 - 0.2 \leq r \leq 0.3 + 0.2$;
- The coefficient of short-term (operative) memory correlation with professional success is 0.2. The selective correlation coefficient is insignificant and random values are uncorrelated.

Comparison of several profiles allows us to see the excess of the optimal level of PDQ development in DAS over the requirements of OPR, which also, in some cases, is a signal that the employee has 'outgrown' his profession, which means that his interest in work and the level of motivation are reduced, and this in certain situations lead to a malfunction.

A differential diagnostic scale has been developed for assessing the results of psychophysiological examinations using a standardized indicator that divides the normal distribution of PDQ measurements into 10 points on a one-dimensional ranking scale from the lowest value equal to 1 point to the highest one, which is 10 points (each point is a certain part in % of a representative sample of the examined environmental engineers. The 'raw' indicator obtained during the examination is translated into normalized scores, which makes it possible to compare the results of the examination of various groups of workers–to develop uniform criteria for assessing optimal readiness for a profession.

Based on the data obtained, a histogram of factor values of the performance efficacy of environmental engineers was constructed, where, in addition to factor values, the possibility of mastering the profession of the environmental engineer was taken into account.

Thus, professional defining qualities (PDQ) for environmental engineers can be considered: psychomotor qualities (readiness for emergency action (REA)), neurodynamic qualities (sensorimotor reactions), and sensory-perceptual qualities (emotional stability and switching of attention). The least pronounced is the dependence of professional success on the mnemonic qualities of environmental engineers of the two groups.

No reliable relationship has been established: between professional success and biological age as well as age from birth certificate; between cognitive potential and biological and age from a birth certificate. This can be explained by compensation for the deterioration of physiological and psychophysiological indicators with age due to the high level of professionalism and motivation of the employee. The correlation coefficient of professional success (Ekho) and calculated cognitive potential (CP) is 0.87.

The results of psychophysiological examinations indicate that the construction of the psychophysiological standard of the profession with sufficient probability predicts the success and reliability of the work of employees and makes it possible to identify contraindications to professional activities.

Obviously, the value (see Formula (6)) should be small (with satisfactory consistency of the expert judgments) for sufficiently significant indicators. Testing this ratio on the experimental results showed that the Δ_i value increases sharply for insignificant indicators. As a result, we obtain a series of significance coefficients of qualities {X} that determine the standard of a professional worker's profile.

The flowchart of the method for constructing the psychophysiological standard of the 'man-nature' type profession is presented in Figure 3.

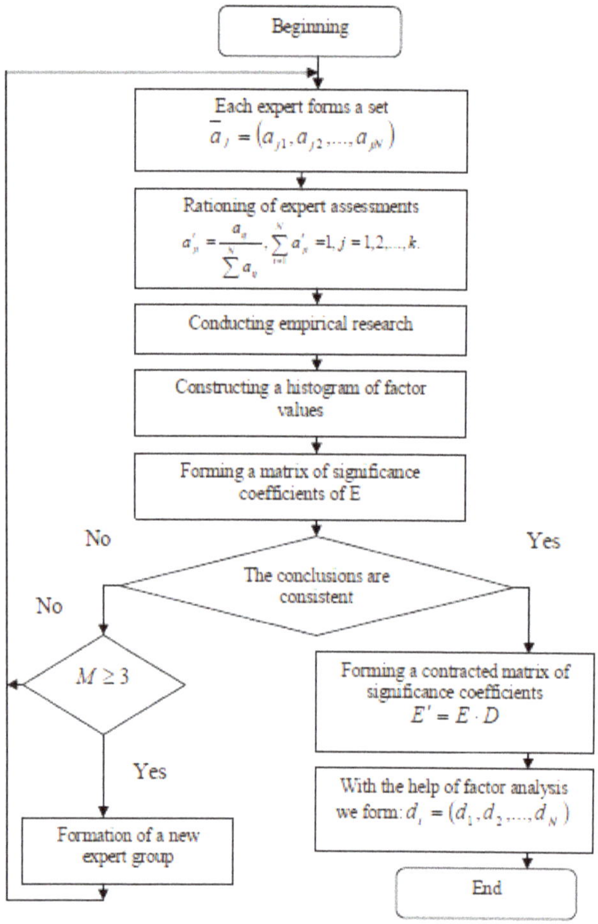

Figure 3. The method of constructing a psychophysiological standard of the 'man-nature' type specialist.

Using the matrix of expert competence coefficients for certain groups of qualities allows improving the quality of the constructed standard model. In addition, the use of factor analysis makes it possible to identify hidden relationships between qualities, which will allow us to monitor complexes of interrelated PDQ groups in the diagnosis of the 'man-nature' type profession to increase the reliability of the results of the professional selection procedure.

Thus, the formalization of the structure of professional selection makes it possible to more clearly represent and formulate requirements for the basic models and procedures used at various stages of professional selection.

Stage IV Feedback and improvement provides a list of key activities aimed at developing and further practical use of the cyber-physical system: determination of the general content, components, and factors of professional development of the specialist; as well as the need for continuous research aimed at tracking changes in psychological and psychophysiological aspects of professional activity in professions of the 'man-nature' type, appropriate consideration of these changes in the activities, obtaining information on the effectiveness of activities to improve them and train professionals for sustainable development.

The developed cyber-physical system contains all the characteristics of cybernetic systems-it is controllable, consists of a complex of subsystems that act as mechanisms for working out and implementing self-preserving reactions. That is, it is effective, characterized by a certain set of parameters. This explains the key role of the system of transmission and processing of information in training specialists for sustainable development, as well as the importance of intellectualization of management processes.

4. Discussion

The theoretical and empirical results obtained during the research provided an opportunity to formulate and substantiate the cyberphysical system of psychophysiological support for becoming a specialist in professions of the 'man-nature' type in the formation of specialists for sustainable development. A similar approach was first used by the authors. The works [39–42] are described as the automation of the decision-making process in the implementation of vocational guidance work. Professions of the 'man-nature' type make up a fairly large group. This includes professions that require significant theoretical training, and hence, higher and secondary specialized education (environmental engineer, agronomist, veterinarian, veterinary technologist, zootechnician, gardener, landscape architect, etc.), and professions that require practical skills, the foundations of which are laid, for example, in a vocational school (livestock breeder, tractor driver of general profile, master of machine milking). Thus, there is a need to study these professions, drawing up professiograms and using them in the organization of practical career guidance work. These professions need special attention because, firstly, they are common in agriculture, which is an important sector of the national economy. Secondly, there is a shortage of qualified professionals. Thirdly, the requirements for the psycho-physiological characteristics of workers are very important, and their decoding is of considerable interest for the practice of career guidance and vocational counseling of school youth [43–45].

5. Conclusions

According to the purpose of the study, the authors have integrated numerous, but rather heterogeneous, mixed studies related to various aspects of the formation of modern specialists of the 'man-nature' type. The authors have implemented a combination of purely psychological and psychophysiological aspects of this problem. An integral system of psychophysiological support has been developed for all stages of the formation of professionals of the 'man-nature' type to train specialists for sustainable development. The model built by the authors of the psychophysiological profile of a specialist is presented in the form of a hierarchical structure, the upper level of which reflects the employer's vision and contains the necessary qualities and their significance coefficients. The lower level of the structure is a set of characteristics that affect the corresponding qualities and the corresponding weight values.

This model of a psychophysiological profile of a specialist in the 'man-nature' type profession allows you to visualize this profile when working on building a specialist's standard using the method of expert assessment and allows you to easily automate the process of creating the standard model of a professional worker for the 'man-nature' type profession, as well as to obtain estimates of the significance of each of the professional characteristics. The novelty of the proposed cyberphysical system lies in the integration of points of view on the standard of a specialist on the part of customers (employers) and on the part of psychologists, in this case, the significance of the top-level elements comes from the expert customers for each profile of a professional worker, while the weights of the relationship between characteristics and certain qualities can be obtained from expert psychologists in advance. A comprehensive assessment of the professional characteristics of a specialist of the 'man-nature' type is formed using methods of hierarchy analysis, taking into account the significance (weights) of experts. Thus, the use of the cyberphysical system proposed by the authors for psychophysiological support of professional self-realization in professions of the 'man-nature' type in the formation of specialists for sustainable development will allow in the future to formalize the structure of professional selection. In the future, developing this approach will make it possible to more clearly represent and formulate the requirements for the basic models and procedures used at various stages of the professional selection of workers.

Author Contributions: Conceptualization, M.R.; methodology, M.R. and T.B.; software, M.R. and T.B.; validation, O.M.; formal analysis, O.K. and T.B.; investigation, O.M. and O.K.; resources, M.R. and O.K.; data curation, T.B.; writing—original draft preparation, M.R.; writing—review and editing, O.K.; visualization, M.R. and T.B; supervision, O.M; project administration, O.K. All authors have read and agreed to the published version of the manuscript.

Funding: This research received no external funding.

Institutional Review Board Statement: The study was conducted according to the guidelines of the Declaration of Helsinki, and approved by the Institutional Review Board of VIACHESLAV CHORNOVIL INSTITUTE OF SUSTAINABLE DEVELOPMENT (protocol code 2 of 17 November 2020).

Informed Consent Statement: Informed consent was obtained from all subjects involved in the study.

Conflicts of Interest: The authors declare no conflict of interest.

References

1. Boholiubov, V.M. *Stalyi Rozvytok Suspilstva: Sotsialno-Ekolohichni Aspekty Formuvannia Profesiinoi Kompetentnosti Mahistriv-Ekolohiv [Sustainable Development of Society: Socio-Ecological Aspects of the Formation of Professional Competence of Masters of Ecology]*; Hrin: Kherson, Ukraine, 2013. (In Ukrainian)
2. Melnyk, L.H. *Fundamentalnыe OsnovыUstoichyvoho Razvytyia [Fundamentals of Sustainable Development]*; Unyversytetskaia Knyha: Sumy, Ukraine, 2003. (In Russian)
3. Skyba, Y.A. *Formuvannia Upravlinskykh Kompetentsii Maibutnikh Ekolohiv na zasadakh Zbalansovanoho Rozvytku: Teoriia ta Praktyka: Monohrafiia [Formation of Managerial Competencies of Future Ecologists on the Basis of Balanced Development: Theory and Practice: Monograph]*; NPU Imeni M.P. Drahomanova: Kyiv, Ukraine, 2013. (In Ukrainian)
4. Mitcham, C. The concept of sustainable development: Its origins and ambivalence. *Technol. Soc.* **1995**, *17*, 311–326. [CrossRef]
5. Klarin, T. The Concept of Sustainable Development: From its Beginning to the Contemporary Issues. *Int. Rev. Econ. Bus.* **2018**, *21*, 67–94. [CrossRef]
6. Pometun, O.I. Pedahohichni zasady osvity dlia staloho rozvytku v Ukrainskii shkoli [Pedagogical principles of education for sustainable development in the Ukrainian school]. *Ukr. Pedahohichnyi Zhurnal* **2015**, *1*, 171–182. (In Ukrainian)
7. Tilbury, D.; Wortman, D. *Enganging People in Sustainability*; IUCN: Gland, Switzerland, 2004.
8. McKeown, R. *Education for Sustainable Development Toolkit*; Printed in UNESCO's Workshop: Paris, France, 2006.
9. Andersone, R. The Content Reform of Education for Sustainable Development. In Proceedings of the Rural Environment. Education. Personality. International Scientific Conference, Jelgava, Latvia, 15–16 May 2015; pp. 75–81.
10. Pometun, O.I. *Pidhotovka Vchyteliv do Vykladannia Pytan Staloho Rozvytku. Navchalno-Metodychni Materialy dlia Vykladachiv Vyshchykh Pedahohichnykh Navchalnykh Zakladiv ta Systemy Pisliadyplomnoi Pedahohichnoi Osvity [Teacher Training for Teaching Sustainable Development Issues. Educational and Methodical Materials for Teachers of Higher Pedagogical Educational Institutions and the System of Postgraduate Pedagogical Education]*; Pedahohichna Dumka: Kyiv, Ukraine, 2015. (In Ukrainian)
11. Vysotska, O. *Osvita dlia Staloho Rozvytku [Education for Sustainable Development]*; Roial Prynt Dnipropetrovsk: Dnipropetrovsk, Ukraine, 2011. (In Ukrainian)
12. Ilchenko, V.R. Modernizatsiia Zmistu Zahalnoi Serednoi Osvity Ukrainy na Zasadakh Osvity dlia Staloho Rozvytku: Nacherk Proektu Eksperymentalnoho Doslidzhennia [Modernization of the Content of General Secondary Education in Ukraine on the Basis of Education for Sustainable Development: An outline of a Pilot Study Project]. *Postmetodyka* **2011**, *5*, 16–17. (In Ukrainian)
13. Shvets, V.Y.; Palekhova, L.L.; Shmidt, M.; Palekhov, D.A. Partnerstvo kak metod prodvyzhenyia modely vыssheho obrazovanyia dlia tselei ustoichyvoho razvytyia [Partnership as a Method to Promote a Higher Education Model for Sustainable Development]. *Rozvyt. Ekon. Osv.* **2015**, *3*, 146–154. (In Russian)
14. Soloshych, I.O.; Pidlisniuk, V.V. Doslidzhennia obiznanosti ideiamy staloho rozvytku suspilstva v konteksti ekolohichnykh problem [Research of awareness of the ideas of sustainable development of society in the context of environmental problems]. *Vypusk* **2011**, *3*, 155–158. (In Ukrainian)
15. Rudyshyn, S.; Koreneva, I. Development of Unsversity Students' Ability to Undestad the World Scientific Pattern. *Adv. Sci. J. US* **2014**, *5*, 7–12. [CrossRef]
16. Azizov, A.A.; Akinshina, N.G. *Obrazovanie v Interesah Ustojchivogo Razvitija: Uchebno-Metodicheskoe Posobie [Education for Sustainable Development: Study Guide]*; NESKO: Tasheknt, Uzbekistan, 2016. (In Russian)
17. Isaienko, V.M. Stratehichne bachennia innovatsiinoi roli osvity dlia staloho rozvytku Ukrainy [Strategic vision of the innovative role of education for sustainable development of Ukraine], Stratehiia staloho rozvytku Ukrainy: Zavdannia osvity shchodo yii realizatsii: Materialy III Vseukrainskoho forumu "Osita dlia zbalansovanoho rozvytku". In Proceedings of the Tsentr Ekolohichnoi Osvity ta Informatsii, Kyiv, Ukraine, 12–13 April 2017; pp. 8–10. (In Ukrainian)

18. Marfenin, N.N. *Osnovnye Problemy i Zadachi Obrazovanija dlja Ustojchivogo Razvitija [The Main problems and Tasks of Education for Sustainable Development]*; Jagodina, G.A., Ed.; Centr «Shkol'naja Kniga»: Moscow, Russia, 2009; pp. 9–31. (In Russian)
19. Campbell, J.S.; Castaneda, M.; Pulos, S. Meta-analysis of personality assessments as predictors of militaryaviation training success. *Int. J. Aviat. Psychol.* **2009**, *1*, 92–109. [CrossRef]
20. Kal'nish, V.V.; Ena, A.M. Sovremennoe sostoyanie professional'nogo psihofiziologicheskogo otbora vUkraine [The current state of professional psychophysiological selection in Ukraine]. *Occup. Med. Ind. Ecol.* **2006**, *3*, 12–17. (In Ukrainian) [CrossRef]
21. Ancyferova, L.I. *Razvitie Lichnosti Specialista kak Subekta Svoej Professional'noj Zhizni [Development of the Personality of a Specialist as a Subject of His Professional Life], Psihologicheskie Osnovy Professional'noj Dejatel'*; PER SJe; Logos: Moskva, Russia, 2007; pp. 499–507. (In Russian)
22. Gulyas, I.A. Determinanty' Samozdijsnennya Osoby'Stosti [Determinants of personality self-realization]. *Pedagog. Proces: Teor. Prakty'ka* **2011**, *17*, 83–92. (In Ukrainian)
23. Dubchak, G.M. *Psy'Xologichni Osnovy' Profesijnoyi Stresostijkosti Majbutnix Faxivciv [Psychological Bases of Professional Stress Resistance of Future Specialists]*; Talkom: Kyiv, Ukraine, 2017. (In Ukrainian)
24. Kokun, O.M. *Psy'xologiya Profesijnogo Stanovlennya Suchasnogo Faxivcya [Psychology of Professional Development of a Modern Specialist]*; Inform.-analit. agenstvo: Kiev, Ukraine, 2012. (In Ukrainian)
25. Korniyenko, I.O. *Psy'Xologiya Opanuval'Noyi Povedinky' Osoby'Stosti: Monografiya [Psychology of Mastering Personality Behavior: A Monograph]*; RVU MDU: Mukachevo, Ukraine, 2017. (In Ukrainian)
26. Kalmy'kovoyi, L.O.; Xomy'ch, G.O. *Rozvy'tok Osoby'Stosti v Rizny'x Umovax Socializaciyi: Kolekty'vna Monografiya [Personality Development in Different Conditions of Socialization]*; Vy'davny'chy'j Dim «Slovo»: Kyiv, Ukraine, 2016. (in Ukrainian)
27. Zeer, E.F. *Psikhologiya Professiy [Psychology of Professions]*; Izdatel'skiy Tsentr «Akademiya»: Moscow, Russia, 2005. (In Russian)
28. Zeer, E.F. *Psikhologiya Professional'nogo Razvitiya [Psychology of Professional Development]*; Izdatel'skiy tsentr «Akademiya»: Moscow, Russia, 2007. (In Russian)
29. Symanyuk, E.E. *Psikhologicheskie Bar'ery Professional'nogo Razvitiya Lichnosti: Praktiko-Orientirovannaya Monografiya [Psychological Barriers to Professional Development of the Individual. Practice-Oriented Monograp]*; Moskovskiy psikhologo-sotsial'nyy institut: Moscow, Russia, 2005. (In Russian)
30. Whitaker, C.A. The Technique of Family Therapy. In *Changing Sexual Values and the Family*; Sholevar, G.P., Ed.; Charles Thomas: Springfield, IL, USA, 1976.
31. Kuzikova, S.B. Samorozvitok osobistosti: Sub'єktnij pidhid [Self-development of the imagination: Subjectiveapproach]. *Psihologiya Osobistosti* **2013**, *1*, 77–86. (In Ukrainian)
32. Pavlenko, M.A. Metody i procedury otbora operatorov ASU pri ispol'zovanii intellektual'nyh sistem podderz-hki prinyatiya reshenij [Methods and procedures for the selection of ACS operators using intelligent decision support systems]. *Collect. Sci. Work. HUPS* **2012**, *4*, 171–177. (In Russian)
33. Kokun, O.M. *Psihologiya Profesijnogo Stanovlennya Suchasnogo Fahivcy [Psychology of Professionalformation of a Modern Specialist]*; DP «Informanalit. agentstvo»: Kyiv, Ukraine, 2012; p. 200. (In Ukrainian)
34. Muchinsky, P.M. *Psychology Applied to Work: An Introduction to Industrial and Organizational Psychology*; Cengage Learning: Boston, MA, USA, 2006; p. 554.
35. Ericsson, K.A.; Charness, N.; Feltovich, P.J.; Hoffman, R.R. *The Cambridge Handbook of Expertise and Expert Performance*; Cambridge University Press: New York, NY, USA, 2018; p. 918.
36. Brown, T. *A Confirmatory Factor Analysis for Applied Research*; Guilford Press: New York, NY, USA, 2015; p. 462.
37. Kahneman, D.; Klein, G. Conditions for intuitive expertise: A failure to disagree. *Am. Psychol.* **2009**, *64*, 515–526. [CrossRef] [PubMed]
38. Evetts, J. Professionalism: Value and ideology. *Curr. Sociol.* **2013**, *61*, 778–796. [CrossRef]
39. Gegenfurtner, A.; Lehtinen, E.; Säljö, R. Expertise Differences in the Comprehension of Visualizations: A Meta-Analysis of Eye-Tracking Research in Professional Domains. *Educ. Psychol. Rev.* **2011**, *23*, 523–552. [CrossRef]
40. Ayaz, H.; Shewokis, P.A.; Bunce, S.; Izzetoglu, K.; Willems, B.; Onaral, B. Optical brain monitoring for operator training and mental workload assessment. *NeuroImage* **2012**, *59*, 36–47. [CrossRef] [PubMed]
41. Dew, N.; Read, S.; Sarasvathy, S.D.; Wiltbank, R. Effectual versus predictive logics in entrepreneurial decision-making: Differences between experts and novices. *J. Bus. Ventur.* **2009**, *24*, 287–309. [CrossRef]
42. Kim, J.O.; Mueller, C.W. *Factor Analysis: Statistical Methods and Practical Issues*; Sage: Newbury Park, CA, USA, 1978; p. 88.
43. Wickens, C.D. *Engineering Psychology and Human Performance*; Psychology Press: New York, NY, USA, 2015; p. 544.
44. Taylor, S.E. Health psychology: The science and the field. *Am. Psychol.* **1990**, *45*, 40–50. [CrossRef] [PubMed]
45. Frank, J.R.; Snell, L.S.; Cate, O.T.; Holmboe, E.S.; Carraccio, C.; Swing, S.R.; Harris, P.; Glasgow, N.J.; Campbell, C.; Dath, D.; et al. Competency-based medical education: Theory to practice. *Med Teach.* **2010**, *32*, 638–645. [CrossRef] [PubMed]

sustainability

MDPI

Article

Building on Strategic eLearning Initiatives of Hybrid Graduate Education a Case Study Approach: MHEI-ME Erasmus+ Project

Radovan Madleňák [1,*], Stephen P. D'Alessandro [2], Agostino Marengo [3], Jenny Pange [4] and György Iván Neszmélyi [2]

1 Department of Communications, University of Žilina, 010 26 Žilina, Slovakia
2 Department of Entrepreneurial Studies, Advenio eAcademy (AeA), 33/4 Abate Rigord Street, XBX 1128 Ta'Xbiex, Malta; sda@aea.academy (S.P.D.); neszmelyi.gyorgy@uni-bge.hu (G.I.N.)
3 Department of Economics and Finance, University Aldo Moro, 70121 Bari, Italy; agostino.marengo@uniba.it
4 Laboratory of New Technologies and Distance Learning, School of Education, University of Ioannina, 451 10 Ioannina, Greece; jpagge@uoi.gr
* Correspondence: radovan.madlenak@fpedas.uniza.sk; Tel.: +421-41-513-3124

Abstract: Online courses are gaining popularity because they provide extensive and varied course material, information, knowledge, and skills, whilst also creating an effective educational online community. This research adopts a case study approach to focus on the teaching method and the manner in which a strategic commitment to eLearning provides scope for the development and implementation of top quality educational online fully accredited programs. Entrepreneurship focuses on developing businesses that add value and create wealth and prosperity in our societies. Therefore, entrepreneurship is a key area of learning for graduate students seeking to set up and operate their own SME organizations. It can serve as a benchmark for the teaching of other graduate subjects that require a sound correlation for the correlation of concepts and theories to the challenging complexities of the real world. The program was developed on the basis of the implementation of a state-of-the-art eLearning platform that allowed for a combination of varied self-learning and collaborative learning elements and activities within a single platform. This enabled students to access the online content material efficiently and effectively. It allows for the development of a program based on the flipped classroom teaching methodology. The underlying concept of the flipped classroom methodology is that effective eLearning should comprise both synchronous and asynchronous learning activities. This combination of self-learning and collaborative learning calls for careful planning by the tutor to ensure that the learning objectives are clearly defined for each activity and that the relevant deliverables are monitored. The content material for each subject course module was designed, developed, produced, and presented by the different project partners in a holistic manner structured to motivate participants to learn. The results of our analysis have shown that students were able to learn, discuss their projects, and cooperate during an online course in an effective and participant-focused manner with their tutors. The feedback given highlights the importance of ongoing communications between students and the tutors who often need to act as mentors to retain student engagement.

Keywords: eLearning; higher education; flipped classroom; ARCS model; teaching method; international cooperation

Citation: Madleňák, R.; D'Alessandro, S.P.; Marengo, A.; Pange, J.; Neszmélyi, G.I. Building on Strategic eLearning Initiatives of Hybrid Graduate Education a Case Study Approach: MHEI-ME Erasmus+ Project. *Sustainability* **2021**, *13*, 7675. https://doi.org/10.3390/su13147675

Academic Editors: Miltiadis D. Lytras, Abdulrahman Housawi and Basim Alsaywid

Received: 22 May 2021
Accepted: 7 July 2021
Published: 9 July 2021

1. Introduction

In our current environment of globalization, accessibility and geographic distances take on different values and meanings more than ever before. For managers or even employees, free time, leisure time, or out-of-work dispensable time have shrunk and become scattered. For efficient career building, including vocational training and post-graduate education, time management has become a key issue. People have less and less free time in their daily schedule. This is particularly relevant to those in leading positions. Managers typically find it difficult to gain new skills and information through classic education systems [1]. Traveling to attend lectures or seminars in pre-fixed time frames—even in part-time learning programs—takes too much time and is difficult to harmonize with busy working schedules. This phenomenon has generated an increasing demand for new ways and methods for education systems that meet quality standards but enable students to save time and eliminate commuting. Thanks to the growing strength of widely used broadband internet, nowadays online learning seems to be a very promising alternative to the classic education model. Although online education has its limitations, it also has a number of advantages. Norman [2] specified five traits as follows: (a) you can learn whatever you want; (b) it is comfortable; (c) online courses look great on a resume; (d) self-paced learning; (e) lower costs.

Moreover, Rakic, Pavlovic, Softic, Lalic, & Marjanovic [3] believe that digital technologies have become an indispensable asset for those involved in online teaching, both at school and university, so much so that they are no longer considered a "disruptive tool", and Chivu, Orza, & Popa [4] say that eLearning is marked as the study applied in a virtual world. Human creativity and technology combine to simplify and accelerate the in-depth knowledge of the studied field. Higher education institutions took part in a mass of technology development processes. Experts assert that eLearning brings with it a group of radical innovations in form, organization, and succession of education. It is easy s to notice that all players in the learning process increasingly use online learning environments.

Additionally, Zimmerman [5], considers that online learning is a demanding process and learners have to develop self-regulating strategies in order to be able to attend online courses. Institutions have included online learning in their teaching methods and approve online learning strategies. Dumford & Miller [6], declare that "many colleges and universities are interested in how to best deliver course content for online learners". In their study, they discuss ways in which taking online courses influences student engagement. They demonstrated that "students taking greater numbers of online courses were more likely to engage in quantitative reasoning", and "they were less likely to engage in collaborative learning, student-faculty interactions, and discussions with diverse others, compared to their more traditional classroom counterparts". Moreover, as technology increases nowadays, flipped classroom teaching (or else inverted classroom or reversed classroom) has been appreciated by many teachers as an effective teaching method in online courses [6]. In flipped classroom teaching, students explore the course material and watch the videos of the courses outside the online classroom. The tasks students have to complete in a flipped classroom according to Bergmann and Sams [7] are performed at home and the homework is completed during the class. Thus, flipped classroom activities support personalized learning and promote student-centered learning activities, outside and inside the classroom [8,9].

Entrepreneurs learn from experience, and ongoing learning plays an important role for them [10]. Additionally, learning is a multidimensional process where different learning theories are applied to teach different subjects [11]. According to many researchers [12–14], flipped classroom activities are supporting the constructivist method of learning, combining online learning with offline activities. According to many studies, it is well documented that teaching entrepreneurship online using MOOCs allows students/entrepreneurs to *"reflect on their own learning activities, so they can make necessary connections to course content"* [15]. Students using online teaching methods are motivated by their professors to learn but what learners want to learn is not always what they need, so flipped classroom activities upgrade students' comprehension, enhance instruction, and construct personalized knowledge [16–18].

Although technology has made encouraging steps forward, there are still some disadvantages in the application of online education, according to [19]. In his recent paper, he states that *"any student seeking to enroll for a distance learning program needs to invest in a range of equipment including computer, webcam, and stable internet connection"*. In the same paper, the author refers to a paper written in 2013 by Nagrale [20], who believed that it was quite dangerous for someone to totally rely on distance education for a degree because an online degree may not go on *"to be recognized by private companies in the job market and the same problem in government jobs"*. So, most employers "prefer a degree from a regular college over online or distance education. They think that distance education is still not a serious form of education". At the present time, online teaching and learning have been recognized more and more and have been accepted from many higher educational institutions worldwide. Students and teachers can interact via internet and there are many online tools for evaluating online courses [11,21,22]. The problem of distance learning is not just a problem of universities but also primary and secondary schools. Poultsakis, Papadakis, Kalogiannakis, & Psycharis [23] realized the research focused on obstacles that teachers at primary and secondary schools face when managing digital learning objects (DLOs) and digital simulation tools (DST) in science. Their study showed that the main reason for refusing to deal with DLOs and DSTs is the technological equipment. Therefore, adequate and user-friendly eLearning environment is essential for a successful and entirely acceptable education process.

Davidson & Goldberg [24] start off their book with the critique of a simple situation, a professor in a university classroom reading aloud a passage from the assigned work met with the inattentive classroom more focused on their laptop screens and their private engagements. This simple situation presented the wide range of questions that drove their research into what tools are best suited for education in a digital world. Should the professor's role be that of an oracle or a guide leading students to their own discoveries? Is reading aloud from an assigned text still the best use of educational technologies available? They argue that modes of learning have changed dramatically in recent years, but have our institutions changed? Do we question *"how we teach, where we teach, whom we teach, who teaches, who administers and who services"*? Davidson & Goldberg contend that in addition to these fundamental questions, new learning institutions "need not be based on the contiguity of time and place". This introduces the concept of virtual learning organizations and the collaborative manner in which these can be created by established traditional organizations. They contend that collaborative, interdisciplinary, multi-institutional learning spaces help transform traditional learning institutions. The authors maintain that central to the effective implementation of such programs is an understanding of digital learning or as it is sometimes referred to, participatory learning. The concept of digital learning differs from traditional instructional methodologies. In digital learning, the outcomes can be customized by the learners who are able to focus the learning activities more closely on their own needs through shared collaborative initiatives where authorship becomes a shared and interactive experience.

The MHEI-ME project is a perfect example of Davidson & Goldberg's citation of the power of virtual learning institutions. It brings together a collaborative consortium to pitch for EU funds and to administer the funds to obtain project completion. It is interdisciplinary, involving academics from different faculties from European higher educational institutions, and most importantly, it utilizes the European Qualifications Framework administered by the Maltese educational regulator to guarantee the quality of the program. The pilot program presented in this article included an international mix of students and these had collaborative projects in each subject model, ensuring that there was a continued sense of shared development and learning experiences.

In the midst of the COVID-19 pandemic, Levy [25] raised the issue of proper tools and planning for online education. He stressed the fact that online classes were not just the same classroom classes conducted through a video conferencing platform. He challenged tutors to decide on what material should be kept in live sessions (synchronously), what material for students to engage with on their own time (asynchronously), and what material to eliminate. He recommends synchronous learning as being more effective when the importance is on the following: exchanging perspectives amongst students, students learning from each other, interactions in which the tutor plays the role of the facilitator or mediator, and opportunities to build community. Conversely, Levy proposes asynchronous learning as the more effective option when the importance is on the following: students developing the necessary groundwork in basic ideas or concepts before classes, an assessment of students' perspectives or background on the subject, students' ability to study and analyze content material at their own pace, and the stimulation of thought and reflection on the content by students. Learners need time to reflect and take on the relevance and significance of the new knowledge they are being exposed to. This ties in with Kolb's framework of learning styles which was used in the design and development of the set of eLearning deliverables that MHEI-ME provided to students [26].

Nowadays, the tools to deliver eLearning courses can facilitate distance-learning activities. Still, learning management systems (LMS) have all the features required to deliver online courses, such as the registration of students, the management of training contents, and the evaluation of the knowledge.

The aim of this study is to examine if the flipped classroom teaching method applies to a Master's course in entrepreneurship. For the analysis of our hypothesis, we used first-hand experiences in using flipped classroom activities in a pioneer-sound Master's degree project that has been launched and implemented online by a multi-HEI consortium. We also applied for the first time in an online Master's course the Keller's ARCS model of motivational design theories, to evaluate the attention, relevance, confidence, and satisfaction (ARCS) of the students [27]. Poór et al. [28] refer to Kaplan and Haenlein [29], who considered that student motivation is absolutely the key which comprises commitment, challenges, control, novelty, and competition [29]. The results of this project are summarized and analyzed with the view of further utilization in similar online education programs.

2. Materials and Methods

2.1. A Strategic Commitment to eLearning: The Erasmus+ MHEI-ME Project

The online Master's program in entrepreneurship was initiated in 2017 and coordinated by Maltese Advenio eAcademy in collaboration with five higher educational institutions (HEIs), University of Žilina, Slovakia, Szent István University (Hungary), University of Ioannina (Greece), University Aldo Moro (Italy), and the University of Poltava (Ukraine). Each of the institutions involved was committed to the continued development of eLearning activities within their organizations, both at an infrastructural as well as at an operational level. The network of HEIs was established specifically to develop an online Master's program in entrepreneurship which would reflect best practices and provide a combination of self-learning and collaborative learning.

Kolb's concept of learning style preferences [26] has become a core pillar of eLearning design and development. His classification of learning styles is based on a cyclical process consisting of four separate activity stages as shown below. The learning experience commences with an individual having a concrete experience; this is reflected upon in the following stage in which the learner learns how to respond to this (reflective observation). This serves as the basis for the learner to develop mental models to integrate and make sense of the experience (abstract conceptualization). These are then used as the basis for decision making and problem solving (active experimentation).

This model links two critical aspects: the manner in which information is acquired during the learning process (vertical axis) and the manner in which knowledge is transformed (horizontal) axis. Kolb focuses on the importance of learning by doing and this is one of the key aspects of the more effective eLearning programs that seek to provide opportunities for reflection and absorption of content material and opportunities for implementing the learning through interactive discussion and debate (see Figure 1).

The MHEI-ME model is based on this approach and includes both self-learning elements that provide students with autonomy in the learning process and time to reflect and take on the learning. It also includes collaborative learning elements that require interaction with faculty or students to review, discuss, debate the learning matter, thus serving as an active learning experience.

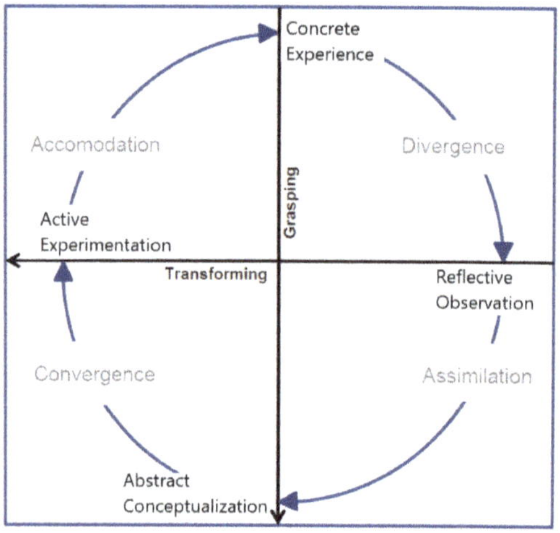

Figure 1. Learning style dimensions and types [26].

The required combination of asynchronous and synchronous learning was possible through the eLearning platform selected, as this provided the required range of online learning options. The MHEI-ME online program was supported through co-funding by Erasmus+ program. A major innovation of the program was its accreditation through the national regulator in Malta, the National Commissions for Further and Higher Education in Malta which confirmed the European Qualifications Framework/Malta Qualifications Framework (EQF/MQF) 90 ECTS credits at Level 7 accreditation. The accreditation of the program was a key element in the sustainability of the project, since it ensured that students have a guarantee of quality and the recognition of their qualification, not only at a national level, but at an international level within all countries that are signatories to the Bologna Process due to their obligations to recognize qualifications awarded within the European Qualifications Framework.

Throughout the discussions and planning of the program, the emphasis was on leveraging best practices to ensure that the new program would provide an engaging and international perspective on entrepreneurship. It was agreed that the "flipped classroom" approach would be adopted with a significant investment in the development and production of pre-recorded video content and other self-learning content. The self-learning content provided flexibility to students to study at their own convenience, since the self-learning content was available 24 × 7. In this way, students were able to study the course material prior to the live online sessions where they could discuss the content with the professor and their peers. The flipped classroom teaching method was evaluated using the attention, relevance, confidence, satisfaction (ARCS) model [27], where students were interviewed on the impact the flipped classroom teaching method had on attention, relevance of interest of course material, confidence of students, and satisfaction with the course.

Throughout the development and production of content material there was a strong sense of collaboration and sharing of knowledge and expertise for developing practical guidelines and templates for faculty in the design, development, and delivery of content. This open and collaborative approach was further enhanced with the ongoing quality assurance maintained by Advenio eAcademy. This QA was an integral part of the program accreditation process, critical to creating a common "feel" throughout all the modules. In this way, whilst each partner HEI was responsible for the design, development, and implementation of one or more course subject module, there was continuity in the style and format of the content. In addition, the ongoing workshop sessions served as a valuable training and development platform for both faculty and support staff.

For the purpose of this study, all comments from participating members of faculty and students during synchronous communications were collected and analyzed. From the Checklists & Rubrics for Evaluating Online Programs [30] we used the questions based on the Blended Course Implementation Checklist [31] to evaluate the appreciation for this program of the students.

The main factor concerning the limitations of the external validity of our results is the size of the group of the participants, as in our group we only had twelve Master's students who completed the program. Another factor affecting the validity of our results is the selection procedure. In this research, we used all participants in the MSc program for the evaluation of the ARCS model, so it can be considered as a convenience sample and not a random sample because in this Master's course we included in our research all the 15 participants.

2.2. The eLearning Platform

One of the fundamental elements for the development and implementation of the project was the commitment to seek best practice at all levels of the design, development, production, and implementation of the program, within the budgetary constraints of the project. The use of a dedicated state-of-the-art eLearning platform that would allow for the development of an online program that includes self-learning, collaborative learning and allows flipped classroom activities and functionality was critical. In addition, the sustainability of the platform post-project was also an important factor. Advenio eAcademy proposed the use of the eLearning platform it had been using for a number of years, which had already been heavily customized to allow for extensive functionality within one eLearning platform. The platform was reviewed by the partners and a decision was taken to develop the program which would leverage the extended functionality of the eLearning platform.

The MHEI-ME program was thus hosted and delivered through a dedicated eLearning platform, eB-Learn. This was developed and offered as software on a services basis by eBusiness Systems. This eLearning platform allows for the organization of a broad range of learning objects and activities generally categorized into asynchronous and synchronous activities. The former is referred to in the system as self-learning objects which users can follow at their own convenience, whilst the latter refers to the collaborative elements usually scheduled for particular time slots. Integrating the self-learning and collaborative elements is an integral part of the flipped or flipped classroom approach adopted for the development and implementation of the online program and was crucial for obtaining EQF certification for the program.

In 2011, eB-Learn was the winner of the Malta Communications Authority national eBiz Award for Best Use of Technology in Education and Training. Later version upgrades of the system improved the security and integrity of the back end as well as added functionality such as auto-issue of course transcripts and tracking of student achievements. The architecture and web-based infrastructure of eB-Learn ensures that it is able to accommodate eLearning operations within different operating scenarios, both within the corporate training and development sector as well as the academic sector. eB-Learn collaborative learning has added functionalities built into the system to allow for online tutoring/mentoring, virtual assistance, online self-assessment, and options for intelligent document handling (see Figure 2). eB-Learn is offered on a software-as-a-solution (SaaS) basis.

The eB-Learn platform is already very well set for mobile access in order to help students and typically more than half of the users access the system through mobile phones and smart phones. The continued growth and expansion of 5G networks in the coming years will provide even more functionality to the networks hosting the eLearning platform, allowing for even broader access and faster communication speeds.

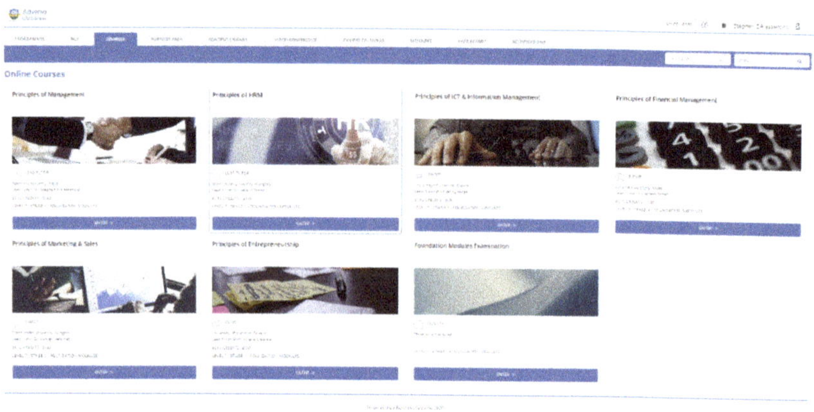

Figure 2. eB-Learn display of foundation modules to MHEI-ME participants [32].

2.3. *The MHEI-ME Process*

This Master's program consisted of four stages: foundation courses x 6, comprising principles of management, principles of HRM, principles of ICT, principles of marketing, principles of financial management, and principles of entrepreneurship. The second stage consisted of three specialist subject modules comprising: SME fundraising and growth, creativity and innovation, and digital marketing. The third stage involved a twelve-week internship in which the work packages assigned in the first stages had to be implemented and reported on. The fourth stage was the research project in which participants had to develop, submit, and present a business plan for a new SME operation or the growth of an existing SME project. The first intake of this joint project, Multiple Higher Educational Institutions–Masters in Entrepreneurship (MHEI-ME), commenced in April 2018 and the participants who successfully completed the program graduated in November 2019. Interviews were held with all students during all stages of their studies. The content material was all online for every course and students had to communicate with their tutors every week in a synchronous and asynchronous method using the flipped classroom method.

Bill Tucker [33] was one of the earlier promoters of the flipped classroom approach, promoting the successful experiment by Jonathan Bergmann and Aaron Sams [7], who as chemistry teachers in Woodland Park High School in 2007 started to record separate class sessions to be available for absentee students on demand. They found that even those students who did not miss class used the pre-recorded content and came to class better prepared and ready to engage in the discussion and review of the relevant content matter.

Jonathan Bergmann at Panopto.com was one of the pioneers of the flipped classroom pedagogy which evolved from an obscure experiment to a widely implemented model for improving the student learning experience in universities and educational institutions. Conceptually, Bergmann defines the flipped classroom as a situation in which "Lectures are recorded outside of class, and then shared with students to review as their "homework". Classroom time is reserved for students to engage in discussions and activities". He contends that by making lecture materials available on-demand, students can learn at their own pace, coming to class better prepared to engage in the review, discussion, and application of the content matter. Bergmann contends that classroom activities may include "group work, comprehension tests, presentations, and other applications of the subject matter". This more engaging approach provides more space for individual queries and personalization of content to the interests of the particular student group [34].

The challenge of adding interactive content meant that each stage of the program, during the course subject stages of the foundation or specialist modules, the internship stage or the final business plan stage, was met at both planning and implementation levels (see Figure 3).

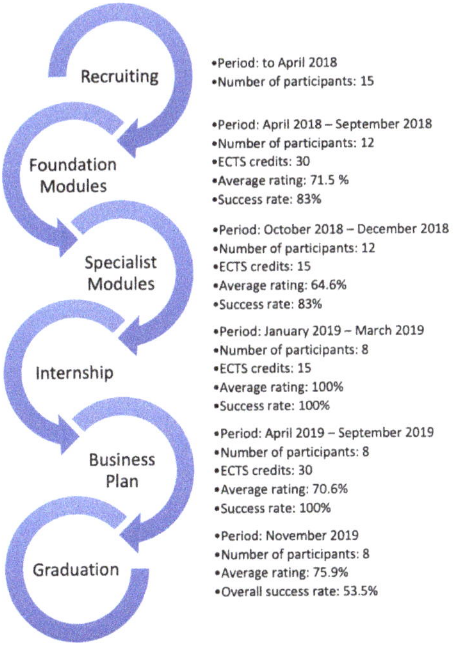

Figure 3. Process of MHEI-ME—facts and numbers.

Students from every economic background and country were able to participate. Students who enrolled and were accepted for the full program were interviewed and replied to our questions concerning the flipped classroom teaching method. The students lived in different geographical areas including Malta, Slovakia, Azerbaijan, Ukraine, Kenya, and Ghana.

The range of the ages of the participants was 36 years with max 59 years and min 23 years of age. Thirteen students were males and two females.

2.3.1. The MHEI-ME Self-Learning Elements

The elements in this category included:

- Pre-recorded audio video sessions carefully planned and prepared to reflect a professional presentation and not merely a reading of the accompanying power point presentation or course notes, to help students read the course material before class tutoring.
- Supporting PowerPoint presentation or notes to facilitate the absorption of the content being delivered during the presentation for the flexibility of the flipped classroom process.
- Reference lists providing the key topics for accompanying live tutorial sessions.
- Reading lists were required for each subject module to encourage students to spend time reviewing the readings and understanding their relevance and correlation with the core course materials before or after the class tutorial.
- FAQs were provided for each subject module to provide an ongoing source of content development and support to the flipped classroom method.
- Online self-assessment facilities allowed for student self-evaluation, enabling them to identify any areas or issues within the course subject module. These self-assessment facilities enabled students to develop skills in the flipped classroom teaching method.

- Glossaries were included, not only to facilitate the precise use of jargon within the particular subject module, but often to ensure that the correct meaning of the English vocabulary used was understood for the flipped classroom process.

2.3.2. MHEI-ME Collaborative Learning Elements

- Calendaring facilities within the eB-Learn platform provided an important link with students as it provided them with auto notification of collaboration events in each module of the program.
- Live video conferencing facilities for online tutorials proved to be one of the most effective collaborative elements in the program. This allowed for inclusion of video clips or other content for discussion. The live interactive aspect of the tutorial/workshop was critical to the overall success of the program. This faculty—student interactivity was crucial to the success of the program, enabling faculty to leverage the technology to bring into play their expertise and experience for the students' benefit.
- Forum sessions proved to be a very powerful learning tool which challenged students to refer to reference material, readings, and other external sources to support their discussion and debate of topics put forward by the forum moderator.
- Individual and/or group assignments were designed to encourage students to collaborate and research both the theory and, most important to the focus of the program, the application, of the various concepts and theories in the context of entrepreneurial activity. The various assignments were discussed during the tutorial/workshop sessions ensuring that students received feedback which enhanced the learning experience of the compilation of the assignment.
- File repository for sharing of documents enabling student-to-group interaction is an important functionality to support groupwork in the flipped classroom process, particularly group assignments.
- Internships were included in the initial MHEI-ME program to provide participants with the opportunity to obtain real-world experience.
- Research project in the MHEI-Me program consisted of the development of a business plan for the launch of a new business or the growth of an existing business. Throughout the research project phase, the weekly online tutorials were maintained in order to support flipped classroom activities, in addition to the scheduled online meetings students had with their supervisors. Again, the weekly tutorial sessions provided the continuity of the flipped classroom process and the ongoing communication with an international dispersed student body.

These elements reflected the manner in which a powerful eLearning platform was used to apply the flipped classroom teaching method and create customized content targeted at the target student audience. It provided the infrastructure for a range of online self-learning and collaborative learning activities that ensured a valid evaluation process which was critical to the EQF certification that was the guarantee of quality promised to students. This model involved more than just turning up for the weekly live tutorial and supported students in the many ad hoc challenges they inevitably faced during an eighteen-month program.

3. Results and Discussions

We had feedback about the flipped classroom teaching method from 12 Master's students who graduated (8 received an MHEI-MI and 4 received a certificate). In particular, two students received an Award in Entrepreneurship, another two students received a Post-Graduate Certificate in Entrepreneurship, and eight students ended up with a Master's in Entrepreneurship.

We used the attention, relevance, confidence, satisfaction (ARCS) model (Keller, 1987) for the evaluation and the analysis of the flipped classroom teaching method. For the evaluation of the ARCS model, we applied qualitative research methods. We interviewed all participants in the course, and according to the results of this model, we found that:

1. Attention was gained by all participants, mainly by posing challenging questions during the online sessions, or by their active participation in problems to be solved. This process also included an active collaboration of students–professors. Students from Malta and Slovakia said in their interviews: *"following the course of Masters in Entrepreneurship was more relaxed to follow because, you can study at your own pace, through a number of videos on each topic. This is easy to follow and with the weekly tutorials with the lecturer and other students gave us the impetus needed to communicate, learn from one another"* and *"I attended a study program MHEI level 7 Master in Entrepreneurship and I authorize that this institution delivered top quality programs with a focus on student needs using ICT. One of the biggest benefits is the support of collaboration between students coming from different countries and cultures"* [35]. According to the views of the students, active participation during the flipped class teaching method was very much appreciated by them according to all their reports.
2. Relevance was well-organized in the courses in order to increase a learner's motivation. Students found the online course material relevant to their interests and they were able to make comments on the course material or search for more information on the internet. Students from Kenya and Slovenia said: *"was an eye-opener for me, far beyond my expectations. Time spent with highly qualified lectures and other students was extremely valuable. The diversity of participants from different nations and the impressive curriculum, has given me inspiration, insight, and enhanced my confidence of start my own business"* and *"as participants, we dealt with multiple issues in particular fields of various subjects and help each other understand the situation in our countries. Not only it helps us strengthen our cooperation but also we gained new views on various topics"* [35]. In all courses we used actual language and examples with which the students were very familiar. During the online meetings of the courses, all students were able to answer questions such as "*What will the subject matter do for me today?*" and "*How can I apply it to a theoretical case tomorrow?*". These questions describe the skills of the students gained through the relevance of the course material and their motivation, because all students replied that they found the course interesting and helpful for their later jobs.
3. Confidence was upgraded in every course because most students met the objectives of learning during the online MHEI-ME. According to the report from a student from Kenya *"I learnt a lot and it gave me a good grounding in the basics of entrepreneurship. I feel that over the duration of the course, I gained a lot of knowledge and practical strategies that will be beneficial to me especially to my future businesses. The course exceeded my expectations significantly"* [35]. Professors were increasing the confidence of students through the online meetings provided to their students, videos, online PowerPoint presentations, list references, and online tests for feedback and support. During the online meetings, students were able to discuss the course material and create a brainstorming list of new ideas. This kind of communication via brainstorming and forums had an added value for the flipped classroom method because students were well prepared and confident to share their ideas with their professors and other classmates. Specifically, a student from Ukraine said that *"Tutors possess deep expertise and are open for communication"* [35]. This opinion confirms confidence gained through class interaction of students and tutors.
4. Satisfaction of this teaching method was evident for all students. They described it in the final evaluation form as a very useful and beneficial process offering them the extra opportunity to be able later to create their own business plan. A student from Slovakia said that: *"Tutors here provided me with much more practical knowledge, all of them spent years working in companies, where did they gain another kind of knowledge and in this study, they shared it with us, what did I miss at my university the most. Studying together with students from others country has helped me to open eyes and to understand different issues from other countries in all topics around business. What do I like the most?"* [35], additionally, a student from Kenya said that *"A very interesting and valuable experience. I learnt a lot and it gave me a good grounding in the basics of entrepreneurship. I feel*

that over the duration of the course, I gained a lot of knowledge and practical strategies that will be beneficial to me especially to my future businesses. The course exceeded my expectations significantly" [35], and the other student from Malta said *"Whether if one seeks knowledge in transforming an existing company, dream of starting an own venture or even simply wanting to expand the career horizon, this course opens a spectrum of paths to pursue. I recommend this course to anyone who wishes to become an entrepreneur for many reasons"* [35]. According to the previously presented opinions of the students, we find that they were satisfied with the online teaching process and the flipped classroom method because they described it as a valuable experience.

Moreover, in comparison with another type of evaluation of this program, and especially according to the evaluation and certification of the program by the national regulator in Malta, the National Commission for Further and Higher Education, the motivation of the Master's students for this course was high. They eagerly commended the innovation and relevance of the program for up-to-date entrepreneurial issues. They also said that one of the innovative aspects of the MHEI program was that each individual subject module was designed and delivered by one of the partners' HEIs, and it was automatically recognized throughout Europe and any country which is a signatory to the Bologna Process. So, this evaluation supports the findings of ARCS evaluation on the flipped classroom teaching method applied during this Master's course.

4. Conclusions

In conclusion, the application of the flipped classroom teaching method on an online entrepreneurship course gave us positive feedback. We can apply it to other online courses for the benefit of students. The flipped classroom helped students to become more engaged in the course material. During this process, students were able to explore the topics in greater depth and collaborate with their peers before the face-to-face meeting. Students upgraded their attention, relevance, confidence, satisfaction (ARCS) pertaining to the course material.

Additionally, this method of teaching may have great impact during Covid-19 because students will be able to study or complete their learning material at home and work faster during class time. Professors, on the other hand, have the ability to offer more guidance and apply various teaching components such as videos, discussion forums, and feedback to all students online. Another positive impact of the implication of flipped classroom is the fact that students had something to do during class and the online delivery of the course was not a boring experience. So, students' main benefits from the flipped classroom include upskilling to all forms of learning, contribution to course material, upgrading online communication skills.

In fact, there is another outcome of our experiences concerning the validity of online learning, thus confirming the findings of Marengo & Pagano [1]. As a result of the COVID-19 pandemic, in most European countries HEIs (and other schools) had to convert to online. As the pandemic hit Europe suddenly, both professors and students had only a couple of days to adapt to the radically changed situation, including all lecturing and tutoring online. For many lecturers and professors—having no experience in the flipped classroom system or other methods of eLearning—it was and still is a challenge as was mentioned in the Rakic study [3]. There are universities where online teaching methods, including online examinations, are still in the embryonic stage, and now some professors use different LMS like MS Teams, Zoom, or CooSpace, while others are not skilled in teaching in LMS and just share their lectures in PowerPoint. However, according to Levy [25], the online courses have to be interesting for the students. In our courses we cooperated with Advenio eAcademy with a well-functioning and solid platform system which could be adjusted to all necessary online functions during the full study program. After all, we believe that our experiences and all these didactic methods for online teaching that we applied throughout our MHEI-ME project can be useful case studies for teaching by other higher educational institutions, especially in these difficult days when online teaching is a must of the HEIs.

Author Contributions: Conceptualization, S.P.D. and J.P.; methodology, S.P.D.; software, S.P.D.; validation, R.M., J.P., and A.M.; formal analysis, G.I.N.; investigation, S.P.D.; writing—original draft preparation, S.P.D. and G.I.N.; writing—review and editing, R.M.; visualization, R.M.; project administration, S.P.D.; funding acquisition, R.M. All authors have read and agreed to the published version of the manuscript.

Funding: This research was funded by ERASMUS+, grant number: 2017-1-MT01-KA203-026960 Multiple Higher Educational Institutions—Master's in Entrepreneurship.

Acknowledgments: This manuscript was created as a result of the ERASMUS+ project: 2017-1-MT01-KA203-026960 Multiple Higher Educational Institutions—Master's in Entrepreneurship.

Conflicts of Interest: The authors declare no conflict of interest.

References

1. Marengo, A.; Pagano, A. Innovative ways to assess soft-skills: The in-basket game online experience. In Proceedings of the European Conference on e-Learning, ECEL, 2020, Berlin, Germany, 29–30 October 2020; pp. 325–334.
2. Norman, S. 5 Advantages of Online Learning: Education without Leaving Home. Available online: https://elearningindustry.com/5-advantages-of-online-learning-education-without-leaving-home (accessed on 12 March 2021).
3. Rakic, S.; Pavlovic, M.; Softic, S.; Lalic, B.; Marjanovic, U. An Evaluation of Student Performance at e-Learning Platform. In Proceedings of the 17th International Conference on Emerging eLearning Technologies and Applications (ICETA), Vysoké Tatry, Slovakia, 2–22 November 2019; pp. 681–686.
4. Chivu, R.G.; Orzan, G.; Popa, I.C. Education Software and Modern Learning Environment: Elearning. In Proceedings of the 15th International Scientific Conference eLearning and Software for Education, Bucharest, Romania, 11–12 April 2019.
5. Zimmerman, B.J. Becoming a Self-Regulated Learner: An Overview. *Theory Pract.* **2002**, *41*, 64–70. [CrossRef]
6. Dumford, A.D.; Miller, A.L. Online learning in higher education: Exploring advantages and disadvantages for engagement. *J. Comput. High. Educ.* **2018**, *30*, 452–465. [CrossRef]
7. Bergmann, J.; Sams, A. *Flip Your Classroom: Reach Every Student in Every Class. Every Day*, 1st ed.; International Society for Technology in Education: Washington, DC, USA, 2012.
8. Abeysekera, L.; Dawson, P. Motivation and cognitive load in the flipped classroom: Definition, rationale and a call for research. *High. Educ. Res. Dev.* **2015**, *34*, 1–14. [CrossRef]
9. Urfa, M. Flipped Classroom Model and Practical Suggestions. *J. Educ. Technol. Online Learn.* **2018**, *1*, 47–59. [CrossRef]
10. Martin, F.; Smith, R. What is it That Entrepreneurs Learn from Experience? *Ind. High. Educ.* **2010**, *24*, 505–512. [CrossRef]
11. Pange, J.; Lekka, A.; Toki, E. Different Learning Theories applied to diverse learning subjects. A pilot study. *Procedia Soc. Behav. Sci.* **2010**, *9*, 800–804. [CrossRef]
12. Zheng, R. *Digital Technologies and Instructional Design for Personalized Learning*; IGI Global, Ed.; IGI Global: Hershey, PA, USA, 2018. [CrossRef]
13. Lobdell, G. Flipped Classroom Model and Constructivism. 2013. Available online: https://ginalobdell.files.wordpress.com/2013/09/globdell_theflippedclassroommodelandconstructivisminsecondaryandhighereducation.pdf (accessed on 23 June 2021).
14. Duan, W. On the "Flipped Classroom" Teaching Model Through Task-Based Language Teaching from the Perspective of Constructivist Learning Theory: A Case of Hotel English Reform in Yinxing Hospitality Management College of CUIT. In *Asian Research on English for Specific Purposes*; Springer: Singapore, 2020. [CrossRef]
15. Strayer, J.F. How learning in an inverted classroom influences cooperation, innovation and task orientation. *Learn. Environ. Res.* **2012**, *15*, 171–193. [CrossRef]
16. Parslow, G.R. Commentary: The khan academy and the day-night flipped classroom. *Biochem. Mol. Biol. Educ.* **2012**, *40*, 337–338. [CrossRef] [PubMed]
17. Alvarez, B. Flipping the classroom: Homework in class, lessons at home. *Educ. Digest* **2012**, *77*, 18–21.
18. Ritter, N. Teaching Psychology in Secondary School: The Flipped Classroom Approach and the ARCS Motivational Model. Doctoral Dissertation, Secondary School, Tokyo, Japan, 2020. Available online: https://scholarcommons.sc.edu/etd/6049 (accessed on 22 June 2021).
19. Sadeghi, M. A Shift from Classroom to Distance Learning: Advantages and Limitations. *Int. J. Res. Engl. Educ.* **2019**, *4*, 80–88. [CrossRef]
20. Nagrale, P. Advantages and Disadvantages of Distance Education. 2013. Available online: https://surejob.in/advantages-and-disadvantages-of-distance-education.html (accessed on 28 February 2021).
21. Vincenza, B.; Trentin, G. The evaluation of online courses. *J. Comput. Assist. Learn.* **2000**, *16*, 259–270. [CrossRef]
22. Baldwin, S.; Trespalacios, J. Evaluation Instruments and Good Practices in Online Education. *Online Learn.* **2017**, *21*. [CrossRef]
23. Poultsakis, S.; Papadakis, S.; Kalogiannakis, M.; Psycharis, S. The management of Digital Learning Objects of Natural Sciences and Digital Experiment Simulation Tools by teachers. *Adv. Mob. Learn. Educ. Res. (AMLER)* **2021**, *1*, 58–71. [CrossRef]
24. Davidson, C.N.; Goldberg, D.T. *The Future of Thinking—Learning Institutions in a Digital Age*; MIT Press: Cambridge, MA, USA; London, UK, 2010.

25. Levy, D. *"The Synchronous vs. Asynchronous Balancing Act—When and How Pre-Work Can Make Your Live Sessions Stronger"*; Harvard Business Publication Education: Boston, MA, USA, 2020; pp. 2–7.
26. Kolb, D.A. *Experiential Learning. Experience as a Source of Learning and Development*; Prentice Hall: Bergen, NJ, USA, 1984.
27. Keller, J.M. Development and use of the ARCS model of instructional design. *J. Instr. Dev.* **1987**, *10*, 2–10. [CrossRef]
28. Poór, J.; Sasvári, P.; Szalay, Z.; Pető, I.; Gyurián, N.; Suhajda, C.J.; Zsigri, F. The Implementation and Management of E-Learning in Companies—The State of E-Learning in Hungary Based on Empirical Research. *J. Eng. Manag. Compet. (JEMC)* **2020**, *10*, 3–14.
29. Kaplan, A.M.; Haenlein, M. Higher education and the digital revolution: About MOOCs, SPOCs, social media, and the Cookie Monster. *Bus. Horiz.* **2016**, *59*, 441–450. [CrossRef]
30. Bauer, S.; Barrett-Greenly, T. Checklists & Rubrics for Evaluating Online Programs. Available online: https://topkit.org/evaluating/checklists-rubrics/ (accessed on 15 March 2021).
31. University of Central Florida. Blended Course Implementation Checklist. Available online: https://blended.online.ucf.edu/files/2011/06/implementation_checklist.pdf (accessed on 15 March 2021).
32. Elearn.aea.academy, MHEI-ME's foundation modules. Available online: https://elearn.aea.academy (accessed on 23 June 2021).
33. Tucker, B. *The Flipped Classroom: Online Instruction at Home Frees Class Time for Learning*; Hoover Institution Press: Stanford, CA, USA, 2012; pp. 82–83.
34. Panopto.com. *The Practical Guide to Flipping Your Classroom.* 2018. Available online: https://www.slideshare.net/adamstepinski/e-book-flipped-classroom (accessed on 18 March 2021).
35. Aea.academy, Level 7 Student Testimonials. 2019. Available online: https://aea.academy/students-testimonials-page/ (accessed on 5 July 2021).

MDPI
St. Alban-Anlage 66
4052 Basel
Switzerland
Tel. +41 61 683 77 34
Fax +41 61 302 89 18
www.mdpi.com

Sustainability Editorial Office
E-mail: sustainability@mdpi.com
www.mdpi.com/journal/sustainability

www.ingramcontent.com/pod-product-compliance
Lightning Source LLC
Chambersburg PA
CBHW061829210926
51850CB00006B/838
* 9 7 8 3 0 3 6 5 2 1 0 1 5 *